P9-DNJ-222
DISCARD

ALLE ZEIT WACH
1842

Sweetness

Edited by John Dobbing

With 33 Figures

Springer-Verlag
London Berlin Heidelberg New York
Paris Tokyo

John Dobbing, DSc, FRCP, FRCPath

Professor of Child Growth and Development, Department of Child Health, University of Manchester, Oxford Road, Manchester M13 9PT

ISBN 3-540-17045-6 Springer-Verlag Berlin Heidelberg NewYork
ISBN 0-387-17045-6 Springer-Verlag New York Berlin Heidelberg

Library of Congress Cataloging-in-Publication Data
Sweetness.
(ILSI human nutrition reviews) Papers presented at a symposium held in Geneva, May 21–23, 1986 and sponsored by the International Life Sciences Institute.
Includes bibliographies and index.
1. Sweetness (Taste) – Physiological aspects – Congresses. 2. Sweetness (Taste) – Psychological aspects – Congresses. 3. Sweetness (Taste) – Social aspects – Congresses. 4. Food – Sensory evaluation – Congresses. I. Dobbing, John. II. International Life Sciences Institute. III. Series. [DNLM: 1. Carbohydrates – congresses. 2. Sweetening Agents – congresses. 3. Taste – congresses. WI 210 S974 1986] QP456.S93 1987 152.1'67 86-24777
ISBN 0-387-17045-6 (U.S.)

Printed in Great Britain

Photosetting by Tradeset, Welwyn Garden City, Herts. AL7 1BH
Printed by Henry Ling Limited, The Dorset Press, Dorchester.

2128/3916-543210

Preface

Very few books, especially when written by many authors, have passed through any serious process of peer review. That this one has done so is due to the special way in which it was produced.

All eighteen authors were asked to submit chapters which were then circulated to each of the others. Everyone was asked to write considered Commentaries on each chapter, with references where necessary, and these too were circulated to all the other authors.

The purpose of the Commentaries was to be constructively critical and, where appropriate, to highlight areas of difference rather than to reach a consensus. Research workers, of course, always strive to reach a common truth in the end, but in the process their findings and ideas pass through a seemingly interminable period of discussion and argument; and during this time their enthusiasm is sharpened by the constructive cut-and-thrust of lively debate. Progress, indeed, comes as much from open discussion as from new discovery. New ideas and new results are regularly dipped in the acid of other people's opinions and findings, so that what remains is refined and likely to be reliably contributory. Our book is meant to be a contribution to that process.

When all this had been done the eighteen authors and the Editor were able to meet together for two long days and finally hammer out those arguments which were much less easy to discuss by correspondence. We met privately and informally, without any audience and with no recording, working through each chapter in turn and all the Commentaries until the finished product emerged.

Each chapter, however, remains its author's responsibility. After reading and listening to everyone else's attitudes to his subject, and after defending his own, an author was left completely free to write the chapter in the way he wanted. It was nevertheless often interesting to see what a substantial modification actually took place as a result of the full and frank discussions we were able to have in a closed, domestic atmosphere.

Those Commentaries which resulted in changes in the chapters have not been published in the book. Those which remain and are reproduced at the end of each chapter are some which address points of scientific interpretation which were not, or did not need to be, resolved. In some cases the original author has contributed his 'reply' and this immediately follows the relevant Commentary.

You, the reader, cannot therefore participate in much of the fascinating 'domestic' debate we had. It was private to us, just as the referees' comments on a paper submitted to a learned journal are private. But in return we can and do present you with chapters which have been peer-reviewed like no other, by seventeen colleagues in the general field; and we hope you will appreciate this and even be able to discern the advantages of their enhanced quality.

Editorially we have tried to respect the differences between American and English English; all non-American chapters of whatever nationality are, as far as possible, expressed in the English variety. The way in which the book was produced, and the way we worked, allowed a much faster publication time, but that did involve everyone in much more work than is usual in an enterprise of this kind, and may also have resulted in a few errors, for which we apologise.

We would like to record our Editorial appreciation of the way authors responded to what on occasions were importunate demands, and we hope the final product will be regarded by them too as the better for all their hard work. We are also grateful to our Publishers for their ready cooperation.

Finally our grateful thanks are due to the International Life Sciences Institute (ILSI), whose Organising Committee selected the participants and who sponsored our endeavour, to my secretary Mrs. Irene Warrington, who bore the brunt of an immense administrative load, and to my wife, Dr. Jean Sands, who helped a great deal in the detailed scientific management. At the end of our deliberations a public Symposium was held in Geneva, on 21–23 May 1986, again sponsored by ILSI, at which the authors presented their finished papers.

St. Julien de Cénac John Dobbing
June 1986

Note

Sweetness is a sensation that all of us experience and which seems to affect a good deal of our eating and our social behaviour. When we invited distinguished scientists from the fields of fundamental research, health and the food industry to join together to write this book they were asked as far as possible to consider *sweetness*, as distinct from *sweeteners*, and we think this is the first time *sweetness* has been specifically treated in this way. However, contributors have naturally not always been able to distinguish between the two because of the obviously indissoluble link between them. Our book therefore does deal, though not exhaustively, with sugar where this seemed relevant.

The Organising Committee

Contents

SECTION I. The Nature and Taste of Sweetness

Chapter 1. Chemical Aspects of Sweetness
Gordon G. Birch 3

Introduction 3
Theories of Sweetness 3
Chemical Classes Which Cause Sweetness 5
Quality, Intensity and Persistence of Sweetness in Relation to Chemical Structure 7
Interactions of Molecules and Tastes 7
The Fundamental Chemoreception Mechanism 9
New Approaches to Sweetness Chemoreception Studies 10
Conclusions 12

Chapter 2. Neurophysiological Aspects of Sweetness
Thomas R. Scott and Barbara K. Giza 15

Introduction 15
The Neural Code for Sweetness 15
Spatial Distribution 15
Temporal Distribution 20
Conclusion 21
Sweetness and the Neurophysiology of Hedonics 21
General Considerations 21
Alterations in the Neural Code for Sweetness 22
Conclusions 28
Summary 28
Commentary 31

Chapter 3. Is Sweetness Unitary? An Evaluation of the Evidence for Multiple Sweets
Linda M. Bartoshuk 33

Introduction 33
Neurophysiological Evidence for Multiple Sweet Receptor Mechanisms 33

Single Fiber Recordings 33
Substances That Compete with or Damage Sweet Receptor Mechanisms 34
Saccharin and Sucrose Mixtures 36
Species Differences in Responses to Sweeteners 36
Psychophysical Evidence for Multiple Receptor Mechanisms 36
Cross-adaptation 37
Sweetness on Various Tongue Loci 41
Sweetness and PTC/PROP Status 42
The Artichoke Effect 42
Correlations Across Sweeteners 42
Sweetness and Diabetes 43
Mixtures 43
Sweetness and Age 44
Sweetness and Reaction Time 44
Summary 44
Commentary 46

Chapter 4. Selected Factors Influencing Sensory Perception of Sweetness
R. M. Pangborn 49

Introduction 49
Sweetness Responses in Relation to Psychophysical Measurements 49
Thresholds and Discrimination 50
Perceived Intensity 50
Descriptive Analysis 51
Affective Tests 51
Interaction of Sweetness with Other Tastes 51
Interactions in Aqueous Solutions 51
Interactions in Foods and Beverages 55
Interaction of Sweetness with Other Sensory Attributes 56
Color 56
Aroma 57
Viscosity 57
Solution Temperature 59
Conclusions 60
Commentary 63

Chapter 5. Sensory Sweetness Perception, Its Pleasantness, and Attitudes to Sweet Foods
Jan E. R. Frijters 67

Introduction 67
Perceived Sweetness Intensity 68
Experienced Pleasantness of Sweetness 69
The Origin of the Pleasantness of Sweetness 73
Attitudes Towards Sweetness and Sweet Foods 74
Commentary 79

SECTION II. The Social Context of Sweetness

Chapter 6. Attitudes Towards Sugar and Sweetness in Historical and Social Perspective
Claude Fischler ... 83

Introduction ... 83
Uses and Perceptions of Sweetness in Western European History: An Overview ... 84
The Early Status of Sugar: Spice and Medicine ... 84
The Emergence of Reservations About Sugar ... 85
"Sweet Revolution" in the Nineteenth Century: the "Vulgarisation" of Sugar ... 86
Contemporary Attitudes ... 87
Modern "Saccharophobia" ... 87
Ambivalent Attitudes and the Social Management of Pleasure ... 89
Fluctuations and/or Trends in Attitudes and Usage ... 91
Sweets and Sugar in the French Press: a Content Analysis ... 93
Medicine, Social Ideologies and Individual Behaviour ... 96
Commentary ... 98

Chapter 7. Sweetness, Sensuality, Sin, Safety, and Socialization: Some Speculations
Paul Rozin ... 99

Sweetness, Sin and Sensuality ... 99
Sweetness and Safety ... 101
Some Semantics: What's in a Name? ... 103
Sweetness and Socialization ... 104
Intrinsic and Extrinsic Motivation ... 104
A Psychological Taxonomy of Food Acceptance and Rejection ... 105
Sweetness as a Socializing Agent ... 106
Sweetness as Reward: Reversing or Preventing Liking ... 107
Acquiring Contexts and Meanings ... 108
Summary ... 109
Commentary ... 110

SECTION III. Inborn and Acquired Aspects of Sweetness

Chapter 8. Opioids, Sweets and a Mechanism for Positive Affect: Broad Motivational Implications
Elliot M. Blass ... 115

Introduction ... 115
Experimental Analyses ... 119
General Discussion ... 121
Commentary ... 124

Chapter 9. Development of Sweet Taste
Gary K. Beauchamp and Beverly J. Cowart 127

Introduction 127
Newborn Infants 128
Maturation and Sensitivity 128
Dietary Experience and Sweet Acceptability 130
Age and Sweet Acceptability 132
Discussion 136
Commentary 138

SECTION IV. Sweetness and Food Intake

Chapter 10. Sweetness and Food Selection: Measurement of Sweeteners' Effects on Acceptance
D. A. Booth, M. T. Conner and S. Marie 143

Objective Acceptability of Sweetness 143
Perception of Sweetness 143
Food Selection: Observed Datum or Hypothetical Process? 144
Objective Influences of Sweetness on Selection 145
Sweetness in the Selection of Familiar Foods 145
Development of Effects of Sweeteners on Food Selection 145
Preference for Sweetness Within the Food Complex 146
The Linear Sweetener/Food-Selection Function 147
Food Sweetness Ideal Points and Rejection Points 147
Objective Sensory Distances from Ideal to Rejection 147
The Tolerance Triangle 148
Characterised Tolerance Functions 149
Differences in Sweetness Acceptance Between Individuals 153
Stability of Measurement 153
Variation Among Individuals in Average Ideal 154
Roles of Sweetness in Food Selection 154
Rapid Objective Measurement of Sweetness Selection 154
Weight Control Problems: Sweet Preference or Abnormal Satiation? 154
Normal Role of Sweetness in Food Selection 156
Commentary 158

Chapter 11. Sweetness and Satiety
Barbara J. Rolls 161

Introduction 161
Alliesthesia: Physiological Usefulness and the Changing Hedonic Response to Sweetness 161
Sensory-Specific Satiety 162
How Specific Is Satiety? 164
Sweetness and Monotony 164
Differences Among Sugars in Their Effects on Satiety 165
Sweetness in Drinks 166
The Sweet Taste Can Override Physiological Satiety Signals 166
Low Energy Sweet Drinks and Body Weight 167

Sweet Foods and Caloric Regulation 169
Conclusions 170
Commentary 172

SECTION V. Sweetness and Obesity

Chapter 12. Sweetness and Obesity
Adam Drewnowski 177

Introduction 177
Perception of Sweetness 179
Hedonic Response 180
Individual Differences 181
Complex Systems 182
Taste and Energy Status 185
Infancy and Childhood 186
Conclusions 187
Commentary 190

Chapter 13. Sweetness and Eating Disorders
Judith Rodin and Danielle Reed 193

Introduction 193
Types of Food Eaten 195
Sweet Sensitivity and Hedonic Valuation 196
Biological Mechanisms Relating Sweet Preference to Disordered Eating 198
Conclusions 201
Commentary 203

Chapter 14. Sweetness and Performance
Edward Hirsch 205

Introduction 205
Methodological and Conceptual Issues 206
Research Strategies 206
Defining the Independent Variable 209
Test Population 211
Prevailing Nutritional State 212
Research Findings 212
Sweetness and Children's Behavior 212
Sweetness and Athletic Performance 216
Sweetness and Antisocial Behavior 218
Commentary 223

SECTION VI. Implications of Sweetness

Chapter 15. Implications of Sweetness in Upbringing and Education
Matty Chiva 227

Introduction 227

The Child 228
Sweetness and Education in General 229
Sweetness and the Social Behaviour of Non-verbal Communication 230
Nutritional Education 232
Sweetness and the Nutritional Education of Children 232
Sweetness and the Nutritional Education of Adults 234
Conclusion 235
Commentary 237

Chapter 16. Sweetness in Marketing
Howard G. Schutz and Debra S. Judge 239

Introduction 239
Consumer Attitudes 239
Sweetness in Contemporary Marketing 240
Methodologies in Marketing 243

Chapter 17. Sweetness in Product Development
P. Würsch and N. Daget 247

Introduction 247
Sweetness in the Food System 250
Choice of Sweetener 252
Product Optimisation 254
Establishment of Measuring Instruments 255
Optimisation Trials 256
Conclusions 258
Commentary 259

Chapter 18. Sweetness in Food Service Systems
Herbert L. Meiselman 261

Introduction 261
Food Attitudes 262
Food Acceptance of Sweet Foods 265
Food Selection and Intake of Sweet Foods 267
Food Waste of Sweet Foods 270
Conclusions 272
Commentary 274

Index 277

Contributors

Dr. Linda M. Bartoshuk
John B. Pierce Foundation Laboratory, 290 Congress Avenue,
New Haven, Connecticut 06519, U.S.A.

Dr. Gary K. Beauchamp
Monell Chemical Senses Center, 3500 Market Street, Philadelphia,
Pennsylvania 19104, U.S.A.

Dr. Gordon G. Birch
National College of Food Technology, Department of Food Technology,
Food Studies Building, University of Reading, Whiteknights,
Reading RG6 2AP, England

Prof. Elliot M. Blass
Department of Psychology, The Johns Hopkins University, Baltimore,
Maryland 21218, U.S.A.

Dr. D. A. Booth
Department of Pathology, The University of Birmingham, PO Box 363,
Birmingham B15 2TT, England

Professor Matty Chiva
Centre de Psychologie de l'Enfant, 200 Avenue de la République,
92001 Nanterre Cedex, France

Dr. M. T. Conner*
Department of Psychology, The University of Birmingham, PO Box 363,
Birmingham B15 2TT, England

Dr. Beverly J. Cowart*
Monell Chemical Senses Center, 3500 Market Street, Philadelphia,
Pennsylvania 19104, U.S.A.

Dr. N. Daget*
Research Department, Case Postate 88, CH-1814 La tour de Peilz, Switzerland

Dr. Adam Drewnowski
The University of Michigan, School of Public Health, Ann Arbor, Michigan 48109, U.S.A.

Dr. Claude Fischler
Association Francaise pour un Science de l'Homme, 3 rue Fessart, 92100 Boulogne, France

Dr. Jan E. R. Frijters
Department of Human Nutrition, De Dreijen 12, 6703 BC Wageningen, The Netherlands

Dr. Barbara K. Giza*
University of Delaware, College of Arts and Science, Department of Psychology, 220 Wolf Hall, Newark, Delaware 19716, U.S.A.

Dr. Edward Hirsch
Human Engineering Branch, Behavioral Sciences Division, Science and Advanced Technology Laboratory, US Army Natick Research and Development Center, Natick, Massachusetts 01760, U.S.A.

Dr. Debra S. Judge*
College of Agriculture, Agricultural Experiment Section, Department of Consumer Sciences, University of California, Davis, California 95616, U.S.A.

Dr. S. Marie*
Department of Physchology, The University of Birmingham, PO Box 363, Birmingham B15 2TT, England

Dr. Herbert L. Meiselman
DOD Food Program, US Army Natick Research and Development Center Natick, Massachusetts 01760, U.S.A.

Prof. R. M. Pangborn
College of Agricultural and Environmental Sciences, Department of Food and Science and Technology, 1480 Chemistry Annex, University of California, Davis, California 95616, U.S.A.

Dr. Danielle Reed*
Department of Psychology, Yale University, 2 Hillhouse Avenue, PO Box 11A Yale Station, Newhaven, Connecticut 06520-7447, U.S.A.

Dr. J. Rodin
Department of Psychology, Yale University, 2 Hillhouse Avenue, PO Box 11A Yale Station, Newhaven, Connecticut 06520-7447, U.S.A.

Dr. Barbara Rolls
Department of Psychiatry and Behavioral Sciences, The Johns Hopkins Hospital, Henry Phipps Psychiatric Clinic, 600 N. Wolfe St., Baltimore, Maryland 21205, U.S.A.

Dr. Paul Rozin
Centre for Advanced Study in the Behavioral Sciences, 2020 Junipero Serro Boulevard, Stanford, California 94305, U.S.A.

Prof. Howard G. Schutz
College of Agriculture, Agricultural Experiment Station, Department of Consumer Sciences, University of California, Davis, California 95616, U.S.A.

Prof. Thomas R. Scott
University of Delaware, College of Arts and Science, Department of Psychology, 220 Wolf Hall, Newark, Delaware 19716, U.S.A.

Dr. P. Würsch
Research Department, Case Postale 88, CH-1814 La tour de Peilz, Switzerland

* Did not attend workshop

Section I

The Nature and Taste of Sweetness

Chapter 1

Chemical Aspects of Sweetness

Gordon G. Birch

Introduction

Chemical aspects of sweetness embrace molecular theories of chemoreception as well as simple structure: activity relationships of defined classes of sweetener. Chemical and physicochemical interactions probably affect both the quality and the intensity of sweet taste response, and the molecular structure of a selected sweetener will govern both its suitability and stability in a particular food system. Much of the commercial interest in the subject of sweetness has centered on finding an alternative to sucrose with a view to matching its ideal taste properties and effort has concentrated on selected and/or modified sugars, proteins and peptides, flavonoids, glycosides, saccharin and acesulphame.

The considerable amount of structural and synthetic chemical work of the past two decades has led to commercial success and further promise and has advanced our understanding of chemoreception mechanisms. The work of neurophysiologists, anatomists, and psychophysicists has benefited from an understanding of the behaviour of molecules and their likely interaction with receptors. The involvement of "active sites" or pharmacophores in the sensory effects of sapid molecules is a modern concept that has stimulated the efforts of many research groups. Both gustatory and olfactory responses may originate in similar pharmacophores but sweetness is generally considered a simple taste phenomenon. This chapter therefore concentrates on chemical aspects of sweet taste.

Theories of Sweetness

The earliest recorded theory of sweetness is probably that of Theophrastus in *De Sensibus* (Stratten 1964). He describes its origin in "small round molecules" and attributes the idea to Democritus the Atomist about 500 BC.

Early this century theories of sweetness centered on solubility, "sapophoric groups" (e.g. OH and NH_2) and other simple chemical explanations (Moncrieff

1967). Cohn (1914) used the ideas of multiple hydroxyl groups and sapophores[1] to explain sweetness and, following this and the chemistry of dyestuffs, Oertly and Myers (1919) developed a theory of auxoglucs[2] and glucophores[3]. This postulated the presence of the two different kinds of group in the same molecule as a prerequisite for the sweetness response. Oertly and Myers (1919) listed putative auxoglucs and glucophores and some of their suggestions do seem to be partly justified by research in the 1970s and 1980s. However, most of these early theories did not explain the sweetening power of intense sweeteners such as saccharin.

Beck (1943) suggested that the sweetness of sugars correlated well with the ratio, sum of atomic volumes : molecular volume. This interesting idea touches on the use of parachors[4] to predict trends of homologous series in a given chemical class and it is both surprising and unfortunate that so little attention has yet been paid to the implications of Beck's suggestion. Some of the very recent structure : activity studies in both sugars and peptides seem to be re-exploring this approach.

Probably all substances with sapid effects must possess some water solubility in order to gain access to receptors. On the other hand it is known that many lipophilic substances are much sweeter than the natural (hydrophilic) sugars, so Deutsch and Hansch (1966) have attempted to correlate sweetness with partition coefficient. The sapid substance is thus imagined as binding with a hypothetical sweet receptor located on the taste cell membrane. The lipid nature of this membrane necessitates a favourable partition coefficient for effective accession. Other theories to explain sweet taste have been electronic vibration (Kodama 1920) and enzyme activity (Baradi and Bourne 1951). The latter theory has not excited much interest but the recent finding of Schiffman et al. (1985) of adenosine involvement in receptor activity may restimulate the search for enzyme intermediation.

Quite the most important hypothesis to explain the chemical basis of sweetness has been that of Shallenberger and Acree (1967), who used a previous observation of hydrogen bonding in relation to sweetness (Shallenberger 1963) to develop the concept of an AH,B glucophore. According to Shallenberger and Acree (1967), all sweet substances possess this unit (Fig. 1.1), in which A and B are each electronegative atoms and AH acts as an acid while B acts as a base. The AH,B unit then forms a doubly hydrogen-bonded complex with a similar AH,B system on the taste receptor. A separation of 2.86 Å between the centres of the atomic orbitals of A and B appears to be ideal for sweetness, and probably the entire molecular geometry, as well as the AH,B system, governs the strength of the hydrogen-bonded complex and both the quality and the intensity of the sweet taste response.

Kier (1972) extended the AH,B theory of sweetness by introducing the concept of a third, lipophilic binding site on the molecule, usually termed 'γ'. This site was supposed also to create a favourable partition coefficient facilitating accession of the sapid stimulus to the lipophilic environment of the taste cell membrane. The idea of a tripartite AH,Bγ glucophore is now often quoted to explain sweetness (Hough

---

1. *Sapophore:* A molecular feature giving rise to a basic taste.
2. *Glucophore:* One of an essential pair of molecular features needed for sweetness (the other is an auxogluc).
3. *Auxogluc:* One of an essential pair of molecular features needed for sweetness (the other is a glucophore).
4. *Parachor:* The parachor is related to the apparent molar volume (ϕV) and the surface tension (γ) by the expression: $[P] = \phi V \cdot \gamma^{1/4}$. In other words the parachor may be viewed as equal to the apparent molar volume if the surface tension were to remain at unity.

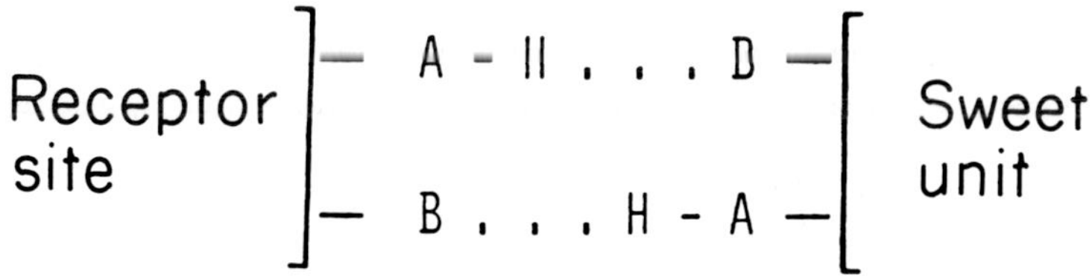

Receptor site — A - H . . . B — Sweet unit
— B . . . H - A —

Fig. 1.1. The Shallenberger and Acree (1967) AH,B glucophore.

1985) but there seems little reason to over-complicate Shallenberger and Acree's (1967) original simple and elegant theory by extending it in this way. The role of lipophilicity in enhancing sweetness is undisputed and γ may be viewed as contributing to the orientation and accession of a sapid stimulus to the receptor. However, the idea of γ as a binding site of the glucophore is unnecessary and, indeed, no such site is obvious in the sugars, inositols or polyols; yet the quality of their sweetness response in unsurpassable.

Chemical Classes Which Cause Sweetness

Chemical classes which cause sweetness (Table 1.1) include sugars, glycosides (and modified sugars, e.g. polyols), amino acids, peptides and proteins (and modified types), coumarins, dihydrochalcones, ureas and other nitrogenous compounds, substituted aromatic substances and certain salts. There is clearly no categorical pattern in the chemical class of molecule able to elicit the sweet response and in some classes, e.g. inorganic salts, there is no apparent glycophore. On the other hand certain broad

Table 1.1. Chemical classes and sweetness[a] (sucrose = 1)

Class	Order of sweetness (Molar basis)
Simple sugars	0.1–2.0
Hydrogenated sugars	0.1–2.0
Chlorodeoxy sugars	0–2600
Terpenoids and their glycosides	0–1000
Dihydrochalcones	0–10 000
Peptides	0–30 000
Proteins	0–300 000
Nitroanilines	0–2350
Sulphamates	0–26
Oximes	0–750
Isocoumarins	0–200
Saccharins	(0–1000)[b]
Acesulphames	(10–250)[b]
Tryptophans	(0–1300)[b]
Ureas	(0–250)[b]

[a] References: van der Heijden et al. (1985a,b), Birch et al. (1971), Compadre et al. (1985), Birch et al. (1980)
[b] These figures are on weight basis

chemical features (e.g. polarity, lipophilicity, chirality, molecular size and shape) are evidently important for many known sweeteners and give strong clues in the search for structure : activity relationships.

Most of the chemical synthesis in sweetness research has centered on modification of sugars, peptides (and proteins) and dihydrochalcones, and a guiding principle throughout has been the Shallenberger and Acree AH,B theory. Probably some hundreds of derivatives of these three types are now available (Birch et al. 1971; Birch and Parker 1982; Grenby et al. 1983).

In almost all classes of sweet molecule possible AH,B glucophores may be recognised but it is rare for them to be unequivocally located. The inorganic salts (e.g. lead and beryllium) provide an interesting exception. However, salts are heavily hydrated in solution and the hydration shells which surround each molecule may themselves constitute suitable AH,B systems.

Chemical synthesis and modification of new sweeteners, principally in the United States, United Kingdom, Ireland, The Netherlands and Japan, has established many structure : activity relationships (de Vos et al. 1985; Soejarto et al. 1982; Dubois et al. 1984; Compadre et al. 1985). The subject has been excellently reviewed by Beets (1978). Although the products of synthetic work are numerous, few have been toxicologically cleared for human consumption. Those which are suitable differ in their chemical and physical properties and are conveniently grouped into those which confer body and viscosity to foods, like the sugars (*bulk sweeteners*), and those which have much more sweetening power than the sugars (*intense sweeteners*). The latter are used at such low concentrations that they confer no body. Table 1.2 lists the new permitted sweeteners in the United Kingdom (HMSO 1983). Quite the most important of these is aspartame (L-aspartyl L-phenylalanine methyl ester), which has already been widely used in food formulation. One problem is its limited storage stability in acid foods and a solution to this problem is of considerable economic importance. Recently Fuller et al. (1985) have in fact published the synthesis of 12 new *N*-(L-aspartyl)-1,1-diaminoalkanes ranging in sweetness from 5–1000 times the sweetness of sucrose, all of which are stable to hydrolysis.

Chemical modification to increase sweetness is often unsuccessful because the products turn out to be bitter. Sometimes molecules are truly bitter-sweet and may elicit such effects because of the proximity of receptor sites (Birch and Mylvaganam 1976). Chlorination of sugars, for example, often leads to intensely bitter products (Dziedzic 1980) but sucrose itself yields a number of deoxy chloro-derivatives (Table1.3) ranging from 5 to 2200 times the sweetness of sucrose (Hough 1985). The explanation of this remarkable enhancement is not easy but may emerge from a study of solution properties (Mathlouthi et al., to be published).

Table 1.2. Permitted sweeteners in the United Kingdom (HMSO 1983)

Intense sweeteners	Bulk sweeteners
Acesulphame K	Hydrogenated glucose syrup
Aspartame	Isomalt
Saccharin	Mannitol
Sodium saccharin	Sorbitol
Calcium saccharin	Sorbitol syrup
Thaumatin	Xylitol

Table 1.3. Taste properties of some chlorodeoxy sugars (Hough 1985)

Compound	Taste[a]
1′Chloro-1′-deoxy sucrose	Sweet (20)
4-Chloro-4-deoxy galactosucrose	Sweet (5)
6-Chloro-6-deoxy sucrose	Bitter
6′-Chloro-6′-deoxy sucrose	Sweet (20)
4,1′,6′-Trichloro-4,1′,6′-trideoxy galactosucrose	Sweet (650)
4,1′,4′,6′-Tetrachloro-4,1′,4′, 6′-tetradeoxy galactosucrose	Sweet (2200)

[a] Figures in parentheses are the times sweeter than sucrose.

Quality, Intensity and Persistence of Sweetness in Relation to Chemical Structure

Although a large number of intensely sweet molecules have now been synthesised, there are many problems about the quality of their sweetness. There is no currently permitted sweetener that possesses the ideal sensory quality of sucrose, though the alleged quality differences among the sugars themselves may be due to impurities. Problems of quality include bitterness, sourness, cooling and menthol or liquorice-like taste. More subtle effects could result from diffferences in surface tension, viscosity, body and tactile response. By far the most important quality effects, however, are the temporal characteristics, which are manifested as both delayed reaction times and prolonged sweet sensations (*persistence*). Every new intense sweetener synthesised appears to suffer the disadvantage of a long persistence, and with the intense protein sweeteners this may extend to 20 or 30 min. It is not known whether the phenomenon is due to a "hysteresis" effect of the receptor or whether persistence of stimulus molecules themselves is responsible, but unless the persistence of sweeteners is properly understood, there seems to be little hope of solving the problem. Delayed reaction time is also a problem and this is difficult to measure accurately, being of the order of 1 s or less, but depending on the number of papillae stimulated (Halpern 1986).

So far there are no clear indications of the effect of chemical structure on temporal characteristics of the sweet response, but a study of taste interactions and solution properties of the sweeteners is an avenue for further exploration (Birch and Shamil, to be published).

Interactions of Molecules and Tastes

The phenomenon of basic taste change is well illustrated in the sugars, which can easily be converted, by substitution, into bitter substances (Clode et al. 1985). Even a simple configurational change at a single asymmetric carbon atom causes the sweet-to-bitter transition. Thus β-D-glucopyranose is sweet whereas β-D-mannopyranose is

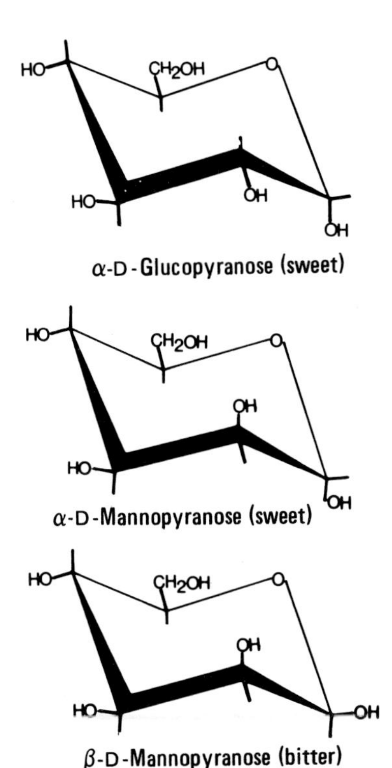

Fig. 1.2. Configurational change and taste.

bitter (Fig. 1.2), so the two basic tastes are fundamentally associated in some way. With bitter-sweet substances (e.g. methyl α-D-mannopyranoside) it has been suggested (Birch and Mylvaganam 1976) that the two types of receptor may be simultaneously spanned by one molecule, which indicates their close proximity.

Although the basic tastes are psychophysically discrete (Bartoshuk 1977; McBurney and Bartoshuk 1973), they interact in mixtures (Bartoshuk 1975 and Chap. 3, this volume) so that the perceived intensity may be different from that of a single component of the mixture. Certainly sweetness lowers the intensity of bitterness and vice versa (Bartoshuk 1975; Birch et al. 1972), but when two sweet substances are mixed the result is less clear. The apparent synergism or suppression of the mixture may be a logical outcome of the two substances' individual intensity power functions (Bartoshuk 1977). If sensorial interaction does occur between sugars, it is certainly not due to interactions between the sugar molecules themselves, as can be seen, for example, by optical rotation measurements. This raises the interesting possibility that the sugar molecules in a mixture may exert different effects on water structure, which in turn affect taste (Mathlouthi et al., to be published). The well reported differences (Bartoshuk 1977) in water taste after adaptation may be related to such effects.

Munton and Birch (1985) have used intensity-time studies of mixtures of two sugars to determine their "effective concentrations". Interestingly the least sweet sugar, in the mixtures tested, appeared to be dominant in the mixture (i.e. the sugar which was the most effective stimulator) and this could be related to its compatibility

Table 1.4. Dominance of sweeteners in binary mixtures (Munton and Birch 1985; Yamaguchi et al. 1970)

Sucrose	: Sorbitol[a]
Sucrose	: Xylose[a]
Sucrose	: Maltose[a]
Sucrose	: Galactose[a]
Lactose	: Glucose[a]
Glucose	: Galactose[a]
Sucrose	: Glucose[a]
Sucrose	: Mannitol[a]
Sucrose[a]	: Saccharin
Fructose	: Glucose[a]
Fructose	: Xylose[a]
Fructose	: Sorbitol[a]
Xylose[a]	: Sodium cyclamate
Sucrose	: Glycine[a]
Mannitol[a]	: Saccharin

[a] Indicates dominant sugar in mixture

with water structure (Birch and Shamil, to be published). A molecule which is highly compatible with water may be viewed as causing the least disturbance or displacement of water molecules and a good measure of this is its apparent molar volume. Yamaguchi et al. (1970) have also observed dominance[5] of one sugar in a mixture but without considering the persistence of response. Table 1.4 lists some results of dominance studies.

The interaction of stimulus molecules with water is of such direct importance to sweet taste response that many related physical parameters should be taken into account (Lawrence and Ferguson 1959). Viscosity, surface tension, and water activity are examples of properties that might affect taste results. The total volume of solution presented to a panellist will affect both the intensity and persistence of response (Birch et al. 1982).

The Fundamental Chemoreception Mechanism

Shallenberger and Acree's (1967) AH,B theory suggests that one molecular feature of a sweet molecule (*a glucophore*) interacts with a sweet receptor. Transduction then occurs and a neural impulse is sent to the brain. If the AH,B concept is correct it ought to be possible to identify the AH,B glucophores in different molecules and to use this information to elucidate similarities and differences of structure : activity patterns among defined classes of sweetener. In practice this is very difficult because most sweet molecules can adopt an almost unlimited number of conformations. Therefore simple cyclic molecules, such as the sugars which adopt a favoured conformation, are the best models in which to conduct a glucophore search. In the sugars, AH,B systems are α-glycol groups, but there are several possible candidates in each molecule. For example, Shallenberger and Acree (1967) assigned the 1,2 α-glycol group as the

5. Operationally, dominance of one or other sugar in a mixture may be viewed as the matchability of its psychophysical power function (measured alone) with that of the mixture. Thus the "effective concentrations" of each of the sugars in a mixture can be calculated.

AH,B system of D-fructopyranose, the sweetest simple sugar, on the basis that OH-2 is the most acidic proton. However, this conclusion is possibly now questionable (Birch et al. 1986, unpublished work) because D-fructopyranose may orientate itself on the taste receptor analogously to D-glucopyranose, in the favoured chair conformation, and its 1,2 α-glycol system would then be unable to bind. The best models for elucidating AH,B systems of sugar molecules are the α-D-glucopyranosides. When oxygen atoms are eliminated stepwise around the ring (Birch and Lee 1974), it is found that only the 3-deoxy derivative exhibits loss of sweetness. Thus OH-3 in glycopyranoside structures must constitute B in the Shallenberger and Acree AH,B system. Substitution of the molecules (Birch 1976) indicates that OH-4 is the primary AH. When the deoxy derivatives of α,α-trehalose and methyl α-D-glucopyranoside are compared they show completely parallel sets of sensory properties in corresponding molecules. This provides a striking illustration of the chemical basis of sweetness as well as a location of the AH,B system.

The binding of a sweet molecule to a receptor via an AH,B glucophore does not in itself explain all the facets of the sweet response. In particular the temporal characteristics of sweetness (reaction time and persistence) probably ensue from other mechanisms. The molecule must accede to the micro-environment of the receptor before it can activate it with an AH,B system, and lipophilic characteristics may facilitate such accession. The great persistence of some of the newly available sweetening agents could be explained by a localised concentration of stimulus molecules at or near to the receptor. This has been clearly established, for example, in insect chemoreception, and persistence can result from prolonged stimulation by a number of molecules in localised concentrations or "stores" until they are depleted. It seems logical to view these localised concentrations as organised for cellular activity. Thus an "orderly queue" hypothesis has been proposed (Birch et al. 1980) to explain how stimulus molecules might be stored and utilised consecutively. Molecules emerging at the head of the queue trigger the ionophore sequentially and the process continues until the queue is emptied of stimulus molecules. Length of queues then governs persistence time whereas the rate of passage through a queue governs reaction time. Accession of stimulus molecules to queues is probably enhanced not only by lipophilicity but also by compatibility with water structure (Birch and Catsoulis 1985; Birch and Shamil, to be published). This introduces an interplay of hydrophilic and hydrophobic forces in the initial stages of taste chemoreception.

The fundamental mechanism of taste chemoreception is still not understood because taste receptors have not yet been isolated and characterised. Considerations of accession of stimuli and activation of receptors suggest that there are at least two stages of the taste chemoreception process and detailed comparisons of stimulus molecules, in homologous series of sweeteners, can indicate how these separate stages might take place.

New Approaches to Sweetness Chemoreception Studies

Probably no substance can be tasted unless it is first dissolved (in water or oral fluid), so water itself is probably involved in a number of different ways in the taste response. Water acts not simply as a vehicle for carrying stimulus molecules to the receptor, but also as an extremely active hydrogen-bonding agent. Sugar molecules are therefore heavily hydrated in solution though they differ in their degrees of hydration due to differences in the equatorial and axial configurations of their hydroxyl groups. The apparent molar volume (ϕV) of a sugar gives a measurement of its apparent displacement of water molecules. Heavily hydrated sugar molecules have low (ϕV) values and are therefore highly compatible with water structure (Birch and Catsoulis 1985; Birch and Shamil, to be published).

A modern approach to the study of sweetness chemoreception is to consider the interactions of physico-chemical variables which can explain molecular structure and solution properties. Both the effect of the solute on the solvent and the effect of the solvent on the solute must be considered (Mathlouthi et al., to be published). Using this approach it may be noted that the dominant sugar, in mixtures of two, is the one with the smaller (ϕV) value (Birch and Shamil 1986). This does not simply mean that the smaller molecule has a size advantage for access to the receptor, but rather that it is more compatible with water structure and is therefore better transported into the micro-environment of the receptor. Over the entire field of basic tastes, a useful index is specific apparent volume (ϕV/mol. wt.). Table 1.5 lists some specific apparent volumes of known tastants which pass from salty (at the top of the table) down to bitter. Since specific apparent volume represents apparent displacement of water by unit of sapid substance, the results in Table 1.5 show qualitative changes of response according to compatibility with water structure. Molecules with good compatibility (e.g. salts) may be transportable to deeper layers of receptors than, say, sweet and bitter molecules, which may reach only shallow receptors. In accordance with this suggestion Hiji and Ito (1977) have shown that sweet response may be selectively eliminated in rat tongues by pronase treatment, leaving the other basic tastes unaffected.

In a profound new study of structure: activity relationships in homologous series of sweeteners, van der Heijden et al. (1985a,b) have calculated minimum, maximum and optimum distances between putative AH,B and γ sites. Their calculations lead them to the conclusion that nitroanilines, sulphamates, oximes, isocoumarins and dipeptides bind with at least three different classes of receptor. A similar study with saccharins, acesulphames, chlorosugars, tryptophans and ureas also suggests more than one type of sweet receptor. However, van der Heijden et al. (1985a,b) have confined their calculations to interatomic distances and bond torsion angles. They

Table 1.5. Specific apparent volumes and taste

Substance	Mol. wt.	Apparent molar volume, ϕV (cm^3/mole)	Sp. app. volume (ϕV/mol. wt.)	Taste
Ferric chloride	162.22	26.01	0.160	Iron-salt
Sodium chloride	58.44	17.21	0.294	Salt
Monosodium L-glutamate	169.1	76.86	0.455	"Umami"
Ortho phosphoric acid	98.0	44.64	0.456	Sour
D-Gluconic acid	196.16	101.1	0.515	Sour
Acesulphame K	201.17	160.0	0.592	Sweet
Sodium Saccharin	241.2	142.8	0.592	Sweet
D-Fructopyranose	180.16	107.2	0.595	Sweet
D-Galactose	180.16	109.0	0.605	Sweet
Sucrose	342.3	208.3	0.608	Sweet
Sorbitol	182.17	116.2	0.638	Sweet
Methyl β-D-xylopyranoside	164.13	116.9	0.713	Bitter-sweet
Quinine, HCl	360.9	277.8	0.768	Bitter

assume that the 3,4 α-glycol group is the sweet glucophore of chlorosugars but take little or no account of the role of water in the sweet response (Mathlouthi et al., to be published).

The behaviour of sugar molecules in water solution can be conveniently followed by optical rotation measurements and Shallenberger (1982) has undertaken chiral computations in the sugars by summing individual contributions of asymmetric carbon atoms and the helical contributions of ring substituents. This approach leads to prediction of favoured conformation in sugar molecules and provides a sensitive probe of sugar–water interaction.

Yet another new approach to the study of sweetness is to observe the effect of taste modifiers. Both gymnemic acid and ziziphin depress sweetness and the effect may be related to their surface active properties (Adams 1985). This, once again, underlines the importance of solution properties.

Conclusions

1. The complete chemical interpretation of sweetness has not yet been achieved.
2. In certain types of conformationally defined molecule, glucophores have been tentatively located. This implies that these molecules are orientated in a particular manner on the receptor.
3. All intensely sweet molecules are more persistent than sugars. An explanation of the phenomenon of persistence is based on an orderly localised concentration of stimulus molecules.
4. All sweet effects are mediated by water. Therefore solution properties and hydration of sweet stimulus molecules are now under study in an atempt to elucidate the mechanism of taste chemoreception.

Acknowledgement. This research has been assisted by a recent travel grant from NATO.

References

Adams MA (1985) Substances that modify the perception of sweetness. In: Bills DD, Mussinan J (eds) Characterization and measurement of flavor compounds. ACS Symposium Series No. 289, pp 13–25

Baradi AF, Bourne GH (1951) Localisation of gustatory and olfactory enzymes in the rabbit and the problems of taste and smell. Nature 168: 977–979

Bartoshuk LM (1975) Taste mixtures: Is mixture suppression related to compression? Physiol Behav 14: 643–649

Bartoshuk LM (1977) Modification of taste quality. In: Birch GG, Brennan JG, Parker KJ (eds) Sensory properties of foods. Applied Science, London, pp 5–26

Beck G (1943) Sweetness and molecular volume. Wien Chem Ztg 46: 18–22

Beets MGJ (1978) Structure–activity relationships in human chemoreception. Applied Science, London

Birch GG (1976) Structural relationships of sugars to taste. Crit Rev. Food Sci Nutr 8: 57–95

Birch GG, Catsoulis S (1985) Apparent molar volumes of sugars and their significance in sweet taste chemoreception. Chem Senses 10: 325–332

Birch GG, Mylvaganam AR (1976) Evidence for the proximity of sweet and bitter receptor sites. Nature 260: 362–364
Birch GG, Parker KJ (1982) Nutritive sweeteners. Applied Science, London
Birch GG, Shamil S (to be published) Structure/activity relationships in sweetness. Food Chemistry
Birch GG, Green LF, Coulson CB (1971) Sweetness and sweeteners. Applied Science, London
Birch GG, Cowell ND, Young RH (1972) Structural basis of an interaction between sweetness and bitterness in sugars. J Sci Food Agric 23: 1207–1212
Birch GG, Latymer Z, Hollaway MR (1980) Intensity/time relationships in sweetness: evidence for a queue hypothesis in taste chemoreception. Chem Senses 5: 63–78
Birch GG, O'Donnell K, Musgrave R (1982) Intensity/time studies of sweetness: psychophysical evidence for localised concentrations of stimulus. Food Chemistry 9: 223–237
Clode DM, McHale D, Sheridan JB, Birch GG, Rathbone EB (1985) Partial benzoylation of sucrose. Res 139: 141–146
Cohn G (1914) Die organischen Geschmacksstoffe. Franz Siemonroth, Berlin. Quoted by Moncrieff RW (1967) The chemical senses, 3rd edn. Leonard Hill, London
Compadre CM, Pezzato JM, Kinghorn AD, Kamath SK (1985) Hernandulcin: an intensely sweet compound discovered by review of ancient literature. Science 227: 417–419
Deutsch EW, Hansch C (1966) Dependence of relative sweetness on hydrophobic bonding. Nature 211: 75
de Vos AM, Hatada M, Van der Wel, Krabbendam H, Peerdeman AF, Sung-Hou K (1985) Three dimensional structure of thaumatin I, an intensely sweet protein. Proc Natl Acad Sci USA 82: 1406–1409
Dubois GE, Bunes LA, Dietrich PS, Stephenson RA (1984) Diterpenoid sweeteners. Synthesis and sensory evaluation of biologically stable analogues of stevioside J Agric Food Chem 32: 1321–1325
Dziedzic S (1980) The intrinsic chemistry of sweetness in sugar and sugar analogues. PhD Thesis, Reading University
Fuller WD, Goodman M, Verlander MS (1985) A new class of amino acid based sweeteners. J Am Chem Soc 107: 5821–5822
Grenby TH, Parker KJ, Lindley MG (1983) Developments in sweeteners—2. Applied Science, London
Halpern BP (1986) Human judgements of MSG taste: quality and reaction times. In: Kawamura Y, Kare MR (eds) Umami: physiology of its taste. Marcel Dekker, New York
Hiji Y, Ito J (1977) Removal of sweetness by proteases and its recovery mechanism in rate taste cells. Comp Biochem Physiol A58: 109–113
HMSO (1983) Sweeteners in foods regulations. S1 1211. London
Hough L (1985) The sweeter side of chemistry. Chem Soc Rev 14: 357–374
Kier LB (1972) A molecular theory of sweet taste. J Pharm Sci 61: 1394–1397
Kodama S (1920) Taste. J Tokyo Chem Soc 41: 495–534
Lawrence AR, Ferguson LN (1959) Exploratory physicochemical studies on the sense of taste. Nature 183: 1469–1471
Mathlouthi M, Seuvre AM, Birch GG (to be published) Relationship between the structure and the properties of carbohydrates in aqueous solutions: sweetness of chlorinated sugars. Carbohydrate Res
McBurney D, Bartoshuk LM (1973) Interactions between stimuli with different taste quantities. Physiol 01–1106
Moncrieff RW (1967) The chemical senses, 3rd edn. Leonard Hill, London
Munton SL, Birch GG (1985) Accession of sweet stimuli to receptors. I. Absolute dominance of one molecular species in binary mixtures. J Theor Biol 112: 539–551
Oertly E, Myers RG (1919) A new theory relating constitution to taste. J Am Chem Soc 41: 855–867
Schiffman SS, Gill JM, Diaz C (1985) Methyl xanthines enhance taste: evidence for modulation of taste by adenosine receptor. Pharmacol Biochem Behav 22: 195–203
Shallenberger RS (1963) Hydrogen bonding and the varying sweetness of the sugars. J. Food Sci 28: 584–589
Shallenberger RS (1982) Advanced sugar chemistry. AVI, Westport, Connecticut
Shallenberger RS, Acree TE (1967) Molecular theory of sweet taste. Nature 216: 480–482
Soejarto DD, Kinghorn AD, Farnsworth NR (1982) Potential sweetening agents of plant origin. III. Organoleptic evaluation of *stevia* leaf herbarium samples for sweetness. J Nat Prod 45: 590–599
Stratten GM (1964) Theophrastus. De Sensibus. Bonset and Schippers, Amsterdam
van der Heijden A, van der Wel H, Peer HG (1985) Structure–activity relationships in sweeteners. I. Nitroanilines, sulphamates, oximes, isocoumarins and dipeptides. Chem Senses 10: 57–72
van der Heijden A, van der Wel H, Peer HG (1985b) Structure–activity relationships in sweeteners. II. Saccharins, acesulphames, chlorosugars, tryptophans and ureas. Chem Senses 10: 73–88
Yamaguchi S, Yoshikawa T, Ikeda S, Ninomiya T (1970) Studies on the taste of some sweet substances. Part II. Interrelationships among them. Agric Biol Chem 34: 187–197

Chapter 2

Neurophysiological Aspects of Sweetness

Thomas R. Scott and Barbara K. Giza

Introduction

In the natural environment, sweetness can nearly always be equated with energy. Consequently its detection is associated with a powerful hedonic appeal which declines only as energy needs are met. In this chapter we will first describe the sensory code by which sweet stimuli are represented and then discuss the neural basis of the hedonic response to sweetness and its malleability by physiological need.

The Neural Code for Sweetness

There are multiple means by which sensory information may be encoded in neural activity. The topographic organization of responsive cells, the profile of activity each cell gives to a stimulus array, and the temporal features of each response may all serve to identify stimulus quality and intensity. Sweetness can clearly be distinguished from other taste qualities on the basis of each of these characteristics of the neural code.

Spatial Distribution

The most obvious spatial feature of the activity evoked by a tastant is the physical location of the responsive neurons. Beginning at the receptor surface and extending through each synaptic relay of the taste system, there is a tendency for cells that are grouped together to have similar response properties. For the sweet quality that grouping is often toward the front. Sweet detection thresholds for humans are lowest on the tip of the tongue (Collings 1974; Hanig 1901). Correspondingly, in the first central relay for taste, the nucleus tractus solitarius (NTS), the largest neural responses evoked by glucose are found anteriorly in the macaque (Scott et al. 1986a). This concordance between human psychophysics and neural records in the macaque implies that these subjects may encode sweetness in part through a spatial distribution of evoked activity which is preserved from receptor to CNS. The anatomical data of Beckstead et al. (1980) suggest that any anteroposterior chemotopic arrangement in the macaque NTS is preserved in a lateromedial dimension in the gustatory thalamus, though there are as yet no electrophysiological data to confirm this. At the following synaptic relay in opercular cortex, however, the spatial distribution of taste activity has been investigated, and glucose responsiveness again predominates toward the front of the area (Scott et al. 1986b).

While rodents have served as the subjects of many more taste experiments than the macaque, the limited extent of their gustatory nuclei has prevented a definitive study of chemotopic arrangements. In rat cortex, however, the extent of the taste area is relatively great. Here, Yamamoto et al. (1985) have analyzed the spatial distribution of responsiveness and have concluded that sensitivity to sugar is also represented anteriorly in this species. Since sugar sensitivity is greatest on the anterior tongue of the rat as well (Zotterman 1935), it is reasonable to infer that this chemotopic arrangement obtains across species and at each relay from receptor to cortex.

Thus the representation of sweetness is distinguished first by the physical location of neurons most sensitive to that quality. The next question is whether these cells which respond well to sweet stimuli are functionally distinct from other taste neurons or whether they simply represent the extreme of a continuum of sweet sensitivity.

Most taste neurons respond to a wide range of sapid stimuli. There is general agreement among researchers that taste cell sensitivity is not randomly or homogeneously distributed across the taste dimensions. Controversy remains, however, concerning whether each neuron's profile of responsiveness can be assigned to either a sweet or non-sweet subgroup to the exclusion of the other. Several investigators promote this notion of a sweet (among others) neuron type (Boudreau and Alev 1973; Boudreau et al. 1982; Frank 1973; Frank and Pfaffmann 1969; Frank et al. 1983; McBurney 1978). Some argue, however, that the classification of neurons into types is an artificial gesture born of a need for categorization as an aid to understanding (Erickson 1985; Erickson et al. 1980; Schiffman and Erickson 1980; Woolston and Erickson 1979). To address this issue one must establish the response profiles of a large sample of taste cells to a representative stimulus array, then determine whether those profiles are separable into sweet and non-sweet groups.

Such investigations have now been undertaken on several occasions. Smith et al. (1983) recorded the activity of 31 individual taste neurons in the hamster parabrachial nuclei in response to the application of each of 18 diverse stimuli. Thus for each cell an 18-point response profile was generated. The functional similarity between any two neurons can then be determined by calculating the Pearson product-moment correlation coefficient between the profiles generated by each. The relative similarity of all 31 cells can be described by calculating coefficients between all possible pairs, i.e. $(31 \times 30)/2 = 465$ correlations. The resulting matrix of coefficients may be represented spatially through multidimensional scaling (Shepard 1980). In such a plot the relative similarity of taste cells, as defined by their response profiles to the 18-stimulus array, is represented by the distance which separates them: the more highly correlated the profiles, the closer the neurons that generated them. The two-dimensional space resulting from this analysis is shown in Fig. 2.1. Neurons are numbered according to the basic stimulus which was most effective in activating them: sucrose-best are neurons 1–9, NaCl-best are 10–20, HCl-best are 21–30, and neuron 31 is quinine-best. One clear distinction in this space is between the cluster of neurons

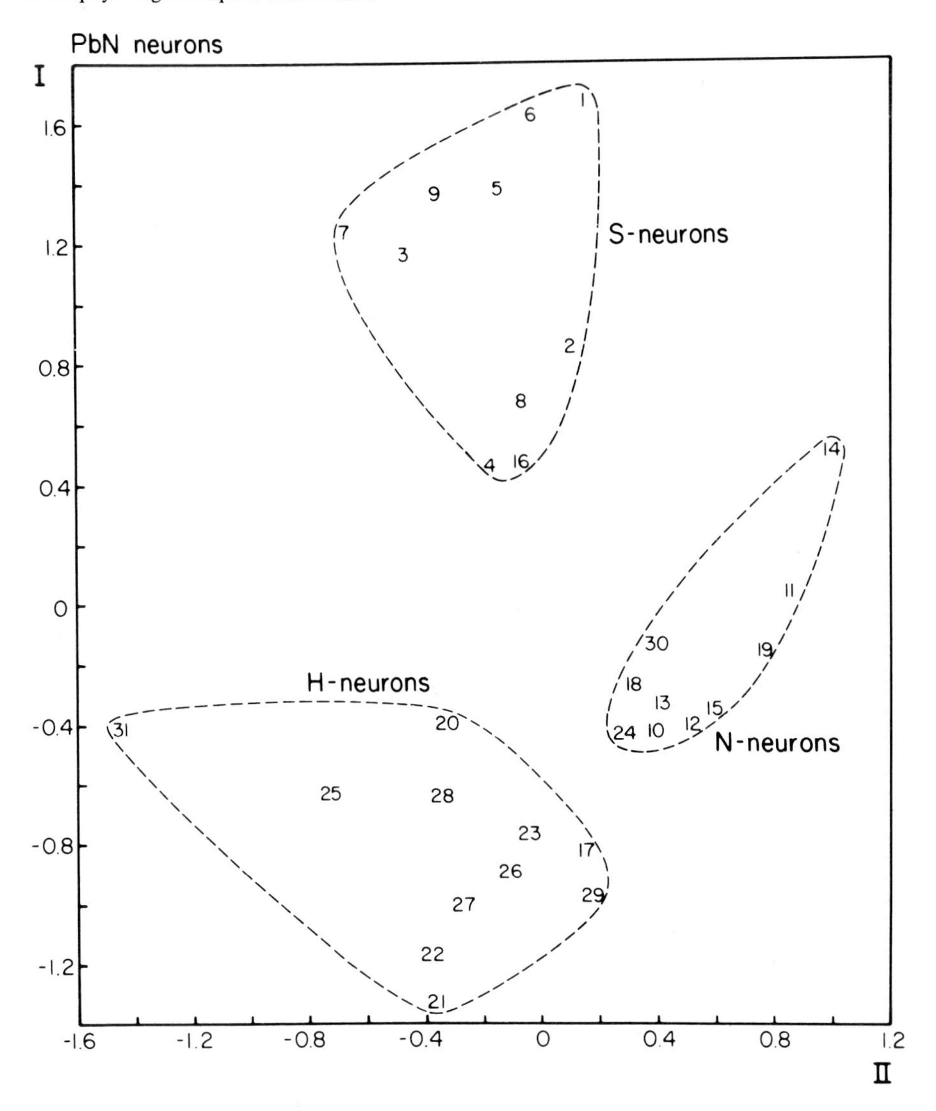

Fig. 2.1. Multidimensional space in which the relative similarities among neuronal profiles are indicated by proximity. Data are from the parabrachial nuclei of the hamster. (Smith et al. 1983)

at the top and the other groups. This proves to be a sweet versus non-sweet dichotomy. The question remains, however, of whether this distinction is sufficiently clear to define sweet and non-sweet sensitive neurons as being qualitatively different.

An objective response to this question may be provided by an analysis of the clusters in Fig. 2.1. Such an analysis is presented in Fig. 2.2 (Everitt 1980). The number of each cell, marked along the right column, corresponds to the number assigned that cell in Fig. 2.1. The position of the line which connects neighboring cells represents the correlation coefficient between the response profiles those two cells generated in response to the array of 18 taste stimuli. Where groups of neurons are connected, the

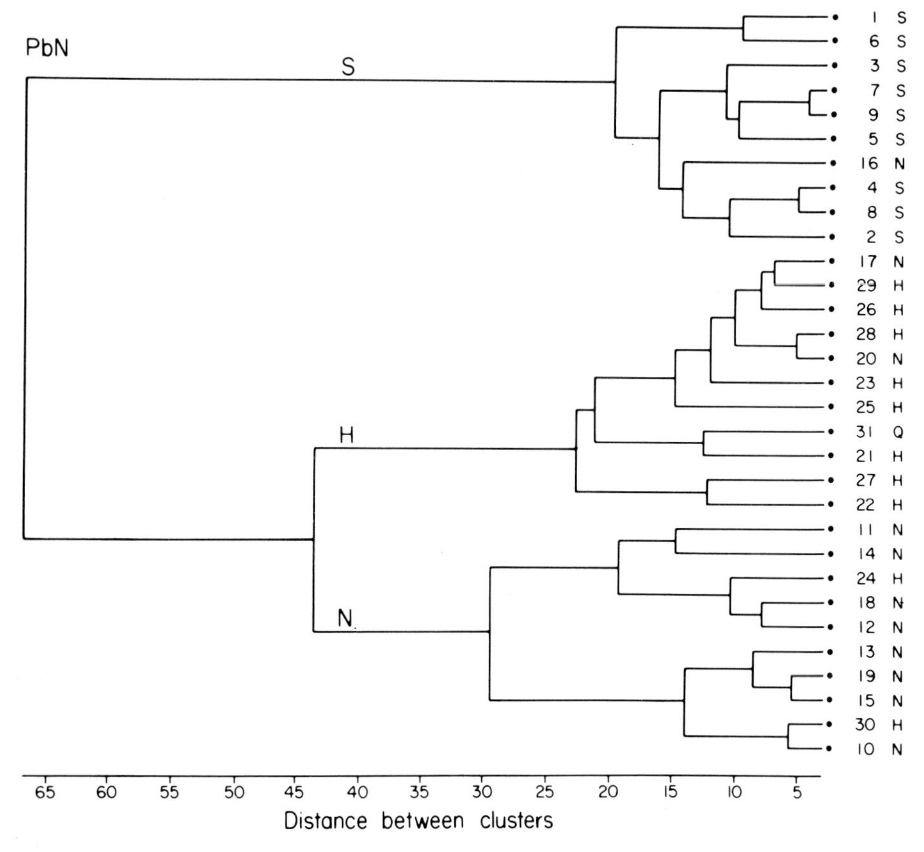

Fig. 2.2. A cluster analysis of the multidimensional space of Fig. 2.1. (Smith et al. 1983)

position of the joining line is set at the mean intercorrelation between two groups. In the resulting "dendrogram" there is the same basic division as was seen in Fig. 2.1 between the ten sweet sensitive neurons[1] at the top and the remaining 21. Thus the most basic functional distinction to be made among taste neurons relates to their degree of responsiveness to sweet stimuli. This is a characteristic which appears to be bimodally distributed, defining the two major classes of taste cells. A similar analysis has been performed by Chang and Scott (1984) with the same results.

Just as each neuron can be characterized by its distinctive response profile across stimuli, so each stimulus evokes its peculiar profile across an array of responsive neurons. It has been suggested that this pattern of activity encodes taste-quality information (Pfaffmann 1941; Erickson 1963); indeed the degree of similarity between two patterns relates well to the perceived similarity between the taste stimuli that generated them as determined by both behavioral (Nachman 1962, 1963; Morrison

1. Since sweet is a human sensation, it cannot technically be referred to in discussing neural responses in the rat. This loose shorthand is used, however, as a widely accepted convenience in describing electrophysiological responses to compounds that humans report as sweet.

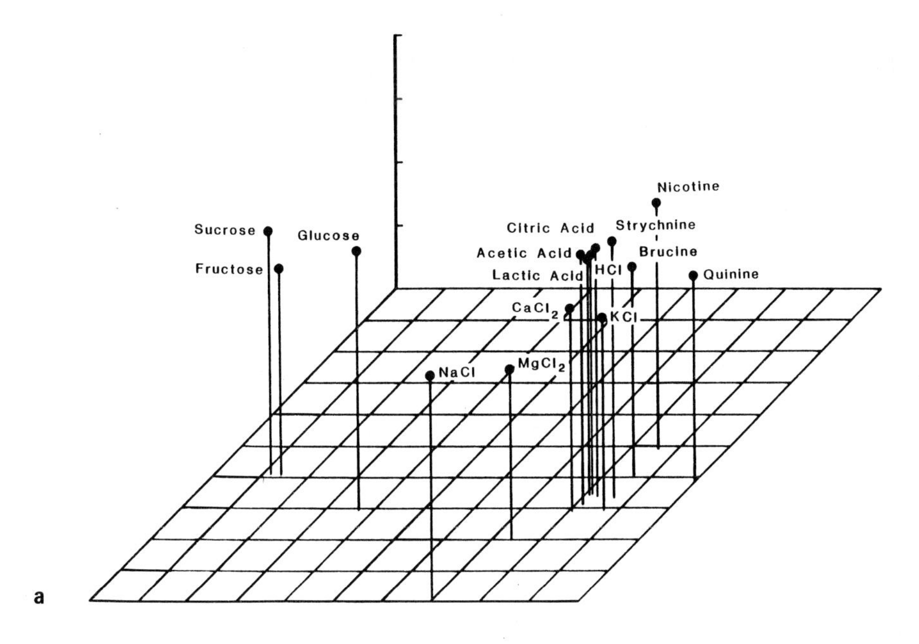

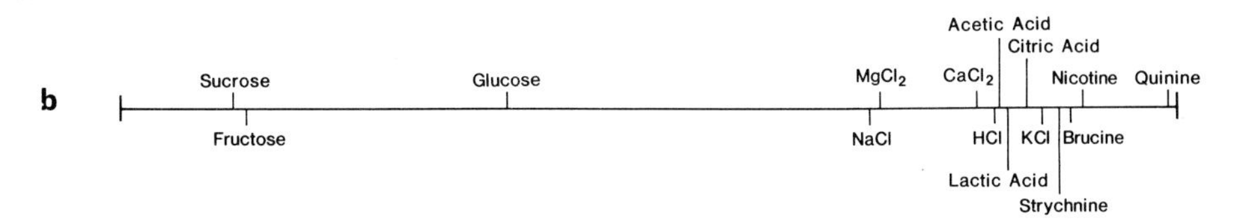

Fig. 2.3a,b. Spatial representations of similarities among 15 diverse taste stimuli based on activity profiles across neurons. **a** Three-dimensional representation accounting for 98% of the data variance; **b** one-dimensional representation accounting for 91% of the variance. The clearest dichotomy is between sweet and non-sweet stimuli in each case.

1967) and psychophysical (Schiffman and Erickson 1971) methods. Thus stimuli can also be characterized according to their profiles.

Several investigators have applied multidimensional scaling routines to similarity measures given by correlations between stimulus profiles (Doetsch and Erickson 1970; Perrotto and Scott 1976; Scott and Erickson 1971; Smith et al. 1983; Woolston and Erickson 1979). The common result of this analysis is that sweet stimuli are rather strictly segregated from non-sweets, with more subtle divisions among non-sweets frequently being possible. This outcome is typified by the data of Mark and Scott (1985), who applied a series of 15 naturally occurring tastants to the rat's tongue while recording evoked single neuron activity in gustatory NTS. For each chemical a profile of activity was generated based on the number of spikes elicited from each of 42 cells during a 5-s poststimulus period. A three-dimensional display of the similarity among these profiles appears in Fig. 2.3a, where the major division between the three sugars and the remainder of the stimuli is apparent. This sweet versus non-sweet dichotomy is shown more clearly when the plot is restricted to the first dimension (which, by itself accounts for 91% of the data variance) as in Fig. 2.3b. Sweet chemicals are to the left and non-sweets, in the general order salt–sour–bitter, are situated to the right.

Temporal Distribution

The preceding analysis recognizes only the spatial distribution of evoked activity: the number of spikes from each cell which accumulate over a fixed period of 5 s. It has been documented that the time course over which this accumulation occurs not only carries reliable information regarding taste quality (Covey and Erickson 1979; DiLorenzo and Schwartzbaum 1982; Nagai and Ueda 1981) but is sufficient to activate appropriate reflexive responses to tastants in behaving rats (Covey and Erickson

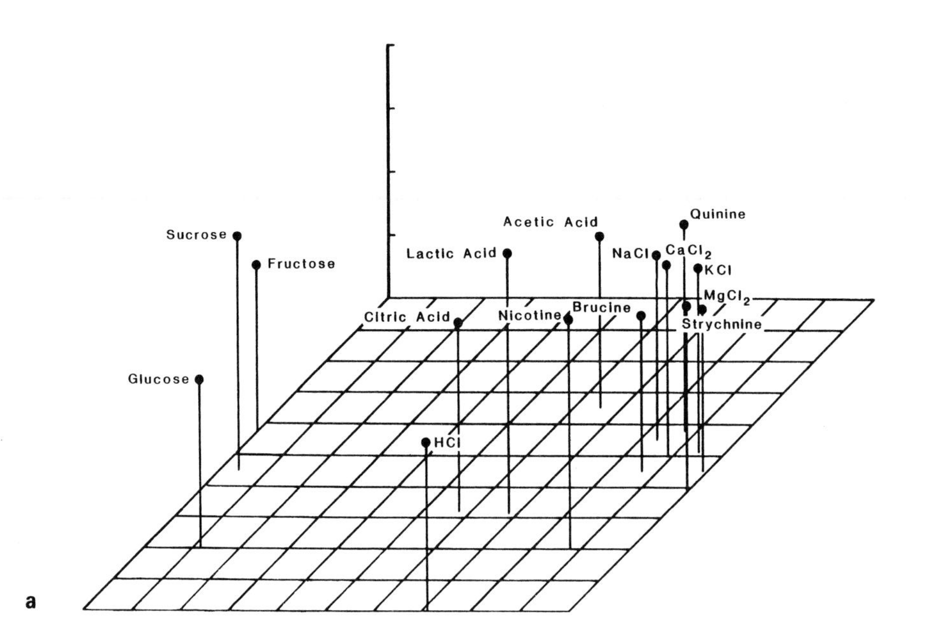

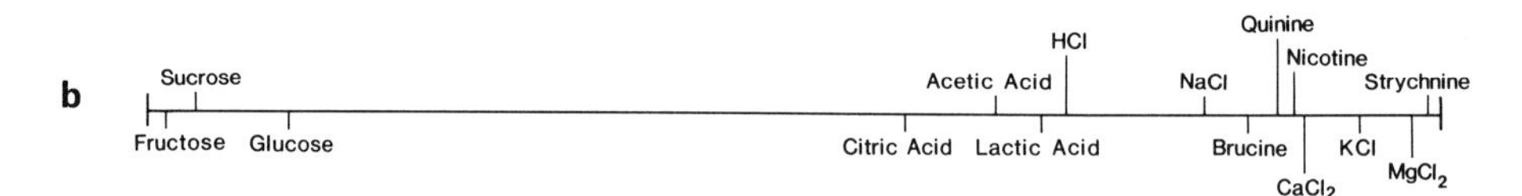

Fig. 2.4. Spatial representations of similarities among the same stimuli based on the time course of evoked activity. **a** Three-dimensional representation accounting for 98% of the data variance; **b** one-dimensional representation accounting for 89% of the variance. The sweet versus non-sweet distinction is clear in this temporal analysis.

1980, unpublished presentation). Thus Mark and Scott (1985) extended the correlational techniques applied to the spatial distribution of activity above, to include a temporal analysis in the NTS. A three-dimensional plot of temporal similarities is shown in Fig. 2.4a. As with the spatial distribution above, the sugars are widely separated from acids, salts, and alkaloids. Collapsing the analysis to a single dimension, which accounts for 89% of the data variance, makes the sweet versus non-sweet dichotomy even clearer (Fig. 2.4b).

The distinctive aspect of the neural time course elicited by sugars in the rat is the lack of a clear phasic component. Whereas salts, acids, and alkaloids typically evoke a sharp burst of activity followed by varying levels of sustained response, sugars elicit little phasic activity and a response which frequently increases over several seconds. It is this reversal of the phasic–tonic time course which sets sugars apart in a temporal analysis.

Conclusion

Information regarding the taste quality of a chemical is carried to and through the nervous system in a spatiotemporal sequence of action potentials. The neural population involved in this signal may include all available cells or be limited to a specialized channel within the taste system. If there are gustatory channels, staffed by functionally distinct neurons, they can be most clearly demonstrated to be specialised for sweet or for non-sweet stimuli (Fig. 2.2). More subtle distinctions among putative cell types specialized for acids or for salts may be possible, but only after the sweet-sensitive neurons have been separated out.

Whether or not an independent sweet channel exists, it is clear that both the spatial (Fig. 2.3) and temporal (Fig. 2.4) aspects of the gustatory neural code emphasize a sweet versus non-sweet dichotomy. Thus the primary characteristic of the neurophysiology of sweetness is its isolation from other taste qualities. The topic of the remainder of this chapter is how this unique quality affects hedonic mechanisms associated with the sense of taste.

Sweetness and the Neurophysiology of Hedonics

General Considerations

Receptors for the sense of taste are situated at the interface between discrimination and digestion. Taste is a chemical gatekeeper of the body, a sense upon whose analysis the decision to swallow or reject potential food is made. Thus the role of gustation transcends that of most sensory systems and extends to involvement in the regulatory processes of feeding. To satisfy the demands of this role the taste sense should be able to identify both nutrients and toxins and have access to mechanisms by which consumption of the former could be encouraged and the latter avoided. At the level of the hindbrain, taste input mediates reflexes through which acceptance or rejection is accomplished (Grill and Norgren 1978b; Steiner 1979). In the diencephalon and telencephalon taste afferents are anatomically associated with areas classically thought to be involved with motivation, emotion, and reinforcement (Norgren 1974, 1976, 1977). These considerations suggest that the afferent code for taste should provide information on the physiological effects of ingesting a chemical with particular reference to the present needs of the organisms. Upon this information, positive or negative hedonic experience would be based.

The spaces of Fig. 2.3 and 2.4 provide a description of how taste qualities are related. Their one-dimensional progression in either the spatial or the temporal representation begins with the sweetness of carbohydrates and ends with the bitterness of alkaloids. This progression corresponds to a physiological dimension of increasing toxicity and to a psychological dimension of decreasing hedonic appreciation. Sweetness, the most neurally distinct of the basic taste qualities, usually represents a nutrient and is associated with positive hedonics. The question to be addressed here is whether the neural activity evoked by a sweet stimulus, and presumably the hedonic value which derives from that activity, is modifiable according to the state of the animal. If a sweet chemical loses its physiological value will its gustatory representation be altered?

Alterations in the Neural Code for Sweetness

Alterations with Experience

An animal's experience may have a pronounced and lasting effect on behavioral reactions to taste stimuli. Preferences may be established by exposure to distinctive tastes during suckling (Capretta and Rawls 1974; Galef and Henderson 1972) or recovery from illness (Revusky 1967, 1974; Young 1966) or even by mere familiarity through regular exposure (Capretta et al. 1973; Domjan 1976). However, the most potent effect of taste experience on subsequent food selection derives from the development of a conditioned taste aversion (Garcia et al. 1955). The aversion established when a novel taste [the conditioned stimulus (CS)] is paired with gastrointestinal distress [the unconditioned stimulus (US)] is so intense and resistant to extinction that the conditioned taste aversion paradigm has itself become a tool for studying taste-related behavior (Bernstein 1978; Smotherman and Levine 1978).

The neural substrates of conditioned taste aversions (CTAs) have been investigated primarily by lesioning brain sites implicated in learning or reinforcement and determining whether subjects retained the capacity to develop aversions. On two occasions, however, recordings were taken from neurons of conditioned rats to determine the effects of a CTA on taste activity evoked by a sweet CS. Aleksanyan et al. (1976) found that the preponderance of hypothalamic activity evoked by saccharin shifted from the lateral to the ventromedial nucleus with formation of a saccharin CTA. They concluded that the taste signal which formerly would have provided the reinforcement associated with lateral hypothalamic activation had, through conditioning, acquired the aversive sensory and motivational properties associated with the ventromedial nucleus.

Chang and Scott (1984) recorded gustatory-evoked activity from single cells of the nucleus tractus solitarius in three groups of rats: unconditioned (exposed only to the taste of the CS, saccharin, with no physiological consequences), pseudoconditioned (experienced only the US, nausea, with no gustatory referent) and conditioned (taste of the saccharin CS paired on three occasions with nausea). They performed comparisons among the responses elicited by an array of 12 stimuli, including the saccharin

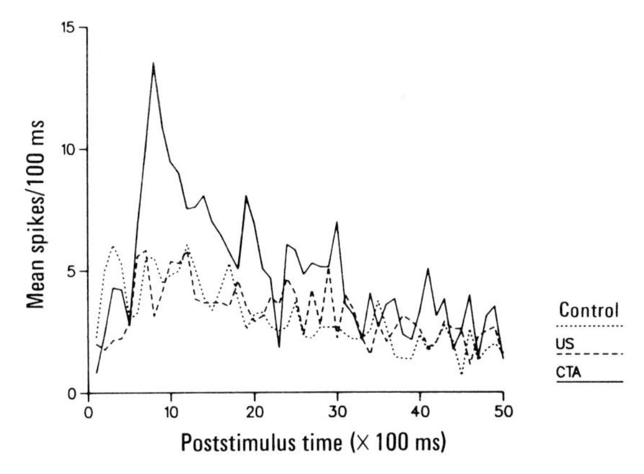

Fig. 2.5. Poststimulus time histograms of the responses of sweet-sensitive neurons in NTS to 0.0025 M sodium saccharin. The activity recorded from unconditioned (*control*) and pseudoconditioned (*US*) rats is typical in response to this weak, sweet stimulus. The peak activity from cells in the conditioned animals (*CTA*) occurs at 900 ms poststimulus, and represents nearly a tripling of control levels. (Chang and Scott 1984)

CS, three saccharides whose taste qualities generalise well to that of saccharin, a more concentrated saccharin solution, and seven salts, acids, and alkaloids through whose responses alteration in the entire system resulting from this taste-learning experience could be evaluated. The primary effect of the conditioning procedure was to increase responsiveness to the saccharin CS through a sharp peak of activity that separated from control recordings at about 600 ms following stimulus onset (Fig. 2.5). Moreover, the effect of this increase was to modify the pattern of activity evoked by the saccharin CS such that it more closely resembled those of non-sweet chemicals, particularly quinine.

Saccharin–quinine proximity may offer a neural concomitant to the increasingly similar behavioral reaction elicited by quinine and sweet chemicals to which an aversion has been conditioned (Grill and Norgren 1978a). Thus in the rat, the neural representation of a sweet stimulus may be altered through experience such that the revised afferent signal is in accord with, and perhaps mediates, the new behavioral reaction.

Alterations with Physiological Need

The detection of a sweet substance, as signalled by the gustatory afferent code, usually elicits a positive hedonic sensation in humans and consummatory behavior in a wide variety of animals. However the hedonic value associated with a taste depends to a considerable degree on the physiological state of the organism. With deprivation, foods become more palatable; with satiety, less so (Cabanac 1971; Richter 1942; Rolls et al. 1981). Can these effects also be attributed to alterations in the gustatory neural code?

There are reports that taste receptor activity in both frogs (Sharma and Doss 1973) and humans (Zaiko and Lokshina 1962) declines as the subject's stomach is distended or as blood glucose levels rise (Budylina and Reztsova 1969). Similarly, peripheral

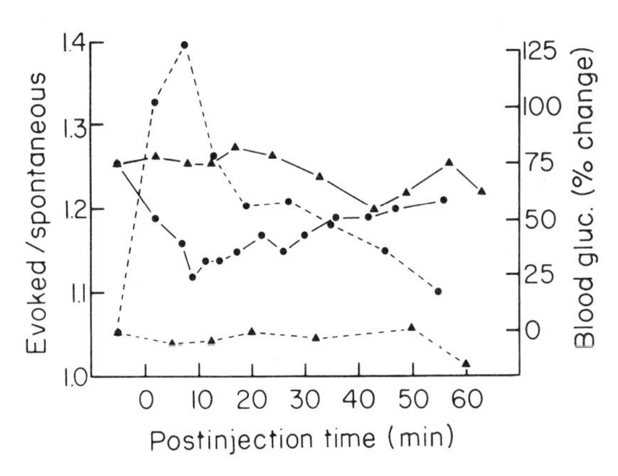

Fig. 2.6. Mean blood glucose levels (*dashed lines*) and gustatory evoked reponses (*solid lines*) to 1.0 M glucose-injected experimental rats (*circles*) and in plasma-injected controls (*triangles*). (Giza and Scott 1983)

taste nerve responses may be altered by the imposition of these same satiety factors in frogs (Dua-Sharma et al. 1973) and toads (Brush and Halpern 1970). These effects extend to the central nervous system. Glenn and Erickson (1976) recorded multiunit activity from the nucleus tractus solitarius (NTS) of freely fed rats under different degrees of gastric distension. They noted a selectively suppressive effect on taste-evoked activity: sucrose responses were most influenced, followed in diminishing order by NaCl, HC1, and quinine HC1, the responses to which were not clearly modified. Relief from distension reversed the effect over a 45-min period.

The administration of other satiety factors has also been shown to influence taste-evoked activity in the NTS. Giza and Scott (1983) infused either 0.5 g/kg glucose (experimental group) or an equivalent volume of the animal's own plasma, withdrawn 2 days earlier (controls), intravenously into rats while monitoring multiunit taste-evoked activity. Elevated blood glucose was associated with a reduction of up to 43% in responsiveness to glucose compared to responses in control rats, with the maximum effect reached 8 min after the intravenous load (Fig. 2.6). As blood glucose levels returned toward baseline over a 60-min period, taste responsiveness to glucose recovered to control levels. Activity elicited by NaCl and HCl was suppressed to a lesser degree and for a briefer period, while quinine-evoked responses were unaffected. Accompanying the reduced gustatory responsiveness in the hindbrain there was a corresponding decrease in perceived intensity for glucose in rats with high blood sugar levels (Giza and Scott 1985a).

Since glucose infusion causes an endogenous release of insulin, which has itself been shown to cause satiety when administered in physiological doses (Anika et al. 1980; VanderWeele et al. 1980; Woods et al. 1979), Giza and Scott (1985b) also studied the effects of hyperinsulinemia on taste responses. The effect of an intravenous injection of 0.5 U/kg regular insulin was similar to that of glucose administration, though less pronounced. Activity evoked by glucose in NTS declined by a maximum 33%, to return to baseline within 35 min. Responses to fructose were similarly affected, but those to NaCl, HCl, and quinine were unmodified.

Therefore gastric distension or injection of glucose or insulin, both of which decrease feeding when administered in the quantities used here, depress taste sensitivity in the rat to sweetness while affecting other taste qualities to a lesser extent or not at all. Neural responsiveness to this representative range of stimuli becomes skewed away from appetitive tastes and toward those which are aversive. Most foods, representing complex mixtures of various taste qualities, would lose hedonic value and termination of a meal would be more likely.

A neural manifestation of the altered hedonic appeal which accompanies changes in the rat's nutritive state has been reported in the hypothalamus, to which taste cells in the caudal brainstem project. Norgren (1970) recorded from lateral hypothalamic neurons under conditions of ad lib feeding or 48-h food deprivation. In the deprived state, cells were responsive to the taste of sucrose, but with satiety this evoked activity declined. Responsiveness to non-sweet taste stimuli was unmodified by deprivation level. Neurons in the same area supported self-stimulation, the rate of which has been shown to be directly proportional to deprivation level (Hoebel and Teitelbaum 1962). It has also been possible to increase the rate of lateral hypothalamic self-stimulation by application of sugar to the rat's tongue (Phillips and Mogenson 1968; Poschel 1968). This implies that the reinforcement derived from electrical activation of the hypothalmus is augmented by an appetitive gustatory input to the same region. Thus it is tempting to conclude that the activity of hypothalamic neurons mediates the reinforcing property of sweet taste, and that suppression of that activity is responsible for the decreased hedonic appeal of sugars which accompanies satiety. The influence of physiological condition on sensory responsiveness may also extend to other senses. Pager et al. (1972) have reported that cells in the rat's olfactory bulb respond to food odors only if the animal is hungry. Burton et al. (1976) have found cells in the lateral hypothalamus which respond to the sight of food only during hunger. Therefore the nutritive state of the rat may affect food-related afferent activity in several modalities, biasing the hedonic value of potential foods and so influencing intake.

If taste activity in the NTS is altered by the rat's nutritive state, then intensity judgments should change with satiety. Just such a result was found in the behavioral test conducted by Giza and Scott (1985a). However, psychophysical studies of human subjects, while not fully consistent among themselves, generally do not support this conclusion. Humans typically report that the hedonic value of appetitive tastes declines with satiety, but that intensity judgments are affected to a lesser extent or not at all (Rolls et al. 1981, 1983; Thompson et al. 1976). Therefore it would be of interest to study the influence of satiety on neural activity evoked at various synaptic levels by sweet stimuli in the human. The closest available approximation to these data may be supplied by subhuman primates. Yaxley et al. (1985) first defined the response characteristics of taste neurons in the NTS of the cynomolgus macaque. Then they monitored gustatory-evoked activity of small clusters of these cells as mildly food-deprived monkeys were fed to satiety with glucose. Taste responsiveness to glucose, NACl, HCl, and quinine HCl was unmodified despite a full behavioural transition from avid acceptance to active rejection upon repletion with the satiating glucose solution. These studies were then extended to the insular and opercular cortices which receive projections from the thalamic taste area and so serve as primary gustatory cortex (Morse et al. 1980; Pritchard et al. 1983, unpublished presentation). While responsiveness to glucose was a bit more prevalent among these highest-order neurons, the effect of feeding the monkey to satiety was the same. As shown in Fig. 2.7, cells maintained their original sensitivity to glucose (and all other basic tastants) despite the total loss of hedonic appeal which accompanied satiety (Scott et al. 1985).

Gustatory projections from primary taste cortex have been traced to the monkey's orbitofrontal region, just anterior to the insular–opercular cortex (E. Rolls, personal

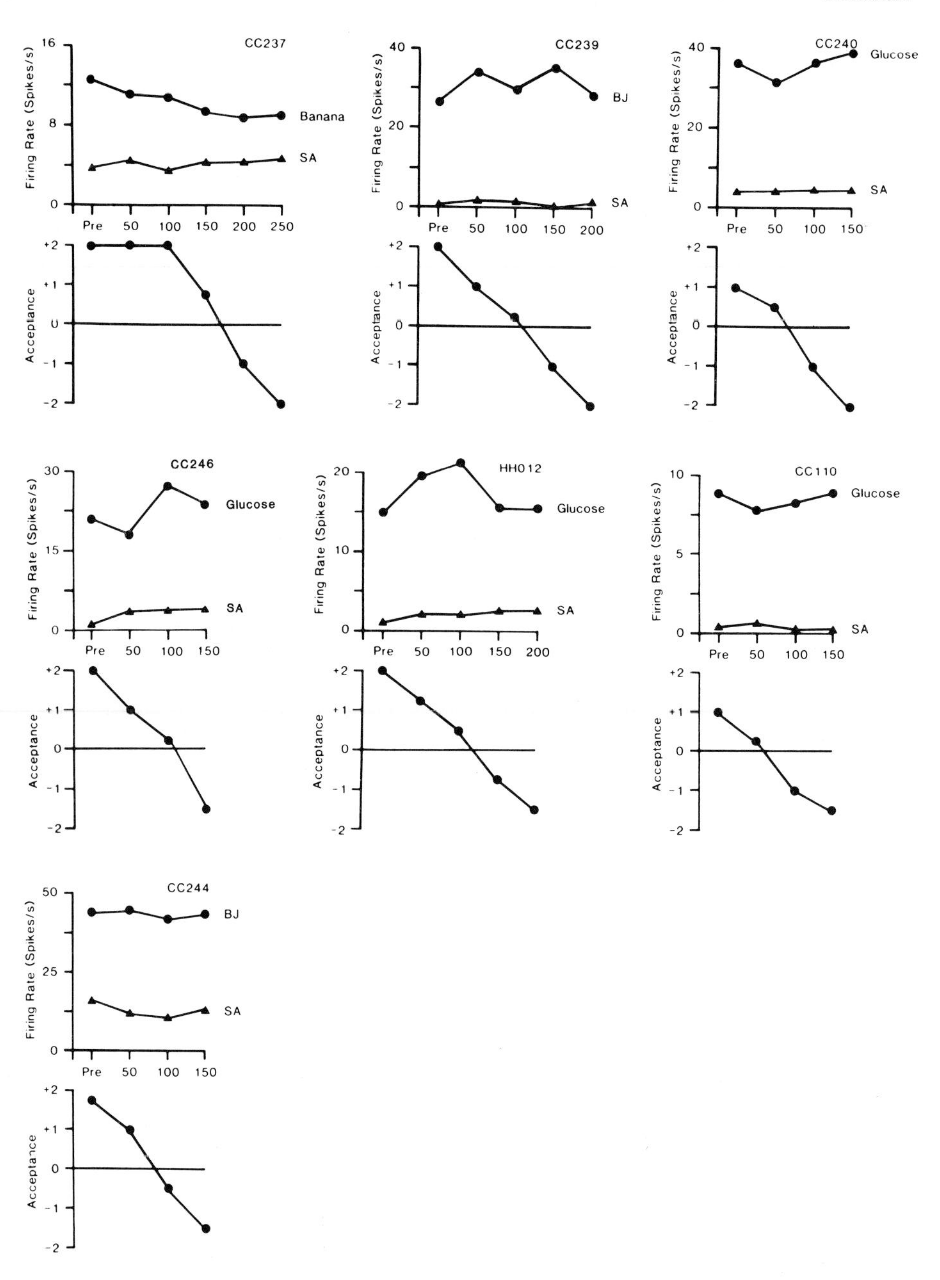

Fig. 2.7. The results of seven independent trials, conducted in insular–opercular cortex, in which a mildly deprived macaque was fed to satiety using either glucose ($n = 5$) or sweet blackcurrant juice (*BJ*; n = 2). Level of acceptance following the consumption of each 50-ml aliquot is indicated on a five-point behavioral scale at the bottom of each frame. The spontaneous activity (*SA*) and evoked response to the satiating stimulus were recorded after each aliquot as well. There was no apparent reduction in sensitivity to the satiating stimulus as satiety was induced. (Scott et al. 1985)

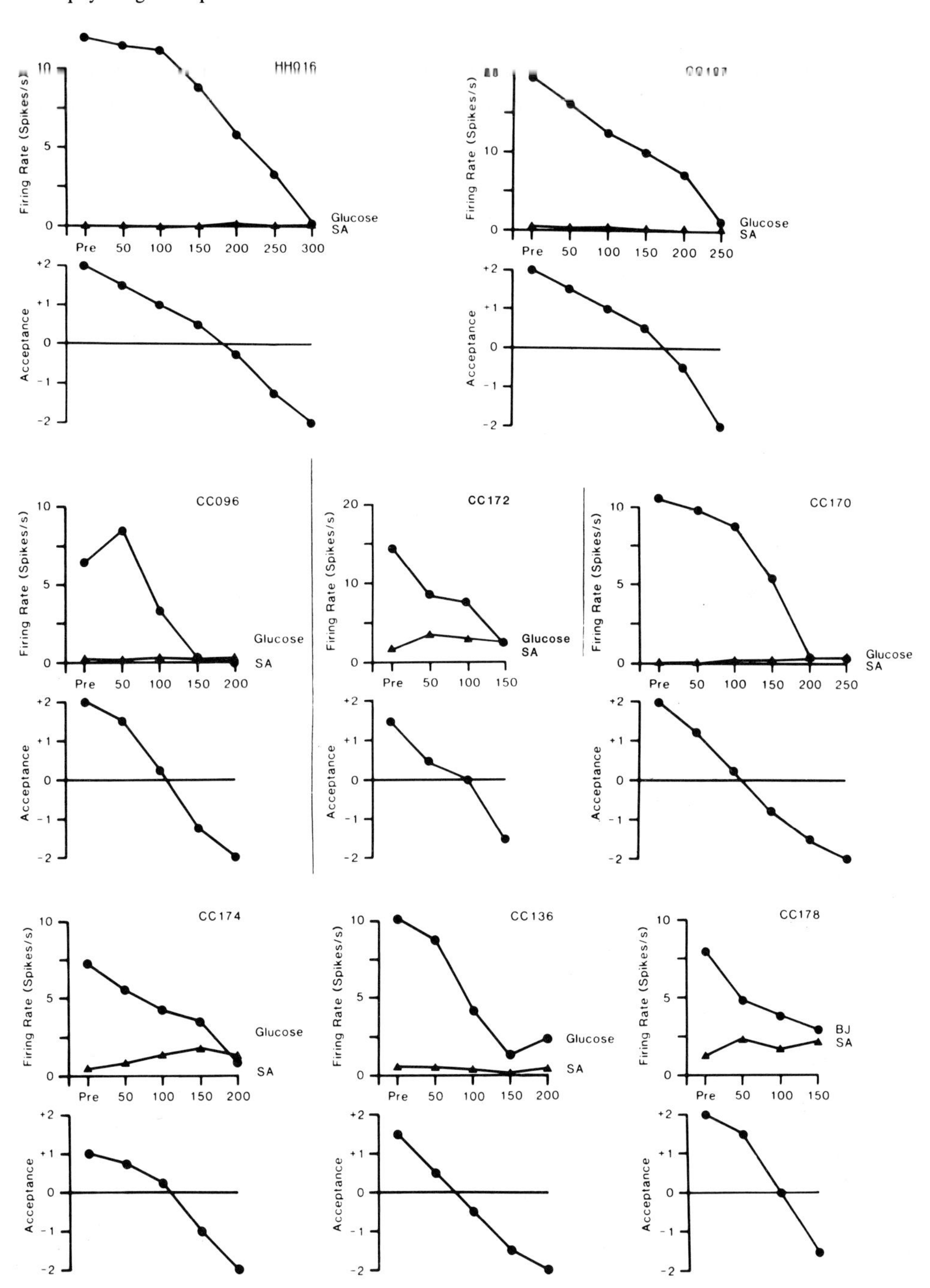

Fig. 2.8. The results of eight independent trials, conducted in orbitofrontal cortex, in which a macaque was fed to satiety using either glucose ($n = 7$) or blackcurrant juice ($n = 1$). All parameters are the same as in Fig. 2.7. The responsiveness of taste cells in the orbitofrontal cortex decreases with increasing satiety. (Rolls et al. 1985)

communication). Electrophysiological investigations of these neurons indicated that they were nearly exclusively sensitive to sweet stimuli and that responsiveness to glucose declined in parallel with behavioral acceptance, as shown in Fig. 2.8 (Rolls et al. 1985). Burton et al. (1976) had previously found corresponding results in the lateral hypothalamus and substantia innominata. At least through primary cortex, then, taste cells respond to intensity independently of hedonic appeal. At subsequent synaptic levels, neurons code exclusively for hedonic value. From these data, it would not be surprising if the macaque joined the human in its capacity to make independent judgments of the intensity and appeal of sweet chemicals.

Conclusion

If the human and macaque can be viewed as representing one phylogenetic level (primates) and the rat another, electrophysiological and behavioral results are consistent. The rat's taste system is influenced by both sweet intensity and hedonics at or before the hindbrain level and its behavior reflects an interdependence of these two variables. In primates intensity and hedonic evaluations show electrophysiological segregation through cortical levels and psychophysical judgments of these variables tend toward independence. Sensory analysis and regulatory physiology may be more closely related in the rat than in the primate.

Summary

Sweetness is the most neurally distinct of gustatory qualities. Its perception is dependent upon neurons which may be physically separable, and clearly are functionally different from cells carrying other tastes, among which less obvious classifications may be possible. The code for sweetness involves afferent activity that is distributed both across neurons and over time in a characteristic pattern sharply different from that representing non-sweet qualities. This pattern initiates acceptance behaviour at nearly any concentration of sugar that occurs in the natural environment, perhaps by activating neurons associated with positive reinforcement in the lateral hypothalamus and elsewhere. Only with clear changes in physiological condition, engendered by experience or satiety, does the reinforcing property of the sweet signal decline. In rats this change appears to involve an alteration in the afferent code for sweet itself; in the macaque it may result from a modification in the properties of neurons associated with reinforcement to which taste axons project.
Acknowledgment. Supported by research grant AM30964 from the National Institutes of Health.

References

Aleksanyan AA, Buresova O, Bures J (1976) Modification of unit responses to gustatory stimuli by conditioned taste aversion in rats. Physiol Behav 17: 173–179

Anika SM, Houpt JR, Houpt KA (1980) Insulin as a satiety hormone. Physiol Behav 25: 21–23
Beckstead RM, Morse JR, Norgren R (1980) The nucleus of the solitary tract in the monkey: projections to the thalamus and brainstem nuclei. J Comp Neurol 190: 259–282
Bernstein IL (1978) Learned taste aversions in children receiving chemotherapy. Science 200: 1302–1303
Boudreau JC, Alev N (1973) Classification of chemoreceptive tongue units of the cat geniculate ganglion.Brain Res 54: 157–175
Boudreau JC, Oravec JJ, Hoang NK (1982) Taste systems of the goat geniculate ganglion. J Neurophysical 48: 1226–1242
Brush AD, Halpern BP (1970) Centrifugal control of gustatory responses. Physiol Behav 5: 743–746
Budylina SM, Reztsova LD (1969) Functional mobility of taste receptors and the blood sugar response. Bull Exp. Biol Med (USSR) 51: 257–262
Burton MJ, Rolls ET, Mora F (1976) Effects of hunger on the responses of neurones in the lateral hypothalamus to the sight and taste of food. Exp Neurol 51: 668–677
Cabanac M (1971) Physiological role of pleasure. Science 173: 1103–1107
Capretta PJ, Rawls LH (1974) Establishment of a flavor preference in rats: Importance of nursing and weaning experience. J Comp Physiol Psychol 86: 670–673
Capretta PJ, Moore MJ, Rossiter TR (1973) Establishment and modification of food and taste preferences: effects of experience. J Gen Psychol 89: 27–46
Chang F-CT, Scott TR (1984) Conditioned taste aversions modify neural responses in the rat nucleus solitarius. J Neurosci 4: 1850–1862
Collings VB (1974) Human taste response as a function of locus of stimulation on the tongue and soft palatc. Pcrccpt Psychophys 16: 169–174
Covey E, Erickson RP (1979) Temporal coding of sensory quality — evidence from single unit taste responses in the rat. Soc Neurosci Abstr 5: 126
DiLorenzo PM, Schwartzbaum JS (1982) Coding of gustatory information in the pontine parabrachial nuclei of the rabbit: magnitude of neural response. Brain Res 251: 229–244
Doetsch GS, Erickson RP (1970) Synaptic processing of taste-quality information in the nucleus tractus solitarius of the rat. J Neurophysiol 33: 490–507
Domjan M (1976) Determinants of the enhancement of flavored-water intake by prior exposure. J Exp Psychol [Anim Behav] 2: 17–27
Dua-Sharma S, Sharma KN, Jacobs HL (1973) The effect of chronic hunger on gustatory responses in the frog. Physiologist 16: 300
Erickson RP (1963) Sensory neural patterns and gustation. In: Zotterman Y (ed) Olfaction and taste. Pergamon, Oxford, pp 205–213
Erickson RP (1985) Grouping in the chemical senses. Chem Senses 10: 333–340
Erickson RP, Covey E, Doetsch GS (1980) Neuron and stimulus typologies in the rat gustatory system. Brain Res 196: 513–519
Everitt B (1980) Cluster analysis. Halstead, New York
Frank M (1973) An analysis of hamster afferent taste nerve response functions. J Gen Physiol 61: 588–618
Frank M, Pfaffmann C (1969) Taste nerve fibers: a random distribution of sensitivities to four tastes. Science 164: 1183–1185
Frank ME, Contreras RJ, Hettinger TP (1983) Nerve fibers sensitive to ionic taste stimuli in chorda tympani of the rat. J Neurophysiol 50: 941–960
Galef BG, Henderson PW (1972) Mother's milk: a determinant of the feeding preferences of weanling rat pups. J Comp Physiol Psychol 78: 213–219
Garcia J, Kimeldorf DJ, Koelling RA (1955) Conditioned aversion to saccharin resulting from exposure to gamma radiation. Science 122: 157–158
Giza BK, Scott TR (1983) Blood glucose selectively affects taste-evoked activity in the rat nucleus tractus solitarius. Physiol Behav 31: 643–650
Giza BK, Scott TR (1985a) Intravenous glucose loads decrease sweet intensity judgments in behaving rats. Chem Senses 10: 449
Giza BK, Scott TR (1985b) The effect of intravenous insulin injections on responsiveness of taste neurons in the rat nucleus tractus solitarius. Chem. Senses 10: 440
Glenn JF, Erickson RP (1976) Gastric modulation of afferent activity. Physiol Behav 16: 561–568
Grill HJ, Norgren R (1978a) The taste reactivity test. I. Mimetic responses to gustatory stimuli in neurologically normal rats. Brain Res 143: 263–279
Grill HJ, Norgren R (1978b) The taste reactivity test. II. Mimetic responses to gustatory stimuli in chronic thalamic and chronic decerebrate rats. Brain Res 143: 281–297
Hanig DP (1901) Zur Psychophysik des Geschmacksinnes. Philos Stud 17: 576–623
Hoebel BG, Teitelbaum P (1962) Hypothalamic control of feeding and self-stimulation. Science 135: 375–377

Mark GP, Scott TR (1985) Taste responses to an extended stimulus array in the rat nucleus tractus solitarius. Chem Senses 10: 440–441
McBurney DH (1978) Pscyhological dimensions and perceptual analyses of taste. In: Carterette EC, Friedman MP (eds) Handbook of perception, vol VIA. Academic Press, New York, pp 125–155
Morrison GR (1967) Behavioral response patterns to salt stimuli in the rat. Can J Psychol 21: 141–152
Morse J, Beckstead R, Pritchard T, Norgren R (1980) Ascending gustatory and visceral afferent pathways in the monkey. Soc Neurosci Abstr 6: 307
Nachman M (1962) Taste preferences for sodium salts by adrenalectomized rats. J Comp Physiol Psychol 55: 1124–1129
Nachman M (1963) Learned aversion to the taste of lithium chloride and generalization to other salts. J Comp Physiol Psychol 56: 343–349
Nagai T, Ueda K (1981) Stochastic properties of gustatory impulse discharges in rat chorda tympani fibers. J Neurophysiol 45: 574–592
Norgren R (1970) Gustatory responses in the hypothalamus. Brain Res 21: 63–77
Norgren R (1974) Gustatory afferents to ventral forebrain. Brain Res 81: 285–295
Norgren R (1976) Taste pathways to hypothalamus and amygdala. J Comp Neurol 166: 17–30
Norgren R (1977) A synopsis of gustatory anatomy. In: LeMagnen J, MacLeod P (eds) Olfaction and taste, vol 6. IRL Press, Washington, DC, pp 225–232
Pager J, Giachetti I, Holley A, LeMagnen J (1972) A selective control of olfactory bulb electrical activity in relation to food deprivation and satiety in rats. Physiol Behav 9: 573–579
Perrotto RS, Scott TR (1976) Gustatory neural coding in the pons. Brain Res 110: 283–300
Pfaffmann C (1941) Gustatory afferent impulses. J Cell Comp Physiol 17: 243–258
Phillips AG, Mogenson GJ (1968) Effects of taste on self-stimulation and induced drinking. J Comp Physiol 66: 654–660
Poschel BPH (1968) Do biological reinforcers act via the self-stimulation areas of the brain? Physiol Behav 3: 53–60
Revusky SH (1967) Hunger level during food consumption: effects on subsequent preference. Psychonom Sci 7: 109–110
Revusky SH (1974) Retention of a learned increase in the preference for a flavored solution. Behav Biol 11: 121–125
Richter CP (1942) Increased dextrose appetite of normal rats treated with insulin. Am J Physiol 135: 781–787
Rolls BJ, Rolls ET, Rowe EA, Sweeney K (1981) Sensory-specific satiety in man. Physiol Behav 27: 137–142
Rolls ET, Rolls BJ, Rowe EA (1983) Sensory-specific and motivation-specific satiety for the sight and taste of food and water in man. Physiol Behav 30: 185–192
Rolls ET, Yaxley S, Sienkiewicz ZJ, Scott TR (1985) Gustatory responses of single neurons in the orbitofrontal cortex of the macaque monkey. Chem Senses 10: 443
Schiffman SS, Erickson RP (1971) A psychophysical model for gustatory quality. Physiol Behav 7: 617–633
Schiffman SS, Erickson RP (1980) The issue of primary tastes versus a taste continuum. Neurosci Biobehav Rev 4: 109–117
Scott TR, Chang F-CT (1984) The state of gustatory neural coding. Chem Senses 8: 97–114
Scott TR, Erickson RP (1971) Synaptic processing of taste-quality information in the thalamus of the rat. J Neurophysiol 34: 868–884
Scott TR, Yaxley S, Sienkiewicz ZJ, Rolls ET (1985) Satiety does not affect gustatory-evoked activity in the nucleus tractus solitarius or opercular cortex of the alert cynopmolgus monkey. Chem Senses 10: 442
Scott TR, Yaxley S, Sienkiewicz ZJ, Rolls ET (1968a) Gustatory responses in the nucleus tractus solitarius of the alert cynomolgus monkey. J Neurophysiol 55: 182–200
Scott TR, Yaxley S, Sienkiewicz, ZJ, Rolls ET (1986b) Gustatory responses in the frontal opercular cortex of the alert cynomolgus monkey. J Neurophysiol 56
Sharma KN, Doss MJK (1973) Excitation and control of gustatory chemoreceptors. Proc 10th Int Conf Med Biol Eng 3: 53
Shepard RN (1980) Multidimensional scaling, tree-fitting, and clustering. Science 210: 390–398
Smith DV, Van Buskirk RL, Travers JB, Beiber SL (1983) Coding of taste stimuli by hamster brain stem neurons. J Neurophysiol 50: 541–558
Smotherman WP, Levine S (1978) ACTH and ACTH4-10 modification of neophobia and taste aversion responses in the rat. J Comp Physiol Psychol 92: 22–33
Steiner JE (1979) Human facial expressions in response to taste and smell stimulation. In: Rese HW, Lipsett L (eds) Advances in child development, vol 13. Academic Press, New York, pp 257–295

Thompson DA, Moskowitz HR, Campbell RG (1976) Effects of body weight and food intake on pleasantness ratings for a sweet stimulus. J Appl Physiol 41: 77–83
VanderWeele DA, Pi-Sunyer FX, Novin D, Bush MJ (1980) Chronic insulin infusion suppresses food ingestion and body weight gain in rats. Brain Res Bull 5: 7–11
Woods SC, Lotter EC, McKay LD, Porte D Jr (1979) Chronic intracerebroventricular infusion of insulin reduces food intake and body weight of baboons. Nature 282: 503–505
Woolston DC, Erickson RP (1979) Concept of neuron types in gustation in the rat. J. Neurophysiol 42: 1390–1409
Yamamoto T, Yuyama N, Kato T, Kawamura Y (1985) Gustatory responses of cortical neurons in rats. II. Information processing of taste quality. J Neurophysiol 53: 1356–1369
Yaxley S, Rolls ET, Sienkiewicz ZJ, Scott TR (1985) Satiety does not affect gustatory activity in the nucleus of the solitary tract of the alert monkey. Brain Res 347: 85–93
Young PT (1966) Influence of learning on taste preferences. Psychol Rep 19: 445–446
Zaiko NS, Lokshina ES (1962) Reflex reaction of the taste receptors of the tongue to direct stimulation of the gastric receptors. Bull Exp Biol Med (USSR) 53: 9–11
Zotterman Y (1935) Action potentials in the glossopharyngeal nerve and in the chorda tympani. Scand Arch Physiol 72: 73–77

Commentary

Beauchamp and Cowart: Running through several chapters, including our own, is the idea that sweetness perception is intimately tied to nutrition through its role as a signal for calories. Scott and Giza make this explicit in their very first sentence ". . . sweetness can nearly always be equated with energy." Yet sugars are not the calorically densest foods, fats are. If sweetness serves principally as a signal for calories, why aren't fats sweet? Is it possible that sweeteners (natural sugars mainly) may be associated evolutionarily with something else in addition to, or instead of, calories? One possibility (J.A. Desor, personal communication) could be that since naturally sweet foods (e.g., fruits) tend to be rich in certain vitamins, minerals, or other nutrients which themselves occur in amounts too small to be easily detected, sweetness may serve as a cue for such micronutrients.

A second issue is raised by the very interesting differences Scott and Giza report between the neurophysiology of rat and monkey responses to sweetness. Do these differences imply that behavioral and neurophysiological studies on rats are of only limited value in advancing understanding of human sensory responses to sweetness?

Würsch: In the study of Giza and Scott (1985) the effect of glucose and fructose were similar, probably because fructose in rat is converted into glucose to a large extent after absorption. A test with a non-glucogenic carbohydrate would have been instructive.

Pangborn: The description of the relationship between taste, nutrition, and hedonics is overly simplistic, and unprovable. It would hold in only a few species, certainly not in humans, who respond favourably to non-sweet nutrients, and to non-nutritive sweeteners.

Chiva: Scott as well as Frijters and Beauchamp elsewhere in this volume keep referring to Steiner's GFR (gustofacial response) as meaning acceptance or rejection of a food, based on taste stimuli. In fact the GFR does not *mean* anything: the real thing is swallowing or spitting out; the GFR is nothing but an accompanying behavior, probably of a reflex nature. The meaning is acquired, in fact constructed, only in a social context on the basis of the interpretations made by the surrounding community. In other words the GFR has to be interpreted, and it is the behavioral modifications it causes which make it effective. The first step is that those who see the child displaying the GFR attribute meaning to the facial expression ("he likes or dislikes") and behave accordingly. Then the subject himself gradually includes the GFR in his system of non-verbal communication and social interaction.

References

Chiva M (1985) Taste. GFR, expression of emotions and non-verbal communication: a developmental study. Presented at the First International Meeting of the International Society for Research on Education (unpublished)

Chiva M (1985) Le doux et l'amer. P.V.F., Paris

Bartoshuk: The achievement of a behavioral generalization gradient based on perceived intensity is of great importance for animal psychophysics. To the best of my knowledge, this is the first time it has been done. I suspect there will be methodological questions because generalization gradients are hard to interpret. I would suggest that the authors try their procedure with a topical treatment that is known to produce reduced perceived intensity to check the methodology. They might paint the rats' tongues with a topical anesthetic (if anyone can work out how to hold the rat's mouth open long enough).

The second half of this chapter provides a synthesis of a variety of papers that was very useful for me and I am sure will be to the readers of our book. The implications for differences between rats and humans in the way the species experience sugars are profound. I would like to add a cautionary note, however. We tend to assume that neurons in the solitary nucleus that respond to chemicals on the tongue are exclusively taste neurons. That may not be true. Some neurons in the solitary nucleus may mediate hedonics in the rat.

Chapter 3

Is Sweetness Unitary? An Evaluation of the Evidence for Multiple Sweets

Linda M. Bartoshuk

Introduction

The nineteenth century concept of sweet as a "basic" taste implied that sweetness was psychologically unitary. But if sweetness is unitary, then why don't all sweeteners taste alike? One possible answer is that various sweetener molecules can be discriminated from one another by non-sweet attributes (e.g., bitterness, slow onset, persistence of sweetness, etc.). Another possible answer is that sweetness itself varies across sweeteners.

Neurophysiological and psychophysical data are relevant to this issue. In both disciplines the key logical approach is to find some treatment or effect that produces differential results across sweeteners because differential effects imply the existence of multiple receptor mechanisms.

The popular view today is that the evidence for multiple sweet receptor mechanisms is overwhelming. Yet a careful reading of much of this evidence leaves lingering doubts and these will be discussed below.

Neurophysiological Evidence for Multiple Sweet Receptor Mechanisms

Single Fiber Recordings

At equimolar concentrations, sugars are not equally effective. There was great interest in this issue in Pfaffmann's laboratory at Brown University in the 1960s and 70s and a variety of Pfaffmann's students worked on this problem (see Pfaffman et al. 1976 for a summary of the neurophysiological studies). Snell found that fructose was a better stimulus than sucrose for the whole nerve of the squirrel monkey, but Bartoshuk (cited in Pfaffmann 1969) found that some single neurons in the squirrel monkey responded better to fructose than sucrose while others did the reverse. These data show that sweeteners interact with more than one kind of receptor mechanism.

In 1974, Frank reported that the neurons that responded better to sucrose than fructose were "sucrose-best" neurons. That is, they responded better to sucrose than to representatives of the other "basic" tastes. On the other hand, the neurons that responded better to fructose than to sucrose were "NaCl-best" neurons. That is, they responded better to NaCl than to the other "basic" tastes.

Behaviorally, squirrel monkeys prefer sucrose to fructose, yet the whole nerve shows fructose to be a more effective stimulus than sucrose. Pfaffmann was interested in how the apparent paradox between these neural and behavioral data could be resolved. If the NaCl-best fibers (those more sensitive to fructose) were to code saltiness while the sucrose-best fibers (those more sensitive to sucrose) were to code sweetness, then fructose would taste salty as well as sweet to the squirrel monkey and thereby be less preferred than the pure sweet sucrose. This led Pfaffmann to replace his earlier "pattern" theory of taste coding with the labeled-line theory of taste coding. Thus the emergence of the labeled-line theory, an event of paramount importance to taste, was stimulated by an effort to understand sweetness.

The labeled-line theory permits the following reasoning. Suppose there were only one type of sweet receptor mechanism present on cells connected to sucrose-best fibers. *Sweetness* would then be mediated by only one kind of receptor mechanism even though *sweeteners* have the ability to stimulate other receptor mechanisms as well. The labeled-line theory would argue that those other receptor mechanisms give rise to non-sweet sensations.

Recent work of Beidler (1983) provides comparisons of the neural responses to six sugars in the rat and hamster. He found not only that there was variability across sugars but also that there was variability across species. He concluded that "more than one specific taste receptor site exists for sugars." Although this is consistent with multiple receptor mechanisms for sweetness, we still cannot eliminate the possibility that the receptor mechanisms that give rise to sweetness are all of the same type.

Substances That Compete with or Damage Sweet Receptor Mechanisms

Sweetness Competitors

Jakinovich and Vlahopoulus (Jakinovich 1983; Jakinovich and Vlahopoulos 1984; Vlahopoulus and Jakinovich 1985) have discovered three substances that, when mixed with sweeteners, inhibit their sweetnesses to various extents. These substances are believed to compete with sweeteners for receptor mechanisms. The effects of these competitors are very different from those of *Gymnema sylvestre* (see below) in that they work only when present simultaneously with the sweetener (i.e., in a mixture). These competitors do not have equivalent effects on all sweeteners. This suggests that the competitors are not interacting with identical receptor populations. Again we must note that this does not demonstrate that *sweetness* is mediated by more than one receptor mechanism. Rather, these results demonstrate that *sweeteners* stimulate more than one type of receptor mechanism. Some of the receptor mechanisms may mediate non-sweet tastes.

Alloxan

Because alloxan destroys pancreatic beta cells, possibly via glucoreceptors in the beta cell membrane, Zawalich (1973) tested its effects on the tongue of the rat to see if it might depress taste responses to glucose. Indeed it depressed neural responses to glucose as well as to xylose, mannose, arabinose, sucrose, and 2-deoxy-D-glucose. However, it failed to depress responses to NaCl, QHCl, sodium saccharin, glycine, and sodium cyclamate. These results are consistent with the existence of multiple receptor sites for sweetness but there are alternate explanations for the results as well. Zawalich notes that the "effects of alloxan can be overcome" if the concentration of glucose is raised. This means that to compare the effects on glucose with those on other sweeteners, all of the sweeteners must be matched for intensity. Since only the responses to glucose were shown, we cannot determine whether or not the stimuli were matched. Thus without this control this study cannot be cited as unequivocal evidence for multiple receptor mechanisms for sweetness.

Proteolytic Enzymes

Proteolytic enzymes digest proteins. Since there is considerable evidence that sweet receptors are proteins, the application of these enzymes should abolish sweet taste. Although some enzymes failed, Hiji (1975) and Hiji and Ito (1977) observed depressed taste responses to sucrose after an application of Pronase E or semialkaline protease to the rat and human tongue. Responses to NaCl, HCl, quinine, and citric acid were unaffected. In the rat, additional sweeteners were tested. Responses to glucose, fructose, sorbitol, and saccharin were eliminated but responses to glycine and DL-alanine were only reduced. Hiji and Ito did not claim that their data demonstrated mutiple sweet receptor mechanisms but others have done so (Faurion et al. 1980; Schiffman et al. 1981a; Lawless and Stevens 1983). In fact, the two sweeteners that were not completely abolished were the most intense (judging by the size of the initial transient response). Thus these data cannot be cited as evidence for multiple receptor mechanisms for sweetness.

Gymnema sylvestre

In human subjects, water extracts of leaves of the plant *Gymnema sylvestre* reduce (or even abolish) the sweet tastes of a variety of sweeteners (Warren and Pfaffmann 1959; Bartoshuk et al. 1969; Kurihara 1971). No precise quantitative studies across a variety of sweeteners have been done. Only one "sweetener", chloroform, has been reported to be unaffected by *Gymnema* (Kurihara 1971) but one might debate whether chloroform is properly considered a sweetener at all. Thus there is no convincing argument to be made from the effects of *Gymnema sylvestre* on human taste for multiple receptor mechanisms for sweetness.

In most other species, including primates (Glaser et al. 1984), *Gymnema sylvestre* is not effective. The exceptions to this are the dog and the hamster (Anderson et al. 1950; Faull and Halpern 1971; Hellekant and Roberts 1983). These results are consistent with the idea that sweetness in humans is mediated by types of receptor mechanisms that most other species lack. However, the mechanism of *Gymnema sylvestre* is not understood. Some years ago, *Gymnema* was thought to compete for sweet receptors but DeSimone et al. (1980) have more recently suggested that its surface-active properties may be responsible for its ability to suppress sweetness in human subjects. If the suppression of sweetness is not receptor specific, then its absence does not imply the lack of receptors.

Miracle Fruit

In human subjects, miracle fruit adds a sweet taste to substances that normally taste sour (Bartoshuk et al. 1969). The sweet taste is believed to result from sugar molecules (arabinose and xylose) on the glycoprotein that is the active constituent of miracle fruit (Kurihara et al. 1969). The presence of acid somehow enables these sugar molecules to come into contact with sweet receptors. Miracle fruit is effective with some primates (Hellekant et al. 1981) but fails to be effective with some lower species (Hellekant et al. 1974). As with *Gymnema*, this could result because the primates have receptors lacked by the lower species. If this were so, then these lower species should lack the ability to taste the actual sweet stimulus in miracle fruit (arabinose and xylose). As far as this author is aware, data are lacking on this point.

Of course, the putative mechanism for miracle fruit could be incorrect. Thus, as with *Gymnema*, we must await an explanation of the mechanism of miracle fruit before we can accept the species differences as evidence for multiple sweet receptor mechanisms.

Saccharin and Sucrose Mixtures

Jakinovich (1981) showed in the gerbil that the addition of saccharin to sucrose produced responses greater than would occur if the saccharin were chemically interchangeable (with regard to sweetness) with sucrose. This implies that the mixture must be stimulating some receptor mechanisms that were not stimulated by sucrose alone.

Species Differences in Responses to Sweeteners

Sugars (especially sucrose) appear to be liked almost universally by members of the mammalian kingdom; however, species vary in their responses to a variety of "artificial" sweeteners. For example, Fisher et al. (1965) noted that rats like saccharin but are indifferent to dulcin while squirrel monkeys like dulcin but dislike saccharin.

Hellekant and his colleagues (Glaser et al. 1978; Hellekant et al. 1981) have evaluated a variety of sweeteners both behaviorally and neurophysiologically in several primates. Their results suggest phylogenetic differences in responses to some sweeteners. In particular only higher primates taste thaumatin (and possibly aspartame) as sweet. This is excellent evidence for the existence of multiple sweet receptor mechanisms.

Psychophysical Evidence for Multiple Receptor Mechanisms

Psychophysics can provide evidence for multiple receptor mechanisms. If multiple receptor mechanisms exist, then the possibility exists that sweetness itself varies across sweeteners. However, multiple receptor mechanisms do not automatically imply that perceived sweetness is not unitary. For example a single taste receptor cell might have more than one kind of receptor site on it, but since excitation of all sites stimulates the same neuron, the sites might not give rise to different sweet sensations. If, on the other hand, the evidence were ultimately to show that all sweetness is mediated by one type of sweet receptor, then sweetness must be unitary. The studies discussed below concern the existence of multiple sweet receptor mechanisms.

Cross-adaptation

The logic underlying cross-adaptation experiments is straightforward. When the tongue is exposed to an adapting solution for a prolonged time, taste sensation fades. When equilibrium is achieved between the stimulus and receptor molecules, sensation has faded to zero. At this point a second solution can be tasted. If it interacts with the same receptor mechanisms as the adapting solution, then the second solution will have no taste (i.e., cross-adaptation has occurred). If the second solution interacts with new mechanisms not stimulated by the adapting solution, then it will have a taste (i.e., cross-adaptation has failed).

Cross-adaptation has been evaluated both across qualities and across substances with a common quality. There is little if any cross-adaptation across qualities (McBurney and Bartoshuk 1973). This is to be expected since different qualities are obviously mediated by different receptor mechanisms. Cross-adaptation occurs for sourness (McBurney et al. 1972) and for saltiness (Smith and McBurney 1969). However, cross-adaptation fails for some bitter stimuli (McBurney et al. 1972), suggesting that the bitter taste is mediated by more than one receptor type. Cross-adaptation among sweeteners has proved to be the most difficult to study. A discussion of the available studies follows.

Self-adaptation

Interpretation of cross-adaptation depends on knowledge about self-adaptation. The earliest systematic studies of self-adaptation are those of Hahn (1949). He found the same classic relation between adapting concentration and threshold for sweeteners that he had earlier discovered for salts. The threshold for sweetness is always located just above the adapting concentration.

Later suprathreshold studies allowed us to examine the effects of adaptation to a sweetener on its psychophysical function. For example, Fig. 3.1 shows self-adaptation to sucrose (Bartoshuk 1968). When the tongue is adapted to water (left function in Fig. 3.1), increasing concentrations of sucrose taste increasingly sweet. When the tongue is adapted to 0.1 M sucrose (center function of Fig. 3.1), that concentration becomes essentially tasteless, the next higher concentration is reduced in sweetness, but lower concentrations take on a new taste quality: bitterness. Similarly, when the tongue is adapted to 0.32 M sucrose (right function in Fig. 3.1), that concentration becomes tasteless and only the 1.0 M sucrose remains sweet. Note that the bitter taste is now much more noticeable.

In summary the adapting concentration determines the bottom of a U-shaped function. Concentrations above the adapting concentration have the typical taste of the solute while those below it take on a different quality called a "water taste".

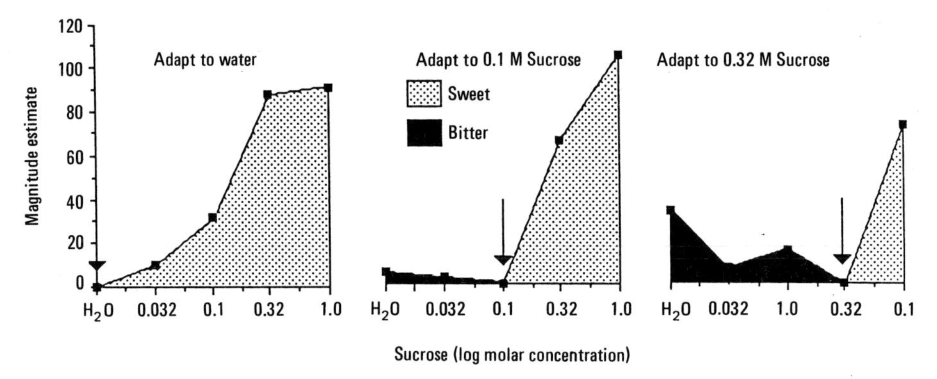

Fig. 3.1. Magnitude estimates of the perceived intensities of sucrose and water under three adaptation conditions: 1. (left function) adaptation to water; 2. (center function) adaptation to 0.1 M sucrose; and 3. (right function) adaptation to 0.32 M sucrose. Arrows indicate the adapting solution.

Adaptation produces a steepening of the sweetness function. For example, note on the center function in Fig. 3.1 that adaptation to 0.1 M sucrose pulls the sweetness to zero at that concentration but by 1.0 M sucrose the sweetness is essentially normal. The adapted function is said to show "recruitment" in analogy with recruitment deafness in the auditory system.

Artifacts in Cross-adaptation Studies

The properties of self-adaptation determine what can be inferred from cross-adaptation. In order to evaluate the meaning of the existing cross-adaptation data, we must consider the kinds of artifacts that complicate the studies.

Unequal Sweetness

First, consider the consequences of testing two concentrations of sucrose for cross-adaptation. If we adapt to the stronger one, then the weaker will be either tasteless or bitter. In either case we will conclude that cross-adaptation failed to occur. However, if we adapt to the weaker, because the adapted function steepens, even a sucrose concentration that is only slightly above the adaptation concentration will taste quite sweet. That is, even when testing two stimuli (i.e., two concentrations of sucrose) that we know stimulate the same receptors, we can observe an asymmetric failure of cross-adaptation. In a true cross-adaptation study the two sweeteners cannot be equated for concentration because we do not know what the actual stimulus is, therefore we equate for perceived sweetness. The logic behind this is that if the two stimulate the same receptors then equi-sweet concentrations will be equivalent. If we make even a small error, then we can expect asymmetric cross-adaptation. That is, adaptation to one substance will completely cross-adapt a second substance but adaptation to the second substance will only partially cross-adapt the first substance.

Incidentally, sweeteners with side tastes (e.g., bitter) may be virtually impossible to equate (for sweetness) with sweeteners with relatively pure sweet tastes. This is because mixture suppression may be operating within the complex sweetener. For example, we know that when sucrose and quinine are mixed, the mixture is less sweet and less bitter than the unmixed components. We also know that at least some of this suppression occurs in the central nervous system (Lawless 1979; Kroeze and Bartoshuk 1985). Thus a bitter–sweet sweetener may have more effective sweet stimulus at the receptors than the perceived sweetness suggests. Equating for perceived sweetness will then produce asymmetric cross-adaptation even if the stimulus for sweetness is identical in the two molecules.

Failure of Self-adaptation

The discussion above is based on complete adaptation which means that the taste of the adapting solution fades to zero during constant stimulation. This requires laboratory conditions. Even with special care, complete adaptation may not occur with all subjects (Meiselman 1975; Meiselman and DuBose 1976). The most successful method has been the use of the McBurney gustometer. This consists of a water bath (to keep solutions at the temperature of the extended tongue, 34°C) and tubing to allow gravity flow of solutions across the extended tongue.

If we produce incomplete adaptation to a given concentration of sucrose, then that concentration will still have a taste. Thus we could appear to get symmetric failure to cross-adapt even when both stimuli affect identical receptors.

Water Taste—Sweet

Adaptation to some sweeteners, especially those with "persistent" sweetness like thaumatin, monellin, and neohesperidin dihydrochalcone, appears to cause sweet water taste (e.g., see Van der Wel 1972). The mechanisms are unknown but the consequences are all too clear. If adaptation to one sweetener causes water to taste sweet, then a second sweetener's taste may be the sum of the solute's sweetness and the water solvent's sweetness. Thus cross-adaptation may appear to fail when in fact it did occur.

Water Taste—Bitter

Adaptation to conventional sweeteners like sucrose tends to make water taste bitter. These water tastes also affect cross-adaptation studies. If adaptation to one sweetener makes water taste bitter then a second sweetener might taste less intense through mixture suppression (Bartoshuk 1975).

Mixture Suppression

Mixture suppression could produce artifacts in another manner. Suppose we have two sweeteners equated for perceived sweetness. One of these (S) is pure sweet while the other is sweet and bitter (SB). Assume that adaptation to S partially cross-adapts the sweetness of SB. When we adapt to S (and remove some of the sweetness from SB), we "release" the bitterness because it was partially suppressed by the sweetness (mixture suppression). The bitterness is then available to suppress the remaining sweetness in SB further. Thus the adapting efficacy of S may appear to be greater than it really is.

Note that these artifacts operate in both directions. The unwary experimenter may be led into concluding that cross-adaptation has occurred when it has not or into concluding that cross-adaptation has failed when it has not. The following discussion examines the published cross-adaptation studies with regard to these artifacts.

Cross-adaptation studies

Three cross-adaptation experiments on sweeteners share a common design (McBurney 1972; Schiffman et al. 1981a; Lawless and Stevens 1983); one concentration of the adapting sweetener is tested on one concentration of the test sweetener. All three studies agreed on one point: sugars are the best cross-adaptors. These studies all found some apparent evidence for some failures of cross-adaptation. However, the amount of this evidence and the interpretations of the data varied.

McBurney (1972)

McBurney found that adaptation to sucrose cross-adapted other sugars as well as a variety of other sweeteners. Adaptation to saccharin was less effective. McBurney concluded that this should not be taken as evidence for failure to cross-adapt because saccharin did not self-adapt as well as sucrose did.

Schiffman, Cahn, and Lindley (1981)

Schiffman et al. (1981a) tested glucose and a variety of other sweeteners. Although as with McBurney's study, the sugar (glucose) was the most effective cross-adaptor, this study reported many apparent failures of cross-adaptation. The authors concluded that more than one receptor mechanism is required to mediate sweetness. Unfortunately, this study failed to demonstrate self-adaptation and also failed to match sweeteners for perceived sweetness (see Lawless and Stevens 1983, for a discussion). Both of these failures would tend to produce apparent failure to cross-adapt even if cross-adaptation occurred. Incidentally, Schiffman et al. observed several cases of "enhancement". That is, after adaptation to one sweetener, another sweetener actually appeared to get sweeter. This unusual result awaits explanation.

Lawless and Stevens (1983)

Lawless and Stevens analyzed and attempted to correct the problems in the study of Schiffman et al. They flowed solutions across the extended tongue (which maximizes self-adaptation) and matched stimuli for perceived sweetness. Lawless and Stevens did not test exactly the same set of sweeteners as Schiffman et al., but there was some overlap. In particular both studies tested aspartame, saccharin, and neo DHC. Schiffman et al. did not observe cross-adaptation between any two of these sweeteners. Lawless and Stevens found mutual cross-adaptation between aspartame and neo DHC and found asymmetric cross-adaptation between the other two pairs. Saccharin cross-adapted aspartame and neo DHC cross-adapted saccharin. Interestingly enough the failures to cross-adapt were actually in the direction of cross-adaptation but did not achieve statistical significance. Note that, as expected, the improved procedure produced less apparent failure to cross-adapt.

Other Cross-adaptation Data

Two additional studies describe cross-adaptation between sucrose and other sweeteners. Van der Wel and Arvidson (1978) observed symmetric cross-adaptation among sucrose, thaumatin, and monellin on small areas of the tongue and on single papillae. This observation is especially important because the use of small areas may reduce problems produced by non-sweet tastes. McBurney and Gent (1978) observed that sucrose cross-adapted fructose, which confirms McBurney (1972). Incidentally, the general effetiveness of sugars as cross-adaptors might be partly an artifact itself (see "Water Taste—Bitter").

Asymmetric Cross-adaptation

Lawless and Stevens observed cases of asymmetric cross-adaptation. For example, in their experiment 1, adaptation to saccharin cross-adapted aspartame but adaptation to aspartame left saccharin unchanged. Lawless and Stevens note that asymmetric cross-adaptation is hard to explain. If cross-adaptation occurs because two stimuli interact with the same receptor sites, then adaptation should be symmetric. Asymmetric cross-adaptation could, of course, be a statistical artifact. In any given experiment, noise in the data may obscure some effects. More subjects and more replications would determine if this were the case. However, as indicated above, asymmetric cross-adaptation could result if stimuli are not precisely equated for sweetness and this is extraordinarily hard to do. When stimuli have side tastes we do not know how to equate for sweetness. In addition, concentrations that appear to be equi-sweet for a group may fail badly for individuals.

Summary and Suggestions

Failure to cross-adapt is the more dramatic result because it implies independent receptor mechanisms. Unfortunately, since errors in cross-adaptation experiments produce, for the most part, apparent failure to cross-adapt, this is also the most uncertain conclusion. McBurney's original caution interpreting his apparent failures to cross-adapt was very well taken. Although the large number of examples of failure to cross-adapt tends to be convincing simply by the weight of numbers, these phenomena need much more attention. One remedy for the problems of interpretation is to test the effects of cross-adaptation on the whole psychophysical function and not just on one concentration. This permits a "strong" test of the failure of cross-adaptation. If the adapting sweetener does not stimulate the same population of receptors as the test sweetener, then even a high concentration of the adapting sweetener should fail to cross-adapt a weak concentration of the test sweetener. Another useful strategy is to use the known genetic difference across individuals (see below) to avoid the problems of bitterness in some sweeteners.

Sweetness on Various Tongue Loci

Van der Wel and Arvidson (1978) noted differences in the spatial distribution of some sweeteners. Sucrose was sweeter at the tip of the tongue than on the lateral edges while thaumatin and monellin were sweeter on the lateral edges.

Similar observations of the sweetness of neo DHC have been made anecdotally by several investigators. Neo DHC is perceived to be sweetest in the rear of the mouth.

This is a strong argument for more than one sweet receptor mechanism and deserves further study.

Sweetness and PTC/PROP Status

The detection thresholds for phenylthiocarbamide (PTC) and other chemically similar compounds like propylthiouracil (PROP) form a bimodal distribution (Harris and Kalmus 1949; Fischer 1971). Those individuals with the highest thresholds (i.e., "taste blind" or "non-tasters") are believed to be homozygous for a recessive gene. Those with lower thresholds (i.e., "tasters") are believed to be heterozygous or homozygous for a dominant gene. Initially, the difference between tasters and non-tasters was believed to be restricted to a single lock-key bitter mechanism. Non-tasters appeared to have a markedly reduced number of receptors for PTC/PROP or none at all. However, later studies showed that PTC/PROP status was also related to the ability to taste chemically unrelated bitter compounds and even sweet compounds (Fischer 1971; Bartoshuk 1979).

Gent and Bartoshuk (1983) examined the sweetness of sucrose and neo DHC in tasters and non-tasters. When the stimuli were delivered through a McBurney gustometer (which stimulates only the front of the extended tongue), sucrose tasted 2.3 times as sweet to tasters as it did to non-tasters. For neo DHC the ratio was 3.2. These ratios were significantly different. In addition, tasters found the highest concentrations of neo DHC to be sweeter than sucrose. Non-tasters found the reverse. These results support the conclusion that neo DHC and sucrose do not stimulate identical receptor populations.

The Artichoke Effect

Some individuals experience a taste-modifying effect from constituents in the globe artichoke (Blakeslee 1935; Bartoshuk et al. 1972). After exposing the tongue to artichoke or the purified active constituents (potassium salts of chlorogenic acid and cynarin), water tastes sweet for several minutes to some individuals. This effect is believed to be genetic but the mode of inheritance is unknown. The possibility exists that this effect may depend on a type of sweet binding site that is present in some individuals and not in others, but there is no direct evidence for this.

Correlations Across Sweeteners

Faurion et al. (1980) looked for correlations between pairs of sweeteners using two taste measures: recognition threshold and suprathreshold matching to a sucrose standard. If there were only one type of sweetness receptor mechanism, then all subjects would show the same relative sensitivities across sweeteners and all sweeteners would be correlated. On the other hand, if there is more than one kind of receptor mechanism and different subjects do not possess equivalent numbers of these, then subjects will not all show the same relative sensitivities to sweeteners. That is, at least some will fail to be correlated.

Faurion et al. found little association among sweeteners at threshold but found some significant positive correlations for suprathreshold sweetness. Surprisingly enough they also found some significant negative correlations. Since a correlation of zero means independence of receptors and a significant positive correlation means that a pair of sweeteners stimulate at least some common receptors, the meaning of a negative correlation is unclear.

One difficulty with this study is the lack of tests of the reliability of the taste measures. An unreliable measure would reduce the correlations. This would appear erroneously to be evidence for independence of sweetener mechanisms. In fact, this study found only 6 (out of 66) statistically significant positive correlations. Some of the pairs of sweeteners that failed to produce significant correlations or that produced negative ones have been shown to cross-adapt. For example, sucrose and thaumatin produced the largest negative correlation (−0.83). Yet Van der Wel and Arvidson found that sucrose and thaumatin cross-adapted one another.

Although the failure of sweeteners to be correlated in this study cannot be taken as evidence of independent receptor sites, the occurrence of a significant positive correlation should still be meaningful. For example, sucrose and saccharin produced a correlation of 0.73. Both McBurney and Lawless and Stevens found some cross-adaptation between the two. The data of Faurion et al. support the conclusion that sucrose and saccharin stimulate some of the same receptor sites.

Sweetness and Diabetes

Settle (1981) tested non-diabetic relatives of diabetics versus controls with no history of diabetes in their families. This design avoids the problem encountered with diabetics of finding taste loss due to pathology produced by the disease. Settle found a subgroup (four out of ten) of relatives of diabetics to have a loss of the ability to taste glucose relative to fructose. This suggests that glucose and fructose do not stimulate identical receptor populations.

Mixtures

If all sweeteners interacted with a single receptor mechanism, then they would be interchangeable in mixtures. This is not the case.

Bartoshuk and Cleveland (1977) selected equally sweet concentrations of sucrose, glucose, fructose, and maltose. These were added to form the six possible two-component mixtures, the four possible three-component mixtures, and the one four-component mixture. The mixtures were compared with the original concentrations and twice, three times, and four times the original concentrations. In addition, the experiment was conducted with two different methods of taste stimulation: dorsal flow (solutions warmed to 34°C) and sip and spit (solutions at room temperature). In general the mixtures were about as sweet as the appropriate multiples of the original concentrations. However, there was one exception. The average of the three-component mixtures (dorsal flow procedure) was significantly higher than the average of the third concentration of each of the sugars, that is, synergism occurred. This suggests that the sugars may not have been acting as simple substitutes for one another. If the populations of receptors stimulated by each of the sugars were not completely identical, then combinations of the sugars could produce sweetness greater than that produced by the appropriate unmixed concentration. However, it should be noted that any superiority of the mixtures was small at best. The more impressive result is how close the sugars come to simply substituting for one another.

Sweetness and Age

Schiffman et al. (1981b) measured taste thresholds and the slopes of psychophysical functions for a variety of sweeteners. Elderly subjects produced significantly higher thresholds for five of the 11 sweeteners. The authors related the degree of elevation of those thresholds to the number of AH,B groups proposed by the Shallenberger-Acree (1967) theory of sweetness. They concluded that the simplest stimuli (those with only one possible type of AH,B system) showed the greatest elevation with age. At suprathreshold concentrations, three of the stimuli (thaumatin, rebaudioside, neo DHC) produced functions with very low slopes (i.e., "flat" functions). The authors concluded that the sweeteners were those "with a considerable number of possible AH,B system types."

Other interpretations of these data are possible. Although the authors conclude that flattened psychophysical functions suggest decreased perceived intensity (i.e., that the function is flattened against the zero taste baseline), this is only an assumption. Flattened functions could also reflect elevated perceived intensities at low concentrations (Bartoshuk et al. 1986). The sweeteners with flat slopes tend to produce sweetness that lingers. In elderly subjects who might have difficulty rinsing as well as younger subjects, this would leave a lingering sweet taste in the mouth that would elevate the lower portion of the psychophysical function, making the whole function appear to flatten.

These considerations make the age data less convincing as evidence for multiple sweet receptor mechanisms.

Sweetness and Reaction Time

Yamamoto et al. (1984) have measured the reaction time necessary to discriminate between pairs of sweeteners. By this measure, sucrose, fructose, glucose, maltose, sorbitol, and aspartame grouped together. Saccharin and cyclamate formed another group. DL-alanine, glycine, neo DHC, and stevioside did not belong to either group nor did they group together. These data demonstrate the temporal differences across sweeteners. If temporal differences reflect events at the sweet receptor site, then this method of grouping should ensure that stimuli in different groups have different mechanisms. However, we might find that sweeteners within a group do not all excite the same receptors.

Summary

The belief that sweetness is mediated by more than one kind of receptor mechanism has been gaining ground. Many different phenomena, neurophysiological as well as psychophysical, seem to support the idea. Yet the weight of all of these studies should not allow us to overlook the weakness of some of the arguments. The discussions above have been directed toward evaluating these weaknesses.

References

Andersson B, Landgren S, Olsson L (1950) The sweet taste fibres of the dog. Acta Physiol Scand 21: 105–119

Bartoshuk LM (1968) Water taste in man. Percept Psychophys 3: 69–72

Bartoshuk LM (1975) Taste mixtures: Is mixture suppression related to compression? Physiol Behav 14: 643–649

Bartoshuk LM (1979) Bitter taste of saccharin: related to the genetic ability to taste the bitter substance 6-*n*-propylthiouracil (PROP). Science 205: 934–935

Bartoshuk LM, Cleveland CT (1977) Mixtures of substances with similar tastes: a test of a new model of taste mixture interactions. Sen Proc 1: 177–186

Bartoshuk LM, Dateo GP, Vandenbelt DJ, Buttrick RL, Long L (1969) Effects of *Gymnema sylvestre* and *Synsepalum dulcificum* on taste in man. In: Pfaffmann C (ed) Taste and olfaction. III: Rockefeller Press, New York, pp 436–444

Bartoshuk LM, Lee CH, Scarpellino R (1972) Sweet taste of water induced by artichoke (*Cynara scolymus*). Science 178: 988–990

Bartoshuk LM, Rifkin B, Marks LE, Bars P (1986) Taste and aging. J Gerontol 41: 51–57

Beidler L (1983) Multiple receptor sites (abstr). Chem Senses 8: 244

Blakeslee AF (1935) A dinner demonstration of threshold differences in taste and smell. Science 81: 504–507

DeSimone JA, Heck GL, Bartoshuk LM (1980) Surface active taste modifiers: a comparison of the physical and psychophysical properties of gymnemic acid and sodium lauryl sulfate. Chem Senses 5: 317–330

Faull JR, Halpern BP (1971) Reduction of sucrose preference in the hamster by gymnemic acid. Physiol Behav 7: 903–907

Faurion A, Saito S, MacLeod P (1980) Sweet taste involves several distinct receptor mechanisms. Chem Senses 5: 107–121

Fischer R (1971) Gustatory, behavioral and pharmacological manifestations of chemoreception in man. In: Ohloff G, Thomas AF (eds) Gustation and olfaction. Academic Press, New York, pp 187–237

Fisher GL, Pfaffmann C, Brown E (1965) Dulcin and saccharin taste in squirrel monkeys, rats and men. Science 150: 506–507

Frank M (1974) The classification of mammalian afferent taste nerve fibers. Chem Senses Flav 1: 53–60

Gent JF, Bartoshuk LM (1983) Sweetness of sucrose, neohesperidine dihydrochalcone, and saccharin is related to genetic ability to taste the bitter substance 6-*n*-propylthiouracil. Chem Senses 7: 265–272

Glaser D, Hellekant G, Brouwer JN, Van der Wel H (1978) The taste responses in primates to the proteins thaumatin and monellin and their phylogenetic implications. Folia Primatol 29: 56–63

Glaser D, Hellekant G, Brouwer JN, Van der Wel H (1984) Effects of gymnemic acid on sweet taste perception in primates. Chem Senses 8: 367–374

Hahn H (1949) Beiträge zur Reizphysiologie. Scherer, Heidelberg

Harris H, Kalmus H (1949) The measurement of taste sensitivity to phenylthiourea (P.T.C.). Ann Eugen 15: 24–31

Hellekant G, Roberts TW (1983) Study of the effect of gymnemic acid on taste in hamster. Chem Senses 8: 195–202

Hellekant G, Hagstrom EC, Kasahara Y, Zotterman Y (1974) On the gustatory effects of miraculin and gymnemic acid in the monkey. Chem Senses Flav 1: 137–145

Hellekant G, Glaser D, Brouwer J, Van der Wel H (1981) Gustatory responses in three prosimian and two simian primate species (*Tupapia glis, Nycticebus, Galago senegalensis, Callithrix jacchus jacchus* and *Saguinus midas niger*) to six sweeteners and miraculin and their phylogenetic implications. Chem Senses 6: 165–173

Hiji Y (1975) Selective elimination of taste responses to sugars by proteolytic enzymes. Nature 256: 427–429

Hiji Y, Ito J (1977) Removal of sweetness by proteases and its recovery mechanism in rat taste cells. Comp Biochem Physiol 58: 109–113

Jakinovich W (1981) Stimulation of the gerbil's gustatory receptors by artificial sweeteners. Brain Res 210: 69–81

Jakinovich W (1983) Methyl 4,6-dichloro-4,6-dideoxy-αD-galactopyranoside: an inhibitor of sweet taste responses in gerbils. Science 219: 408–410

Jakinovich W, Vlahopoulos V (1984) Inhibition of the gerbil's electrophysiological sucrose taste response by para-nitro-phenyl-α-D-glucopyranoside and chloramphenicol. Paper presented at the 6th Annual Meeting of the Association for Chemoreception Sciences, Sarasota, Florida

Kroeze JHA, Bartoshuk LM (1985) Bitterness suppression as revealed by split-tongue taste stimulation in humans. Physiol Behav 35: 779–783
Kurihara K (1971) Taste modifiers. In: Beidler LM (ed) Handbook of sensory physiology, vol IV. Chemical senses, part 2, Taste. Springer, New York Berlin Heidelberg, pp 363–378
Kurihara K, Kurihara Y, Beidler LM (1979) Isolation and mechanism of taste modifiers; taste-modifying protein and gymnemic acids. In: Pfaffmann C (ed) Olfaction and taste III. Pergamon Press, New York, pp 451–469
Lawless HT (1979) Evidence for neural inhibition in bittersweet taste mixtures. J Comp Physiol Psychol 93: 538–547
Lawless HT, Stevens DA (1983) Cross adaptation of sucrose and intensive sweeteners. Chem Senses 7: 309–315
McBurney DH (1972) Gustatory cross adaptation between sweet-tasting compounds. Percept Psychophys 11: 225–227
McBurney DH, Bartoshuk LM (1973) Interactions between stimuli with different taste qualities. Physiol Behav 10: 1101–1106
McBurney DH, Gent JF (1978) Taste of methyl-α-D-mannopyranoside: effects of cross adaptation and *Gymnema sylvestre*. Chem Senses Flav 3: 45–50
McBurney DH, Smith DV, Shick TR (1972) Gustatory cross adaptation: sourness and bitterness. Percept Psychophys 11: 228–232
Meiselman HL (1975) Effect of response task on taste adaptation. Percept Psychophys 17: 591–595
Meiselman HL, DuBose CN (1976) Failure of instructional set to affect completeness of taste adaptation. Percept Psychophys 19: 226–230
Pfaffmann C (1969) Taste preference and reinforcement. In: Tapp JT (ed) Reinforcement and behavior. Academic Press, New York, pp 215–241
Pfaffmann C, Frank M, Bartoshuk LM, Snell TC (1976) Coding gustatory information in the squirrel monkey chorda tympani. In: Sprague JM, Epstein A (eds) Progress in physiological psychology, vol 6. Academic Press, New York, pp 1–27
Schiffman SS, Cahn H, Lindley MG (1981a) Multiple receptor sites mediate sweetness: evidence from cross adaptation. Pharm Biochem Behav 15: 377–388
Schiffman SS, Lindley MG, Clark TB, Makino C (1981b) Molecular mechanism of sweet taste: relationship of hydrogen bonding to taste sensitivity for both young and elderly. Neurobiol Aging 2: 173–185
Settle RG (1981) Suprathreshold glucose and fructose sensitivity in individuals with different family histories of non-insulin-dependent diabetes mellitus. Chem Senses 6: 435–443
Shallenberger RS, Acree TE (1967) Molecular theory of sweet taste. Nature 216: 480–482
Smith DV, McBurney DH (1969) Gustatory cross adaptation: does a single mechanism code the salty taste? J Exp Psychol 80: 101–105
Van der Wel H (1972) Thaumatin, the sweet-tasting protein from *Thaumatococcus Daniellii* Benth. In: Schneider D (ed) Olfaction and taste. Wissenschaftliche Verlagsgesellschaft, Stuttgart, pp 226–233
Van der Wel H, Arvidson K (1978) Qualitative psychological studies on the gustatory effects of the sweet tasting proteins thaumatin and monellin. Chem Senses Flav 3: 291–297
Vlahopoulus V, Jakinovich W (1985) Inhibition of the gerbil's electrophysical sweetener response by methyl 4,6-dichloro-4,6-dideoxy α-D-galactopyranoside, *p*-nitrophenyl α-D-glucopyranoside and chloramphenicol. Chem Senses 10: 432
Warren RM, Pfaffmann C (1959) Inhibition of the sweet taste by *Gymnema sylvestre*. J Appl Physiol 14: 40–42
Yamamoto T, Kato T, Kawamura Y, Yoshida M (1984) Gustatory reaction time to various sweeteners. Chem Senses 9: 79
Zawalich WS (1973) Depression of gustatory sweet response by alloxan. Comp. Biochem Physiol 44A: 903–909

Commentary

Birch: I do not accept the conclusion, arrived at from Bartoshuk's psychophysical data, that there are multiple sweet receptor sites. The AH,B concept leads chemists to the view that there is, in principle, one type of receptor that fits this glucophore.

Presumably all sweet molecules first accede to, and then interact with receptors. The differences which are observed psychophysically between sweetness may therefore arise from their intrinsically different accession efficiencies as well as the nature of the tissue that surrounds the receptor.

Bartoshuk: I believe that we can suggest a resolution to the apparent conflict between Birch's conclusions from the studies I cited that there must be more than one receptor site. According to Birch's view sweet molecules must pass through an orderly queue before they can reach the receptor site. The length of the queue and the rate of passage through the queue can vary. This means that we have a mechanism for handling different sweeteners in different ways. But this is just what we mean when we say that there are multiple sweet receptor mechanisms. In Birch's terminology the orderly queue plus the receptor site (complementary to the AH,B glucophore) *is* the receptor mechanism.

Blass: Bartoshuk's review suggests multiple sweet receptors but also reveals the limitations of human psychophysics in revealing their properties. Bartoshuk discusses at length some of the artifacts that have marred the field but does not focus sufficiently on the intrinsic limitations of stimulus selection for cross-adaptation studies (i.e., should it be perceived intensity or actual concentration). The field does not appear to have advanced significantly beyond the pioneering studies of Hahn (1949) and Blakeslee (1935). Perhaps different psychophysical approaches (as described by Frijters and by Booth et al. in this volume) or a greater focus on animal psychophysics would be more revealing. Included in this approach should be efforts relating different genetic profiles to appetite and feeding habits.

References

Blakeslee AF (1935) A dinner demonstration of threshold differences in taste and smell. Science 81: 504–507

Hahn H (1949) Beiträge zur Reizphysiologie. Scherer, Heidelberg

Rozin: Are there no data on specific localization of sweet tastes on the tongue which might show site-specific differences for different sweeteners? If such data do not exist, why not? How about binding studies for different sweeteners on different places on the tongue or different types of receptors? Is all adaptation peripheral? Is there any cross site adaptation?

Frijters: In addition to differences between individuals with respect to equi-sweet concentrations of different sweeteners it should be remembered that these concentrations are determined by some variant of the method of constant stimuli. Equi-sweetness is therefore a statistical concept: two equi-sweet stimuli may differ in sweetness in a particular trial.

Chapter 4

Selected Factors Influencing Sensory Perception of Sweetness

R. M. Pangborn

Introduction

Taste interactions have long interested psychologists, physiologists, and food scientists because the phenomena are stumbling blocks that must be considered before a functional theory of taste perception can be evolved (Gregson 1968).

The perceived sweetness of a compound is often modified by alteration of the physical or chemical composition of the medium in which the sweetener is dispersed. Usually the modification is one of depression of apparent sweetness intensity, along with qualitative changes. However, the effects of some combinations may be unpredictable because responses depend on the methods of measurement, subjects produce highly individual responses, and results from model solutions may not extrapolate to complex food systems.

The present review explores gustatory responses obtained when additional tastes (sour, salty, bitter) and other sensory attributes (color, aroma, and solution temperature and viscosity), are combined with the primary sweet stimulus. In the interest of simplicity and experimental control most research on these topics has been limited to aqueous solutions. Also most approaches have been empirical in nature, with much room remaining for development of theories to explain observed behavioral phenomena and possible physiological mechanisms.

Sweetness Responses in Relation to Psychophysical Measurements

Responses to sweetness, as well as to other stimuli, are context-dependent. Therefore to interpret the salience of a behavioral response and to estimate its reliability, the psychophysical method employed must be considered. Table 4.1 gives a simplified summary of the types of sweetness information derived from four main measurements: sensitivity, quantity, quality, and affect.

Table 4.1. Examples of sweetness information obtained from diverse measurements

A. *Sensitivity tests*
   1. Thresholds
      a) Absolute: minimum amount of sweetener detected from background
      b) Difference: minimum amount of change in sweetness to be detected
      c) Recognition: minimum amount of sweetener identified as "sweet"
   2. Discrimination
      a) Single sample: Is the sample sweet or not sweet?
      b) Pairs: Which of the two is sweeter?
      c) Duo-trio: Which of the two matches the standard?
      d) Triangle: Which of the three is the odd sample?

B. *Quantitative tests*
   1. Scaling of sweetness intensity
      a) Ranking: place samples in order of sweetness
      b) Category: rate sweetness in terms of adjectives (slight, moderate, etc.), numbers, or lengths of lines
      c) Ratio: e.g., magnitude estimation of how many times lesser or greater the sweetness is compared to a standard
   2. Duration: recording length of time sweet sensation persists.

C. *Qualitative tests*
   1. Use of adjectives to describe, e.g., compound tastes (bitter–sweet), textures (viscosity, smoothness), aromas (caramelized)

D. *Affective tests*
   1. Acceptance: accept/reject the available sweetness
   2. Preference: select one sweetener over another, or one level over another
   3. Hedonic: degree of like/dislike of a sweetness level

Thresholds and Discrimination

The threshold is a fixed value at any brief moment, but over time thresholds are fluctuating, statistically determined end points along a stimulus continuum (Bartoshuk 1978; Pangborn 1980, 1984). For example, two sweeteners with identical thresholds can increase in perceived sweetness intensity with concentration at different rates. At higher levels, perhaps the level at which they occur in the product under study, two sweeteners can differ considerably in intensity. Thresholds can change from trial to trial, reflecting the subject's performance, which can improve with practice or decline from adaptation or fatigue. Threshold values also reflect the a priori theoretical assumptions and mathematical treatments by which the behavior is translated into a single number. In summary, classical methods do not permit measurement of an absolute response independent of the procedure.

Perceived Intensity

Psychophysicists have waged battles for years over the merits of scaling procedures. Ranking is useful for grouping similar products, but simply places stimuli in relative order with no indication of degree of difference among them. Category scales of numbers and/or descriptors are simple to administer and understand but the categories do not represent equal intervals and they suffer from end effects: subjects can run out of categories upon which to express their responses. Ratio scales, such as magnitude estimation, are theoretically infinite scales and represent equal intervals. However, magnitude estimation is too complex for many subjects, is prone to round-number bias, and is inappropriate for bidirectional (like-dislike) scaling (Giovanni and Pangborn 1983). Another alternative has been graphic scaling (visual analog) wherein estimated intensity is indicated by placing a tick on a line anchored with end terms, such as "none" and "very strong." For numerical analysis the tick is converted to cm.

Descriptive Analysis

Description of the qualitative characteristics of sweeteners and sweetened products is critical to product development, product matching, and quality assurance (Larson-Powers and Pangborn 1978; Pangborn 1984) but is beyond the scope of the present review.

Affective Tests

Although not discussed here, affectivity is included in Table 4.1 to emphasize its distinction from the previous measures.

Interaction of Sweetness with Other Tastes

The tissues and receptors in the mouth, throat, and nasal cavity are so situated in relation to one another and so innervated, that all operate simultaneously or sequentially in response to an oral stimulus.The oro-nasal-pharyngeal area integrates sensations of texture (roughness, tingle, astringency), thermal and chemical burning, cooling, moisture, and olfaction, as well as taste. In addition, saliva varying in amount and composition modifies the physical and chemical nature of the stimulus. The fore-going emphasize the many avenues by which sweetness can be modified by secondary stimuli.

Interactions in Aqueous Solutions

In order to partition gustation from other oral sensations and reduce complexity, most research on taste interactions has been conducted with simple, aqueous solutions. Table 4.2 presents an overview of results reported in selected early literature relative to interaction of sweetness with other taste sensations. Five major observations are noteworthy:

1. Contradictions abound, attributable primarily to use of different psychophysical methods. For example, whereas Beebe-Center et al. (1959) used very sensitive paired comparison and direct matching methods, only two subjects participated in the experiment. On the other hand, Kamen et al. (1961) tested 960 subjects in a half-replicate design where half the samples were judged by one group of subjects and the remaining

Table 4.2. Summary of early literature on interaction of sweetness with other taste in aqueous solutions

Year	Investigator	Observation
1892	Zuntz	NaCl increased sweetness of sucrose.
1894	Kiesow	NaCl decreased sweetness of sucrose.
1917	Kremer	NaCl increased sweetness of sucrose.
1937	Cragg	Sucrose decreased sourness of HCl.
1943	Fabian and Blum	NaCl increased sweetness of sucrose, dextrose, fructose.
		HCl and acetic acids decreased sweetness of dextrose but had no effect of sweetness of sucrose.
		HCl and citric acids had no effect on sweetness of fructose.
		Lactic, malic, and tartaric acids increased sweetness of sucrose but decreased sweetness of fructose.
		All sugars decreased sourness of acids and saltiness of NaCl.
1947	Cameron	Urea decreased sweetness of sucrose.
1953	Sjöström, Cairncross	0.5% NaCl increased sweetness of 5%–7% sucrose, but 1% NaCl decreased sweetness of 3%–10% sucrose.
		1%–10% sucrose decreased sourness of 0.04–0.06% acetic acid.
		0.04%–0.06% acetic acid decreased sweetness of 6% sucrose and above, but had no effect on sweetness of 1%–5% sucrose.
1959	Beebe-Center et al.	Mutual masking of NaCl and sucrose except for sweetness enhancement of weak solutions.
1961	Kamen et al.	Sucrose decreased the bitterness of caffeine but had no effect on the saltiness of NaCl. Sweetness of sucrose was reduced by NaCl, increased by citric acid, and unaffected by caffeine.
1960a	Pangborn	Sucrose decreased sourness of citric acid, saltiness of NaCl, and
1961	Pangborn	bitterness of caffeine.
1962	Pangborn	Citric acid and caffeine decreased sweetness of sucrose.
		Low concentration of NaCl enhanced low concentration of sucrose; all NaCl levels depressed the highest sucrose (20.25%) level.
1963	Gregson and McCowen	In weak solutions, some subjects perceived an enhanced and others a decreased sweetness of sucrose by citric acid.
1965	Pangborn	Citric, acetic, tartaric, and lactic acids decreased sweetness of sucrose, fructose, glucose, and lactose.

half by another. Intervals between the two replications ranged from 1 week to 16 months, and a relatively insensitive single sample method was used with a nine-interval intensity scale.

2. Enhancement or depression varies with the specific sweetener, salt acidulant, or bitter compound. Hence it is advisable to specify the compound being tested rather than use the generic term of "sweetness."

3. The concentration of both members of the binary mixture influences responses, with a greater variability at near-threshold concentrations. Using direct-comparison, difference intensity scaling (+3 to −3), Gregson and McCowen (1963) noted marked contradictions among 72 subjects, probably because of differing sensitivity to the low concentrations tested (2, 3.5, and 5 just-noticeable-difference units above threshold).

4. Even within an experiment, subjects do not always agree on enhancement as opposed to depression. Apparently some subjects cannot judge sweetness and other tastes separately but coalesce them into one bipolar sensation (Gregson 1968). Kroeze (1982a) observed "side tastes" (sensations that were qualitatively distinct from the main taste of a stimulus) in mixtures by some, but not all subjects tested. Specific side taste sensations noted were NaCl-sweetness, quinine-saltiness, and

NaCl-bitterness. The addition of the side tastes to the suppressed main tastes tended to give a curvilinear, not a linear function. Diametrically opposite responses were noted by Pangborn and Trabue (1967), wherein seven subjects found depression and eight noted enhancement of the saltiness of NaCl by citric acid. The effect was robust, showing up for the same subjects in water solutions, in green bean and lima bean purées, and in tomato juice.

5. Although not evident from Table 4.2, an additional variation among investigations is the aqueous medium itself, e.g., distilled water, deionized water, "spring" water, and tap water (composition not specified).

Publications utilizing more sophisticated procedures were consulted in an attempt to draw conclusions about sweetness interactions. Using the τ scale (approximately a logarithmic function of stimulus concentration, where one unit of $\tau \simeq 4$ decilogs), Indow (1969) reported large compound × concentration interactions. For example: (a) the sweetness of sucrose was masked by high levels of quinine sulfate and NaCl; (b) there was slight enhancement of sweetness by lower levels of quinine sulfate and higher levels of tartaric acid; and (c) the most marked enhancement of sweetness occurred with the lower levels of sucrose at the intermediate levels of NaCl. Using parotid salivary gland flow, Feller et al. (1965) reported that the addition of citric acid or sodium citrate to sucrose produced responses equal to the sum of the separate constitutents.

Bartoshuk (1975) tested two-, three-, and four-component mixtures using magnitude estimation of perceived taste intensity of solutions applied to the tongue at 34°C by means of a gravity-flow system. Mean sweetness intensities for equally intense concentrations of sucrose (S), sodium chloride (N), hydrochloric acid (H), and quinine hydrochloride (Q) were:

S	SN	SH	SQ	NSH	SHQ	NSQ	NSHQ
21.8	15.8	22.5	15.4	15.3	17.8	11.2	9.0

Except for SH, sweetness of the mixtures was less than for sucrose alone, and three- and four-component mixtures were even less sweet. Bartoshuk (1975) suggested that antidromic inhibition might mediate the observed compression. She further suggested that additivity in taste mixtures is predictable not on the basis of the qualitative differences among the components of a mixture, but on the basis of the manner in which perceived intensity of the components increased with concentration. McBride and Anderson (to be published) disagree, stating, "Mixing two components, both of which have perfectly linear psychophysical functions, is no guarantee that activity will result."

In a subsequent study, Bartoshuk and Cleveland (1977) found that simple taste additivity accounted for sweetness intensity responses for two-, three-, and four-component mixtures of sucrose, fructose, glucose, and maltose when the flow procedure was utilized, but not when tasted by a "sip and spit" procedure. The simple sum of the sweetness of the four unmixed sugars was 40; that of the mixture of the four sugars was 29 by the flow procedure and 123 by sip and spit procedures. Temperature effects may be involved, as solutions were at 34°C in the flow, compared with at room temperature in the sip and spit procedure. McBride's (1984) experiments on binary mixtures of sucrose, fructose, and glucose showed additivity of sweetness up to moderate levels, after which a subadditive trend appeared. This is consistent with the sucrose–fructose study of Curtis et al (1984).

Earlier, Moskowitz (1974) noted general synergism for sugar mixtures, where the perceived intensity of the mixture was greater than would be predicted from the sum of the intensity of the components. Frijters and Oude Ophuis (1983) evolved an equiratio model where the sweetness of glucose–fructose mixtures always fell between the intensities of the unmixed compounds when the total concentration of the mixture (in M) was equal to the concentrations of the unmixed compounds. The validity of the model was confirmed in subsequent experiments with sucrose, sorbital, and three equiratio sucrose–sorbitol mixtures (Frijters et al. 1984). McBride and Anderson (to be published) challenged these conclusions, based on fitting of power functions to magnitude estimations. According to McBride (1983), magnitude estimation is subject to non-linear bias and power functions often provide poor descriptions of sweetness data.

In studies reported by O'Mahony et al. (1983), mixtures of fructose and other taste compounds were judged as tasting more mixed than single, but the three- and four-component mixtures did not differ in their "mixedness." They intepreted this as indicating a degree of synthesis, i.e., the mixtures were perceived as having singular tastes, distinct from their components.

Gillan (1983) noted greater within-modality (sweetness of sucrose and saltiness of NaCl) than between-modality suppression (taste–odor). He concluded that individual stimuli were more intense than taste–odor mixtures, which were more intense than within-modality mixtures.

A recent report based on the use of generalized Procrustes analysis (Francombe and Whelehan 1986) noted that quinine reduced sweetness of sucrose, NaCl combined with citric acid enhanced sweetness, and sucrose and quinine, separately or together, reduced apparent saltiness.

Why are mixtures generally considered to have a lesser taste intensity than unmixed solutions? Several explanations have been proposed by Lawless (1979) and others:

1. *Molecular interactions.* Interactions of molecules in solution could cause mixture suppression through weak physical attraction which could reduce the effective concentration available for binding to receptors compared with equimolar unmixed solutions.
2. *Receptor binding interference.* One compound could inhibit the taste of a second by interfering with the binding of the primary compound to the taste receptor.
3. *Antidromic inhibition.* Taste stimuli induce conduction of action potentials in the conventional orthodromic manner up the sensory axon. In addition, the potentials can be conducted back antidromically, producing inhibition by hyperpolarizing the receptor membrane (Bartoshuk 1975).
4. *Convergence resulting in occlusion.* Both of these possibilities are reasonable if the fibers of the human taste nerve are non-specific with respect to taste quality, as has been found in some other species (Lawless 1979).
5. *Simple adaptation effects.*

As expressed for sweet–sour interactions by Kuznicki and McCutcheon (1979), through their capacity to bind protons, sugars may decrease proton availability to acid receptors, suppressing sourness. Removal of protons from acid receptors also could open up sites, resulting in enhancement of sourness by subsequent acid stimuli.

Evidence advanced by Lawless (1979) implied that mutual suppression of bitter and sweet tastes was due to neural inhibition rather than chemical interactions in solution or competition of molecules for common receptor sites. Gregson and Paris (1967) noted that interaction can arise from adaptation or frame-of-reference effects in the judgment process as from any receptor cell event. Results from Kroeze's (1979) studies of sucrose–sodium chloride mixtures indicated no relation between masking and cross-adaptation. Following adaptation to sucrose and to the mixture of sucrose and NaCl, sweetness intensity decreased by 54.2 and 54.6%, respectively. A significant reduction (87.9%) of the saltiness of NaCl also occurred. No indication of cross-adaptation was observed. Later, Kroeze (1982b) reported that the amount of NaCl–sucrose suppression depended on the relative frequencies of mixed and unmixed stimuli in the series, i.e., a contextual factor.

Applying magnitude estimation to assess cross-adaptation and synergism among selected combinations of nine sweeteners, Gardner (1984) reported significant reciprocal cross-adaptation only with sodium saccharin and Acesulfame K. Synergism was noted at all concentrations between xylitol and D-glucose, sodium cyclamate, sorbitol, and D-glucose, and D-glucose and sodium cyclamate. No synergism occurred between D-fructose and L-sorbose, nor between sucrose and sodium saccharin. Tasteless levels of L-threonine enhanced sweetness of sucrose and D-alanine, had no significant effect on the taste of L-alanine, and tended to decrease sweetness of D-fructose. Ethyl maltol increased perceived sweetness of sucrose only when olfaction was operative. The combined results indicated multiple sweet receptor sites and the existence of primary and secondary sites for sweeteners.

Kroeze and Bartoshuk (1985) tested two hypotheses which could explain taste mixture suppression: (a) peripheral phenomena due to affinity of taste molecules for common receptor sites leading to competition between mixture components, and (b) central phenomena residing in the neural system. Using a box in which the left and right sides of the tongue could receive separate stimuli, these investigators concluded that bitterness suppression of quinine hydrochloride by sucrose occurred centrally. In comparison, bitterness suppression by NaCl was about two-thirds peripheral and only one-third central. This contrast merits additional testing. For a lucid, stimulating history of the search for the key to taste mixtures, see Bartoshuk and Gent (1984).

According to Gregson (1968), there is no support for a scaling model which makes provision for a specific kind of mutually inhibitory or facilitatory interaction between compounds. He advanced a provocative opinion that psychophysically, taste interactions reported in the literature are artifacts of the response measure selected. Thus, an "equality" response would be due to simple averaging of negative and positive individual responses because time-order effects make physically equal stimuli appear different when experienced sequentially. While applicable to subtle differences, surely robust observations require other interpretations.

Interactions in Foods and Beverages

It is risky to generalize taste relationships from aqueous media or model systems to actual foods and beverages. Unfortunately, the literature on taste interactions in foods is very sparse and unsophisticated. In lima bean purée (Pangborn and Trabue 1964) taste results were compatible with those reported previously for aqueous media in that (a) sucrose and citric acid exhibited mutual masking and (b) sucrose and NaCl produced mutual depression except for low levels of NaCl (0.05% and 0.10%), which slightly enhanced the apparent sweetness of 0.4% and 0.8% sucrose. Comparable results were obtained for canned tomato juice (Pangborn and Chrisp 1964). Again, sucrose and citric acid exhibited mutual masking effects, as did sucrose and NaCl, except for a slight enhancement of sweetness by low levels of NaCl.

Studies with white table wines (Pangborn et al. 1964) demonstrated that tartaric acid had a strong depressing effect on apparent sweetness of 0.2%–1.4% added sucrose. Low levels of added caffeine (0.01% and 0.02%) had little effect, whereas a high level (0.04%) significantly reduced sweetness; bitterness was enhanced slightly by sucrose. Later Martin and Pangborn (1970) observed that ethyl alcohol generally enhanced the sweetness of sucrose in aqueous solutions. Vermouths in which viscosity was increased by a non-sweet oligosaccharide mixture (Polycose) were sweeter and less bitter than samples with the same sucrose concentration but lower physical viscosity (Burns and Noble 1985). Viscosity alone contributed 20%–30% of the perceived increase in sweetness due to sucrose addition. More investigations such as the foregoing carefully designed study are needed across a wide variety of foods and beverages.

Although untested in actual food products, Gardner (1984) noted that the perceived sweetness of sucrose was significantly reduced by pregelatinized potato starch but not by pregelatinized tapioca and waxy corn starches. The latter starches may have had lower apparent viscosities due to differential breakdown by salivary alpha-amylase.

Interaction of Sweetness with Other Sensory Attributes

Color

Alteration of the appearance, particularly the color, of model systems and of foods can modify perceived taste quality and intensity, primarily due to expectation. For example, although green coloring did not affect perception of sweetness or of peppermint flavoring in aqueous solutions (Pangborn 1960b), greater sweetness and flavor were ascribed to orange-colored, apricot-flavored, and red-colored, cherry-flavored solutions than to uncolored samples. Maga (1974) observed that sucrose thresholds were decreased by yellow and increased by green coloring added to the solutions. Johnson et al. (1982) and Johnson and Clydesdale (1982) reported that unflavored solutions with greater amounts of red coloring were judged 2%–10% sweeter than were less red samples.

Green-colored pear nectar was considered to be least sweet while colorless samples were judged sweetest, when in fact both had the same sucrose content (Pangborn and Hansen 1963). A second group of judges demonstrated no color–sweetness associations in the nectar. Subsequently, when a dry white wine was miscolored to simulate sauterne, sherry, rose, claret, and burgundy wine types, naive judges showed no color–sweetness associations (Pangborn et al. 1963). Experienced wine tasters, however, ascribed greatest sweetness to the pink-colored samples, probably through their familiarity with rosé wines, which usually have a high residual sugar content. Among small groups of consumers, Lundgren (1976) noted that red-colored raspberry sweets were judged to taste sweeter and more raspberry-like than did uncolored candy.

Aroma

Some sweeteners can influence the intensity of selected aromas and flavors-by-mouth by modifying the concentration of volatile compounds in the headspace above the medium. Wientjes (1968) reported that addition of high concentrations of fructose (79%) or invert sugar (73%) to aqueous solutions of natural and synthetic strawberry volatiles increased headspace concentrations of some components while decreasing others. According to Maier (1970), additions of 1% sucrose, fructose, or sorbitol suppressed the volatility of acetone and acetaldehyde, while increasing the volatility of ethyl acetate. The volatility of ethanol was increased by fructose and sorbitol, but not by sucrose. Nawar (1971) found that 20%, 40%, and 60% sucrose increased the headspace concentration of acetone but decreased that of 2-heptanone and heptanal. As reported by Ahmed et al. (1978), the flavour threshold, of *d*-limonene, the major volatile constituent of orange juice, was increased slightly in water by the addition of 4.9% sucrose, 2.4% glucose, and 2.6% fructose.

The mechanism by which sugars alter the volatility of selected compounds is unknown, but may be related to a simple salting-out effect. Some compounds have limited solubility in water, so that the addition of other more polar solutes such as sucrose further decreases solubility, shifting the equilibrium towards the vapor phase, resulting in an increase in apparent aroma intensity. Nawar (1971) observed that addition of sucrose to some compounds decreased headspace concentration, while addition of the volatile to the sucrose solution increased headspace concentration. He explained that the result was due to interaction of the sugar with water molecules rather than of the sugar with the volatiles.

Considerably more collaborative investigations between physical chemists and psychophysicists are necessary to define the interactions of solvents and solutes in model systems before food scientists can explain comparable phenomena in foods.

Viscosity

The physical properties of a stimulus can influence the rate with which a sweet compound reaches gustatory receptors. Conversely, taste compounds can alter perceived oral viscosity by changing the rheological behavior of the viscous material and/or by inducing saliva to dilute the medium. As early as 1926, Skramlik reported that taste intensity was greater in aqueous media than in paraffin oil. Crocker (1945) speculated that the physical state of a food influenced taste by controlling the quantity of sapid material reaching the taste receptors in a given time. Mackey and Valassi's (1956) often-cited study indicated that taste thresholds for sucrose were lower in water solutions than in tomato juice, each prepared as liquids, gels, and foams. Later (1958), Mackey reported that saccharin was more easily detected in water than in mineral oil, and hypothesized that the lipid inhibited the solubility of the taste compound in the saliva. When methylcellulose was added to water to the same viscosity as that of the oil, ease of detection of the taste compound was intermediate between water and oil. Ohta et al. (1979) confirmed Mackey's findings, as taste intensity of sucrose was highest in water, lowest in water/oil emulsions, and intermediate in oil/water emulsions.

Wick (1963) observed that the accuracy of ranking sweetness intensity of aqueous solutions, sucrose–starch powder mixtures, gelatin gels, and whipped gelatin gels averaged, respectively, 75%, 47%, 43%, and 33%. In a limited study with five subjects, Stone and Oliver (1966) found that the sweetness of 1%, 2%, 5%, and 10% solutions was enhanced somewhat by the viscosity imparted by 2% cornstarch or by 1% gum tragacanth. Using sucrose in combination with cornstarch, guar, and carboxymethylcellulose (CMC), Vaisey et al. (1969) measured the rates of sweetness recognition, matching of equisweetness, apparent sweetness intensities, and ranking of sweetness. CMC, with less drop in viscosity with increasing rates of shear, tended to mask sweetness perception throughout. Subsequently, Marshall and Vaisey (1972) reported that the sweetness of sodium sucaryl was greatest in carrageenan gels and least in cornstarch gels, while intermediate in low methoxy pectin, agar, and gelatin gels. Gels which took more effort to disintegrate in the mouth limited sweetness perception. The sweetness of glucose solutions was also depressed by viscosity imparted by CMC (Moskowitz and Arabie 1970), as was the sweetness of both sucrose and sodium saccharin solutions (Arabie and Moskowitz 1971). In contrast, Pangborn et al. (1973) observed that low concentrations of five hydrocolloids generally reduced sweetness of sucrose but enhanced sweetness of sodium saccharin. Figure 4.1 illustrates the degree of enhancement of saccharin by low-viscosity CMC. The same observation was made with medium-viscosity CMC, where enhancement was more pronounced at higher saccharin levels, and with sodium alginate, where enhancement was independent of the concentration of saccharin.

With high-viscosity CMC, Christensen (1980) reported significant decreases in perceived sweetness intensity of lower concentrations of sucrose, but low-viscosity

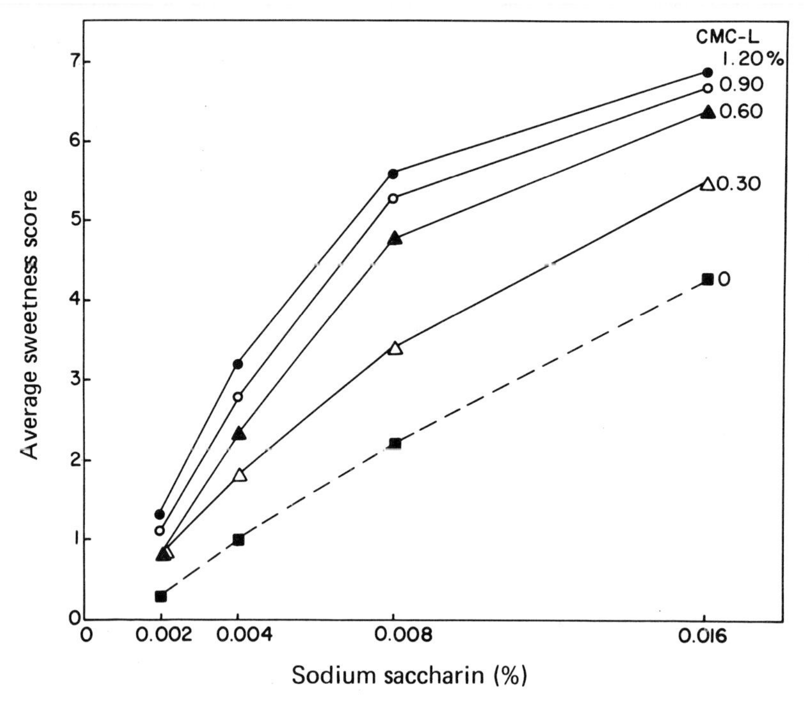

Fig. 4.1. Influence of low-viscosity carboxymethylcellulose on average sweetness intensity of solutions of sodium saccharin. Values for the 0 sample represent 16 subjects × 8 replications; those for the CMC-L samples represent 16 subjects × 2 replications. (Pangborn et al. 1973)

CMC produced little or no sweetness suppression. Izutsu et al. (1981) found that increasing CMC with viscosities from 1 to 100 poises[1] decreased perceived sweetness intensity of 4%–16% sucrose, as expressed by the regression equation:

$$\text{Sweetness intensity (Y)} = 0.86A^{1.04}B^{-0.08}$$

where A is % sucrose concentration of test solutions A and B is viscosity in poise. The equation was applicable to low-, medium-, and high-viscosity CMC.

Increasing viscosity imparted by methyl cullulose, guar seed flour, carob seed flour, and tara seed flour significantly increased the stimulus and recognition thresholds for sucrose and other compounds (Paulus and Haas 1980). In agreement with others, they noted that the thickeners had differential effects on individual taste intensities. With suprathreshold levels of sucrose, Launay and Pasquet (1982) found that sweetness was greater in aqueous solutions than in guar gum solutions.

To explain better the effect of solution viscosity on perceived sweetness intensity of glucose, Cussler et al. (1979) applied diffusion theory. Subsequently they reported that diffusion coefficients and sweetness intensity of sucrose and fructose were decreased 22% and 42%, respectively, by addition of 1.3% and 3.6% insoluble tomato solids (Kokini et al. 1982).

Relationships noted in aqueous systems do not necessarily hold true for more complex food systems. For example, five hydrocolloids (xanthan, hydroxypropylcellulose, sodium alginate, and CMC of low and medium viscosity) significantly reduced aroma, flavor, and sourness of orange drink, but had little effect on perceived sweetness (Pangborn et al. 1978). It was speculated that the reduced sourness permitted better detection of sweetness.

Solution Temperature

Only brief mention will be made of taste–temperature interactions. The variegated history of the topic has been reviewed by Sato (1967), Laffort (1969), and Pangborn et al. (1970). Representative of results cited are those of Shimuzu et al. (1959), who found that half of the young females tested were more sensitive to sweet and salty stimuli at solution temperature between 32°C and 38°C and less sensitive at 15°C and 55°C. The remaining half demonstrated decreasing sensitivity with increasing temperature from 15°C to 55°C. In data reported by Pangborn et al. (1970) and Pangborn and Bertolero (1972), both chilling (0°C) and heating (55°C) reduced taste intensity of solutions of NaCl and seven other salts, as well as of six natural drinking waters. Generally, maximum taste intensity was ascribed to solutions at 22°C, followed by those at 37°C.

Similarly, sensitivity to several sugars at threshold and low supra-threshold concentrations has been reported to be greatest at a stimulus temperature between 20°C and 40°C (Stone et al. 1969; McBurney et al. 1973; Moskowitz 1973; Bartoshuk et al. 1982). Recent intriguing data on thermal and chemical "hotness" showed that 0.01 – 0.29 M sucrose masked pungency of 0.06 to 0.7 mg/liter of capsaicin (Sizer and Harris 1985). Pungency was enhanced by high (60°C) and masked by low (2°C) solu-

---

1. The poise (after Poiseuille) is that viscosity of a flowing liquid in which a velocity gradient of 1 cm s^{-1} is obtained when a force of 1 dyne is applied to two surfaces 1 cm apart (Bourne 1982).

tion temperatures. Unlike sucrose, NaCl and citric acid had no effect on the threshold for capsaicin. Additional studies are needed to elucidate the underlying mechanisms of thermal effects in model systems, and in foods to establish more pragmatic relationships.

Conclusions

In general, perceived sweetness intensity of most sweeteners is reduced by the addition of taste, texture, or other substances in model systems as well as in foods. Heating and chilling solutions generally decreased perceived taste intensity compared to solutions between 20°C and 40°C. As concluded in the review by Izutsu and Wani (1985), the wide divergence of results from studies on mixtures can be attributed mainly to differences in the psychophysical methods employed by various investigators. The physical–chemical interactions of the individual components in a mixture preclude generalizations based on sensations, e.g., sweetness. Rather the specific compounds, the medium of dispersion, the psychophysical methods, and other procedural variables must be considered in the evolution of theories and/or equations to explain the observed relationships.

References

Ahmed EM, Dennison RA, Dougherty RH, Shaw PE (1978) Effect of nonvolatile orange juice components, acid, sugar, and pectin on flavour threshold of *d*-limonene in water. J Agric Food Chem 26: 192–194

Arabie P, Moskowitz H (1971) The effects of viscosity upon perceived sweetness. Percept Psychophys 9: 410–412

Bartoshuk LM (1975) Taste mixtures: Is mixture suppression related to compression? Physiol Behav 14: 643–649

Bartoshuk LM (1978) The psychophysics of taste. Am J Clin Nutr 31: 1068–1077

Bartoshuk LM, Cleveland CT (1977) Mixtures of substances with stimilar tastes. A test of a psychophysical model of taste mixture interactions. Sensory Proc 1: 177–186

Bartoshuk LM, Gent JF (1984) Taste mixtures, an analysis of synthesis. In: Pfaff DW (ed) Taste, olfaction, and the central nervous system, Rockefeller University Press, New York, pp 210–232

Bartoshuk LM, Rennert K, Rodin J, Stevens JC (1982) Effects of temperature on perceived sweetness of sucrose. Physiol Behav 28: 905–910

Beebe-Center JG, Rogers MS, Atkinson WH, O'Connell DW (1959) Sweetness and saltiness of compound solutions of sucrose and NaCl as a function of concentration of solutes. J Exp Psychol 57: 231–234

Bourne MC (1982) Food texture and viscosity: concept and measurement. Academic Press, New York, p 200

Burns DJW, Noble AC (1985) Evaluation of the separate contributions of viscosity and sweetness of sucrose to perceived viscosity, sweetness and bitterness of vermouth. J Text Stud 16: 365–381

Cameron AT (1947) The taste sense and the relative sweetness of sugars and other sweet substances. Sci Rept Series No 9, Sugar Research Foundation, New York

Christensen CM (1980) Effects of solution viscosity on perceived saltiness and sweetness. Percept Psychophys 28: 347–353

Cragg LH (1937) The sour taste: threshold values and accuracy, the effects of saltiness and sweetness. Trans R Soc Can 31: 131–140

Crocker EC (1945) Flavor. McGraw-Hill, New York

Curtis DW, Stevens DA, Lawless HT (1984) Perceived intensity of the taste of sugar mixtures and acid mixtures. Chem Senses 9: 107–120
Cussler EL, Kokini JL, Weinheimer RL, Moskowitz HR (1979) Food texture in the mouth. Food Technol 33: 89–92
Fabian FW, Blum HB (1943) Relative taste potency of some basic food constituents and their competitive and compensatory action. Food Res 8: 179–193
Feller RP, Sharon IM, Chauncey HH, Shannon IL (1965) Gustatory perception of sour, sweet, and salt mixtures using parotid gland flow rate. J Appl Physiol 20: 1341–1344
Francombe MA, Whelehan OP (1986) Taste perception of mixtures (abstr). Chem Senses 11: 162
Frijters JER, Oude Ophuis PAM, (1983) The construction and prediction of psychophysical power functions for the sweetness of equiratio sugar mixtures. Perception 12: 753–767
Frijters JER, de Graaf C, Koolen HCM (1984) The validity of the equiratio taste mixture model investigated with sorbitol–sucrose mixtures. Chem Senses 9: 241–248
Gardner BJ (1984) Investigation of sweet taste mechanism by taste interactions. Diss Abstr Int B 1985 45:7
Gillan DJ (1983) Taste–taste, odor–odor, and taste–odor mixtures: greater suppression within than between modalities. Percept Psychophys 33: 183–185
Giovanni ME, Pangborn RM (1983) Measurement of taste intensity and degree of liking of beverages by graphic scales and magnitude estimation. J Food Sci 48: 1175–1182
Gregson RAM (1968) Simulating perceived similarities between taste mixtures having mutually interacting components. Br J Math Stat Psychol 21: 117–130
Gregson RAM, McCowen PF (1963) The relative perception of weak sucrose–citric acid mixtures. J Food Sci 28: 371–378
Gregson RAM, Paris GL (1967) Intensity–volume interaction effects in gustatory perception. Percept Psychophys 2: 483–487
Indow T (1969) An application of the τ scale of taste: interaction among the four qualities of taste. Percept Psychophys 5: 347–351
Izutsu T, Wani K (1985) Food texture and taste: a review. J Text Stud 16: 1–28
Izutsu T, Taneya S, Kikuchi E, Sone T (1981) Effect of viscosity on perceived sweetness intensity of sweetened sodium carboxymethylcellulose solutions. J Text Stud 12: 259–273
Johnson J, Clydesdale FM (1982) Perceived sweetness and redness in colored sucrose solutions. J Food Sci 47: 747–752
Johnson JL, Dzendolet E, Damon R, Sawyer M, Clydesdale FM (1982) Psychophysical relationships between perceived sweetness and color in cherry-flavored beverages. J Food Protect 45: 601–606
Kamen JM, Pilgrim FJ, Gutman NJ, Kroll BJ (1961) Interaction of suprathreshold taste stimuli. J Exp Psychol 62: 348–356
Kiesow G (1894) Ueber die Wirkung des Cocain und der Gymnemasaure auf die Schleimhaut der Zunge und des Mundraums. Philosoph Studien 9: 510–527
Kokini JL, Bistany K, Poole M, Stier EF (1982) Use of mass transfer theory to predict viscosity–sweetness interactions of fructose and sucrose solutions containing tomato solids. J Text Stud 13: 187–200
Kremer JH (1917) Influence de sensations de goût sur d'autres specifiquement differentes. Arch Neerl Physiol de l'Homme Animaux 1: 625–634
Kroeze JH (1979) Masking and adaptation of sugar sweetness intensity. Physiol Behav 22: 347–351
Kroeze JHA (1982a) The relationship between the side tastes of masking stimuli and masking in binary mixtures. Chem Senses 7: 23–37
Kroeze JHA (1982b) The influence of relative frequencies of pure and mixed stimuli on mixture suppression in taste. Percept Psychophys 31: 276–278
Kroeze JHA, Bartoshuk LM (1985) Bitterness suppression as revealed by split-tongue taste stimulation in humans. Physiol Behav 35: 779–783
Kuznicki JT, McCutcheon NB (1979) Cross-enhancement of the sour taste on single human taste papillae. J Exp Psychol 108: 68–89
Laffort P (1969) Temperature et qualites organoleptique. Annal Nutr Aliment 23: 63–77
Larson-Powers N, Pangborn RM (1978) Paired comparison and time-intensity measurements of the sensory properties of beverages and gelatins containing sucrose or synthetic sweeteners. J Food Sci 43: 41–46
Launay B, Pasquet E (1982) Sucrose solutions with and without guar gum: rheological properties and relative sweetness intensity. Prog Food Nutr Sci 6: 247–258
Lawless HT (1979) Evidence for neural inhibition in bittersweet taste mixtures. J. Comp Physiol Phychol 93: 538–547
Lundgren B (1976) The taste of decolored foods. SIK Information, Göteborg, Sweden 6: 6–9
Mackey A (1958) Discernment of taste substances as affected by solvent medium. Food Res 23: 580–583

Mackey AO, Valassi K (1956) The discernment of primary tastes in the presence of different food textures. Food Technol 10: 238–240
Maga JA (1974) Influence of color on taste thresholds. Chem Senses Flav 1: 115–120
Maier HG (1970) Volatile flavoring substances in foodstuffs. Angew Chem Internat Edit 9: 917–926
Marshall SG, Vaisey M (1972) Sweetness perception in relation to some textural characteristics of hydrocolloid gels. J Text Stud 3: 173–185
Martin SL, Pangborn RM (1970) Taste interactions of ethyl alcohol with sweet, salty, sour and bitter compounds. J Sci Food Agri 21: 653–655
McBride RL (1983) Taste intensity and the case of exponents greater than 1. Aust J Psychol 35: 175–184
McBride RL (1984) The sweetness of binary mixtures of sucrose, fructose, and glucose (abstr). Sixth Ann Meeting Assoc Chemoreception Sci, Sarasota FL
McBride RL, Anderson NH (to be published) Integration psychophysics in the chemical senses. Chem Senses
McBurney DH, Collings VB, Glanz LM (1973) Temperature dependence of human taste responses. Physiol Behav 11: 89–94
Moskowitz HR (1973) Effects of solution temperature on taste intensity in humans. Physiol Behav 10: 289–292
Moskowitz HR (1974) Models of additivity for sugar sweetness. In: Moskowitz HR, Scharf B, Stevens JC (eds) Sensation and measurement: papers in honor of S S Stevens. Reidel, The Netherlands, pp 379–388
Moskowitz HR, Arabie P (1970) Taste intensity as a function of stimulus concentration and solvent viscosity. J Text Stud 1: 502–510
Nawar WW (1971) Some variables affecting composition of headspace aroma. J Agric Food Chem 19: 1057–1059
Ohta S, Sakamoto Y, Kondo D, Kusaka H (1979) Influence of oil and fats in foods on five tastes (in Japanese). J Jap Chem Soc 28: 321–327
O'Mahony M, Atassi-Sheldon S, Rothman L, Murphy-Ellison T (1983) Relative singularity–mixedness judgements for selected taste stimuli. Physiol Behav 31: 749–755
Pangborn RM (1960a) Taste interrelationships. Food Res 25: 245–256
Pangborn RM (1960b) Influence of color on the discrimination of sweetness. Am J Psychol 73: 229–238
Pangborn RM (1961) Taste interrelationships. II. Suprathreshold solutions of sucrose and citric acid. J Food Sci 26: 648–655
Pangborn RM (1962) Taste interrelationships. III. Suprathreshold solutions of sucrose and sodium chloride. J Food Sci 27: 494–500
Pangborn RM (1965) Taste interrelationships of organic acids and selected sugars. In: Leitch JM (ed) Proceedings of the first international congress of food science and technology Gordon and Breach, New York, pp 291–305
Pangborn RM (1980) A critical analysis of sensory responses to sweetness. In: Koivistoinen P, Hyvönen L (eds) Carbohydrate sweeteners in foods and nutrition. Academic Press, London, pp 87–110
Pangborn RM (1984) Sensory techniques of food analysis. In: Gruenwedel DW, Whitaker JF (eds) Food analysis: principles and techniques. Marcel Dekker, New York, pp 37–93
Pangborn RM, Bertolero LL (1972) Influence of temperature on taste intensity and degree of liking of drinking water. J Am Water Works Assoc 64: 511–515
Pangborn RM, Chrisp RB, (1964) Taste interrelationships. VI. Sucrose, sodium chloride, and citric acid in canned tomato juice. J Food Sci 29: 490–498
Pangborn RM, Hansen B (1963) The influence of color on discrimination of sweetness and sourness in pear-nectar. Am J Psychol 76: 315–317
Pangborn RM, Trabue IM (1964) Taste interrelationships. V. Sucrose, sodium chloride and citric acid in lima bean purée. J Food Sci 29: 233–240
Pangborn RM, Trabue IM (1967) Detection and apparent taste intensity of salt–acid mixtures in two media. Percept Psychophys 2: 503–509
Pangborn RM, Berg HW, Hansen B (1963) The influence of color on discrimination of sweetness in dry table wine. Am J Psychol 76: 492–495
Pangborn RM, Ough CS, Chrisp RB (1964) Taste interrelationship of sucrose, tartaric acid, and caffeine in white table wine. J Am Soc Enol Vit 15: 154–161
Pangborn RM, Chrisp RB, Bertolero LL (1970) Gustatory, salivary, and oral thermal responses to solutions of sodium chloride at four temperatures. Percept Psychophys 8: 69–75
Pangborn RM, Trabue IM, Szczesniak AS (1973) Effect of hydrocolloids on oral viscosity and basic taste intensities. J Text Stud 4: 224–241
Pangborn RM, Gibbs ZM, Tassan C (1978) Effect of hydrocolloids on apparent viscosity and sensory properties of selected beverages. J Text Stud 9: 415–436

Paulus K, Haas E (1980) The influence of solvent viscosity on the threshold values of primary tastes. Chem Senses 5: 23–32
Sato M (1967) Gustatory response as a temperature-dependent process. In: Neff WD (ed) Contributions to sensory physiology. Academic Press, New York, pp 223–251
Shimizu M, Yanase T, Higashihira K (1959) Relations between gustatory sense and temperature of drinks. Kaseigaku Kenkyu 6: 26–28
Sizer F, Harris N, (1985) The influence of common food additives and temperature on threshold perception of capsaicin. Chem Senses 10: 279–286
Sjöström LB, Cairncross SE (1953) Role of sweeteners in food flavor. Adv Chem Series, pp 108–113
Skramlik E von (1926) Handbuch der Physiologie der Neiderensinne, Band 1: Die Physiologie des Geruchs und Geschmackssinnes. Thieme, Leipzig
Stone H. Oliver S (1966) Effect of viscosity on the detection of relative sweetness intensity of sucrose solutions. J Food Sci 31: 129–134
Stone H. Oliver S, Kloehn J (1969) Temperature and pH effects on the relative sweetness of suprathreshold mixtures of dextrose and fructose. Percept Psychophys 5: 257–260
Vaisey M, Brunon R, Cooper J (1969) Some sensory effects of hydrocolloid sols on sweetness. J Food Sci 34: 397–400
Wick EL (1963) Sweetness in fondant. Paper presented at the 17th Ann Prod Conf, Penna Mfg Conf Assoc, Lancaster PA
Wientjes AG (1968) The influence of sugar concentrations on the vapor pressure of food odor volatiles in aqueous solutions. J Food Sci 33: 1–2
Zuntz N (1892) Beitrag zur Physiologie des Geschmacks. Arch Anat Physiol, Physiol Abst, p 556

Commentary

Beauchamp and Cowart: Two other studies are relevant to the topic of factors influencing perception of sweetness. Pangborn emphasizes that there is a great deal of individual variation in the perceptual effects of mixing taste stimuli, with some subjects even reporting enhancement of tastes in mixtures. Although these individual differences appear to be stable (e.g., Pangborn and Chrisp 1964), it is difficult to determine whether they arise from perceptual differences or simply from differences in habits of response. Results of a study by Kroeze (1982) suggest perceptual differences may indeed play a role in the observed variation. Specifically, individuals differ in the extent to which they experience side tastes when presented with stimuli traditionally used to represent each of the four basis tastes. Apparently, these side tastes are taken into account in rating the intensity of tastes in mixtures such that an appropriate side taste of a masking stimulus may be added to the predominant quality of the other taste in a binary mixture resulting in low masking or even apparent enhancement of the second taste quality. Finally, in the discussion of possible mechanisms for taste mixture interactions and of studies that have attempted to distinguish among those mechanisms for specific taste combinations, the work of Kuznicki and McCutcheon (1979) is relevant. This paper reports an elegant series of experiments whose results point to molecular interactions as the basis for both mixture suppression and sequential enhancement in sweet–sour interactions. It is concluded that sugars, through their capacity to bind protons, may (a) act to reduce the availability of protons to acid receptors when in an acid mixture, thereby suppressing sour taste, and (b) act to remove protons from acid receptors when presented as simple stimuli, thereby opening up sour taste receptor sites and enhancing the perceived sourness of any subsequent acid stimulus.

References

Kroeze JHA (1982) The relationship between the side tastes of masking stimuli and masking in binary mixtures. Chem Senses 7: 23–37

Kuznicki JT, McCutcheon NB (1979) Cross-enhancement of the sour taste on single human taste papillae. J Exp Psychol 108: 68–89

Pangborn RM, Chrisp RB (1964) Taste interrelationships VI. Sucrose, sodium chloride and citric acid in canned tomato juice. J Food Sci 29: 490–498

Bartoshuk: Pangborn is very charitable about my mixture theory (1975). Unfortunately it turned out to be a disaster. For reasons that none of us understand yet, the exponents of sucrose and QHCl functions predict the amount of suppression in sucrose-QHCl mixtures very well. However, the exponents of other functions (e.g., NaCl and QHCl) fail miserably. This failure was what led Kroeze and me to start to explore differences between these mixtures in the CNS. It's always a shame when lovely theories turn out to fail to fit the real world but I have had to recant.

References

Bartoshuk LM (1980) Sensory analysis of the taste of NaCl. In: Biological and behavioral aspects of salt intake. Academic Press, New York, pp 83–98

Bartoshuk LM, Gent JF (1985) Taste mixtures: an analysis of synthesis. In: Taste, olfaction, and the central nervous system. Rockefeller University Press, New York, pp 210–232

Booth: The complexity of frequency-range effects and the restrictions on a Helson model have been illustrated by McKenna (1984). We call such effects 'bias' in our paper (Chap. 10) and we review procedures that are necessary to reduce the resulting distortions in psychophysical measurement and consumer testing. For example, earlier stimuli may make a later stimulus seem more like them than would have been the case without this context. McKenna (1984) replicates this common "assimilative" effect but also shows that an opposite "contrast" effect can occur under some conditions. Therefore we cannot assume any definite contextual effect on the pattern of responses. We must design our experiments to measure and correct for whatever such biasses occur.

Reference

McKenna FP (1984) Assimilation and contrast in perceptual judgments. Q J Exp Psychol 36A: 531–548

Scott: As with a few other chapters in this group, I am troubled by the state of gustatory psychopyhsics. My inclination is to accept rather little of what the psychophysical literature says without a very careful scrutiny of the particular protocols involved.

Pangborn: I share your concerns. Taste responses to both model systems and foods are highly context dependent, and classical measurements of discrimination, intensity, and hedonics are very sensitive to numerous methodological manipulations. *Caveat emptor.*

Rolls: What knowledge have we gained about sweetness interactions in foods by studying aqueous solutions? Are real foods too complex for this type of analysis? How can we learn about complex foods?

Rodin: Why is the literature on taste interactions in actual foods and beverages so sparse and unsophisticated? Is it just too hard to do? Do the developments need to be methodological or are there conceptual problems that need to be ironed out?

Pangborn: Only selected taste relationships, e.g., sweet–sour interactions, seem to generalize from distilled water systems to foods. Although useful behavioral data can be obtained from model systems relative to discriminability and perceived intensity, measurement of hedonic responses to or ingestion of aqueous solutions can be an exercise in futility. Very few reliable data are available for "real" food for several reasons:

1. There are relatively few active investigators in the field of sensory analysis of food.
2. Most industrial researchers do not publish their findings due to proprietary or other reasons.
3. It is unusual for sensory analysts to use good psychophysics, and for psychophysicists to be well-informed about the multiple functional properties of food.
4. Both sensory analysts and psychophysicists are justifiably intimidated by complex stimuli wherein it may be difficult or impossible to disperse additives uniformly, and where a change in one ingredient can cause multiple physical–chemical interactions which alter several sensory attributes simultaneously — appearance, aroma, texture, taste, etc.
5. Reliable experimentation is time consuming and labor intensive. The typical large variability in human responses requires replicate testing of large numbers of individuals.
6. All of the above lead to the need for research funds, which are sparse.

Because of the increasingly expressed need and interest, we should gradually learn more about sensory interactions in and acceptance of complex foods. The effort could be expedited significantly if:

1. There were more collaboration between food scientists and psychophysicists.
2. Academic institutions provided better education in sensory analysis (including data analysis and interpretation).
3. Food processors provided sensory analysts with formulated foods varying in only one ingredient across a reasonable concentration range.
4. More researchers could participate in disciplinary cross-fertilization at workshops such as these.

Booth: It is not too hard to look at taste interactions in real foods and drinks, if you pick your cases and use methods that measure contextual effects objectively.

It has not been done because of the assumption that stimuli have to be physically simple to investigate sensory systems effectively that exclusive presupposition has proved false in the visual system which is built to analyse complex natural stimuli. It could prove equally productive to investigate some of the simpler "natural models" in taste.

Rozin: It is puzzling from an adaptive viewpoint that viscosity suppresses sweetness, since I would think that the two would be correlated in nature. Of course, on adaptive grounds there may be no reason for the sweetness of a ripe fruit to be enhanced...it *is* already sweeter.

Booth: I was surprised to find an almost total reduction in the sweetness of saccharin when I tasted it in cold, evaporated milk.

Pangborn: Your experience parallels our recent result with cold (~10°C) chocolate milk drinks (Pangborn and Wang 1986, unpublished data). We observed that the sweetness of 0.14% aspartame (but not that of the equivalently sweet 10% sucrose samples) was dramatically reduced by increasing additions of fat (0%–36% milk fat). How much of this decrease can be attributed to the fat and how much to the beverage temperature remains to be established.

Chapter 5

Sensory Sweetness Perception, Its Pleasantness, and Attitudes to Sweet Foods

Jan E. R. Frijters

Introduction

The word "sweet" is used in many languages and cultures to denote an object or experience that is pleasant, nice or very much liked. In a metaphorical sense "sweetness" is traditionally closely associated with pleasantness. In the literal sense it refers to gustatory perception, or to taste impressions that result from stimulation of the taste receptors located primarily on the tongue. Although we are used to speaking about "sweet" foods, sweetness does not refer to the physicochemical composition of the foods themselves, nor to molecules of particular substances or sensory physiological processes. "Sweetness" is an attribute of perceived taste and therefore it is a mental concept. It is generally considered to be an element of a complex taste percept, and because it is an element it is called a sensation.

Most investigators in taste research acknowledge that "sweetness" is very important for the regulation of food intake behaviour and for the food choices individuals make. The basic question to be resolved is how sweetness operates in the eliciting and control of intake behaviour and what mechanisms and processes are involved. At present the answer is far from clear. Most of the research carried out on animals has addressed questions related to the effects of physiological factors, such as bodily or metabolic state, on the intake of sweet foods or substances (Weiffenbach 1977). In human studies the psychophysics and hedonics associated with sweetness perception have received considerable attention. There are very few studies of attitudes to sweetness and the pleasantness of sensory sweetness, or of attitudes to sweetness and the behavioural consequences for food selection, despite growing markets for low calorie alternative sweeteners and increased applications of these substances in foods and beverages. The main difficulty in the study of the effect of sweetness on human behaviour is disentangling the various psychological factors associated with sweetness. At least four different aspects which are frequently confused should be considered: the perceived sweetness intensity of a stimulus as a pure sensory phenomenon; the sensory affect (pleasure, displeasure) associated with perceived sweetness intensity; the motivational state in relation to the pleasantness of sweetness as caused by, for example, bodily and nutritional state and experiential factors such as habitation; and the learned attitude to sweetness or to sweet foods and the related like or dislike of sweetness.

In this paper, a few aspects of these factors are discussed in an effort to contribute to a systematic approach to the study of sweetness in relation to perception, food intake and food choice behaviour.

Perceived Sweetness Intensity

Sweetness is perceived when a sufficiently large amount of a particular substance, usually a carbohydrate, is placed on the tongue or at some locations on the palate or throat. This is a psychophysical phenomenon: a stimulus results in a sensation. The sweetness intensity perceived when eating a food or drinking a beverage can easily be manipulated by changing the concentration of the sweet-tasting substance. The fundamental psychophysical rule is that increasing the concentration up to a certain physical upper limit increases the intensity of the sweetness perceived by the observer. In general, doubling the concentration does not automatically imply that the perceived sweetness intensity becomes twice as strong. The precise relationship between the concentration of a compound and its perceived sweetness intensity has been the subject of a great number of studies, not only in psychophysical taste research (Stevens 1969). To summarise this work very briefly, it may be stated that at present two main psychophysical "laws" are used to describe the relationship between concentration of a sweet-tasting substance and perceived intensity corresponding to that concentration (Wagenaar 1975). The first and oldest of these is the Fechner "law" which is given in its general form by the equation:

$$I = \mathrm{k}\log S + \mathrm{C}, \qquad (1)$$

in which I is the perceived sweetness intensity of a certain concentration S of a particular sweet-tasting substance (McBride 1983a,b). C is a constant determined when fitting the equation. Initially this law was developed with I expressed in JNDs (just noticeable differences) but later it was used in combination with I, the perceived intensity, as assessed with category or visual analog scales (Generalised Fechnerian Functions). The second "law", developed as an alternative to that of Fechner, was developed by Stevens. It is called the psychophysical power law and in its general form it is given by:

$$I = \mathrm{k}S^{\mathrm{n}}, \qquad (2)$$

in which I again denotes the perceived intensity of a particular concentration S (Moskowitz 1970). The perceived intensity is assessed by some form of ratio scaling, mostly magnitude estimation. Stevens claimed that this power law holds for all senses and that the constant n in Eq. (2) was the most critical parameter. Its value was, he said, specific for a particular sensory modality. In addition, however, it is also specific for the particular substance under consideration when the gustatory modality is concerned. For example, the value of the exponent for the psychophysical function of sucrose was claimed to be 1.3 (Stevens 1969, 1975), but in various studies values ranging between 0.46 and 1.80 have been reported for this substance (Meiselman 1972).

At present it is generally accepted that the value of the exponent in any psychophysical power function also depends on a host of experimental factors. The range of concentrations used, the stimulus delivery and presentation procedure, and the instructions given to the subjects are all factors that may affect the exponent. It is thus impossible to determine *the* psychophysical function for a particular compound because these kinds of experimental condition always play a role one way or another. However, despite the observed variability in psychophysical functions determined for the same substance in different experiments, it may be concluded that the psychophysical part of sweetness perception can be studied adequately with the available scaling procedures and concomitant methodology.

Experienced Pleasantness of Sweetness

Psychophysical functions describe stimulus–sensation relationships and are usually established in human experiments. One basic assumption (rarely tested) underlying such functions is that the quantitative verbal response given by the observer is a direct and unbiased estimate of the taste intensity perceived as a consequence of stimulation. This means that a one-to-one relationship between the perceived sweetness (internal) and the rated sweetness intensity (external) is assumed to exist. The rated sweetness increases in magnitude in a linear way with the perceived intensity, and the perceived sweetness intensity increases with increased sweetener concentration, rarely in a linear way. From a sensory psychological point of view sweetness is a sensation, but from a cognitive point of view rated sweetness is a type of information ascribed to or associated with the stimulus. It is a cue that enables or facilitates the identification and categorisation of foods and beverages. In the latter view, sweetness as a sensory phenomenon derives its meaning from the "food context" in which it appears.

Sweetness has another feature which is important from a behavioural point of view: the hedonic aspect. Sensorily perceived sweetness elicits or is associated with sensory affect. A sugar stimulus on the tongue gives rise to both perceived intensity and the pleasantness or unpleasantness of sweetness. Other terms used in the literature for sensory affect are hedonic value and affective value. In a number of studies it has been shown that the pleasantness of sweetness depends on the level of sweetness intensity and not on the concentration used to obtain that sweetness intensity, *ceteris paribus* (e.g. Moskowitz 1971; Moskowitz et al. 1974). The pleasantness of sweetness first increases with increasing sweetness intensity and then decreases with further increments of perceived sweetness intensity (cf. Pfaffmann 1980). Coombs (1979) has called this type of function, which is also typical of other stimuli and sensory modalities, a single-peaked preference function. The peak of the function represents the most liked sweetness intensity and is called the ideal point. Another name for the same function is the Wundt curve. In this paper the term preference function is used for a function that relates sweetener concentration to the sensory pleasantness of sweetness, and the term psychohedonic (Frijters and Rasmussen-Conrad 1982) is used for a function describing the relationship between perceived sweetness intensity and the sensory pleasantness of sweetness.

Fitting a single-peak function to data is technically complicated (Coombs and Avrunin 1977), although Moskowitz et al. (1974) have proposed a general equation which can be determined quite easily on an ad hoc basis.

The affective value associated with a particular type of stimulus, such as sucrose-adulterated lemonade, is frequently assessed using a well-known hedonic scale (Peryam and Girardot 1952). This is a simple category scale mostly with ordered categories from 1 to 9. Each category is verbally anchored (Amerine et al. 1965). The ninth category stands for "like extremely" and the first for "dislike extremely". Two arbitrary examples of the use of this type of scale are studies by Pangborn (1970) and Riskey et al. (1979). In these studies the relationship between the concentration of sucrose in water or in lemonade was plotted against the mean judged pleasantness of sweetness. No formal statistical procedure was used to establish these relationships. Fitting such a function between the concentration of a sweetener and the pleasantness of sweetness is based on the reasoning that the latter is a direct function of physical concentration. However, Moskowitz et al. (1974) have shown, in sugar solutions and also in foods, that there is an invariable relationship between the pleasantness of sweetness and sweetness intensity independent of sugar concentration. For example, the ideal sweetness intensity, that is the most preferred sweetness intensity, corresponds to 1.0 M glucose and to 0.21 M sucrose, which are about equal in sweetness intensity according to Moskowitz et al. (1974). From this finding it can be concluded that separate investigation of the psychohedonic function in addition to the psychophysical function is more appropriate than following the approach in which concentration is directly related to the pleasantness of sweetness. The two-step approach enables a distinction to be made between sensory-perceptual and sensory-affective factors when explaining the origin of or change in the pleasantness of sweetness.

Frijters and Rasmussen-Conrad (1982) followed the proposed approach in their study of sweetness liking in relation to overweight (see also Frijters 1983). Essential to their experiment was the use of a response scale to assess the pleasantness of sweetness which differs from the traditional "hedonic scale". The scale used was a so-called unidimensional ideal-point scale, an idea developed independently by McBride (1982). Later, Booth et al. (1983) showed how a simplified stimulus presentation procedure could be used in combination with this type of scale.

The affective taste response to six different sucrose solutions was assessed on an unstructured bipolar scale. A 170-mm line was bisected at the midpoint, which represented the ideal point of sweetness pleasantness. If a particular stimulus was considered not to be sweet enough by the subject, he marked the line to the left of this ideal point. On the other hand if a solution was judged to be too sweet, the line was marked to the right of the subject's ideal point. The distance between a mark and the ideal point was considered to represent the difference in the pleasantness of sweetness of the rated solution and the most preferred sweetness intensity of the individual.

This approach made it possible to determine the psychohedonic sweetness function for each individual. The general form of such a function is given as:

$$A = \mathrm{c} \log I + \mathrm{m}, \tag{3a}$$

in which I represents the perceived sweetness intensity as assessed by magnitude estimation or by any other "ratio" scaling procedure, A is the experienced pleasantness of sweetness, and c and m are constants to be estimated when fitting the function. It appeared from several analyses that this type of function could easily be determined by using simple linear regression analysis. Since $A = 0$ represents the ideal-sweetness point, the value of the intensity corresponding to this point can be calculated by setting Eq. (3a) equal to 0, solving for log I, and then taking the antilogarithm of the value obtained. This approach can also be used when the perceived sweetness intensity is assessed with category or unstructured scales. The relationship between the pleasantness of sweetness and sweetness intensity then becomes:

$$A = \mathrm{c}I + \mathrm{m}, \tag{3b}$$

in which I represents the perceived sweetness intensity assessed by a category or unstructured bipolar scale, A is the experienced pleasantness of sweetness, and c and m are constants to be estimated when fitting the function.

In this two-stage approach it is possible to find the sweetener concentration that corresponds to the most preferred sweetness intensity. This is achieved by substitution of the psychophysical function into the psychohedonic function. Let us assume that in a particular experiment the psychophysical power function that relates sugar concentration to perceived sweetness intensity has been determined according to Eq. (2). Since in this case Eq. (3a) is applicable, the following expression can be derived:

$$A = \mathrm{c}(\mathrm{n}\log S + \log \mathrm{k}) + \mathrm{m}. \tag{4}$$

Thus, substitution of the psychophysical function for the psychohedonic one produced the preference function (this is the function which relates sugar concentration *directly* to the pleasantness of sweetness) as given by Eq. (4). Following this substitution method, Frijters and Rasmussen-Conrad (1982) determined the sucrose concentration corresponding to the most preferred sweetness intensity for each of a number of subjects. After calculating an individual's Eq. (4), A was set equal to 0 in order to obtain the log value of the sucrose concentration corresponding to the most preferred sweetness intensity. The antilogarithms of the values are given in Table 5.1. As stated earlier, it is obvious that the two-stage approach encompassing the separate determination of the psychophysical and psychohedonic function has theoretical advantages over the one-stage approach in which the preference function is determined. The reason is that in the latter case differences between individual functions are difficult to explain because sensory-perceptual and sensory-affective factors are confounded. For example, in a study of the relationship between body weight and sweetness preference, Thompson et al. (1976) distinguished two types of subject on the basis of the shape of their sugar preference function. Subjects with peak pleasantness ratings between 0.06 and 0.6 M sucrose were classified as type I, and those with peak pleasantness ratings above 0.6 M as type II. It is impossible to decide from this type of data whether the difference between the two types of individual is caused by different functioning of the taste perception system (i.e. they had different psychophysical functions for sucrose, cf. Bartoshuk 1979) or by a difference in the psychohedonics of sweetness (i.e. they had different psychohedonic functions). In order to show that the difference between the two types of individual was due to differences in psychohedonic factors, Thompson et al. (1976) determined the psychophysical functions for sucrose. These appeared to be weight independent, something that has often been shown (e.g. Mower et al. 1977), so that it may now be logically concluded that the differences were caused by different sensory-affective processes in the two types of subject.

The two-step approach has been illustrated with studies using sugar solutions. It will be obvious that the same method can be applied to the cases of real foods or beverages. The use of real foods increases the ecological validity of the study on the

Table 5.1. Maximally preferred sweetness intensities (0.2089 M = 10) and corresponding sucrose concentrations (in M). BMI, body mass index (weight/height^2). (Frijters and Rasmussen-Conrad 1982)

Subject No.	Age	BMI	Maximally preferred intensity	Maximally pref. sucrose concentration
		Overweight group		
1	53	26.5	7.9	0.327
2	42	26.4	0.5	0.054
3	38	26.3	7.3	0.224
4	42	25.2	10.2	0.231
5	38	26.2	0.8	0.050
6	42	27.7	14.3	0.237
7	43	26.7	28.5	0.270
8	33	27.9	130.5	0.902
9	43	25.1	6.4	0.217
10	39	27.9	5.8	0.139
11	24	26.5	0.0	0.000
12	37	27.2	5.8	0.127
13	44	25.2	11.3	0.237
Mean	39.8*	26.5*	7.4**	0.191**
		Normal weight group		
14	45	20.8	2.7	0.047
15	46	20.5	3.8	0.139
16	35	20.2	8.0	0.232
17	31	21.6	4.6	0.109
18	41	21.8	10.6	0.201
19	38	21.0	28.3	0.438
20	37	22.0	5.1	0.156
21	42	21.1	6.2	0.081
22	39	21.8	7.9	0.136
23	26	21.1	16.2	0.242
24	32	22.1	6.6	0.159
25	32	21.9	16.6	0.186
Mean	37.0*	21.3*	7.8***	0.154***

* Arithmetic mean.
** Geometric mean, subject 11 excluded.
*** Geometric mean.

one hand, but on the other hand it is more difficult to establish exactly what is the role of sweetness intensity in relation to the pleasantness of sweetness. Foods are perceived through the synesthetical action of a number of senses, not only through the gustatory system. Moreover, factors that inhibit or stimulate the release of molecules, the amount of salivation produced when actually eating and the way of eating are very product specific. Mackay (1985) has pointed out that gelatin desserts have about 16% sugar content, ice creams 15%–20%, canned fruits 25%, jam and fruit preserves 56% (by U.S. law), chewing gum 50%–60% and boiled sweets even 98%. However, according to this author there is reason to believe that the saliva produced on eating is sufficient to decrease the sugar concentration *as eaten* down to 10%. This is the concentration manufacturers use as corresponding to the maximally preferred sweetness intensity. Thus, although it seems that the maximally pleasant sweetness intensity is highly food specific, as judged from the sugar content of the food item, the additional dilution with saliva causes the peak pleasantness not to be different from what has been found in beverages and simple sucrose solutions.

The Origin of the Pleasantness of Sweetness[1]

There is ample evidence that the sensory pleasantness of sweetness has an innate basis (e.g. Crook 1978; Desor et al. 1977; Lipsitt 1977; Nowlis and Kessen 1976; Maller and Desor 1973; Steiner 1973, 1979). Newborn babies ingest more of solutions of fructose and sucrose than of glucose and lactose, which are less sweet sugars on an equal concentration basis. They also ingest a greater quantity of higher than of lower concentrations of the same sugar (Beauchamp and Cowart 1985; Desor et al. 1973), although no entire (single-peaked) relationship between concentration of a particular sugar and ingested quantity has ever been established in newborn babies or young children.

Post-natal dietary experience with sweet-tasting substances or foods can modify the initial sweetness pleasantness, a change that has been shown to be food specific (Beauchamp and Moran 1984). These findings suggest the existence of a learned and stable change in the psychohedonic sweetness function; more specifically, they suggest that the peak of the function has been shifted to the right. In this study testing for the pleasantness of sweetness was done between two feeds so that it can be assumed that the babies were in a need-free situation (Young 1977). It therefore seems highly likely that the shift of the peak of the psychohedonic function cannot be attributed to a change in the motivational state of the subjects. This does not mean that motivations cannot influence the psychohedonic sweetness function. When an individual has been deprived of food for some time the pleasantness of sweetness of a particular sugar solution changes, and so does the psychohedonic sweetness function according to the set-point theory of Cabanac (Cabanac 1971, 1979a; Cabanac and Duclaux 1970; Mower et al. 1977). According to this theory, odour and taste pleasantness are not only a function of sensory quality and intensity but also of physiological usefulness of the stimulus for the organism. In the case of underweight, undernourishment or hypoglycaemia, intense sweetness is extremely pleasant, whereas ingestion of a meal in a normal stable-weight individual, or overfeeding, results in a lower pleasantness of the same sweetness intensity (e.g. Cabanac and Fantino 1977; Cabanac and Rabe 1976; Fantino et al. 1983). According to Cabanac (1979b) two kinds of signal are necessary to produce alliesthesia, his term for an internal, physiologically determined changed in the sensory affective value of a taste or odour stimulus. The first kind is a peripheral signal emitted by peripheral sensors and responsible for the quality and intensity of the sensation. The second is an internal signal responsible for the affective value. Because internal signals determine the sensory pleasantness of the taste and odour of a food, sensory perception indirectly controls the food intake/rejection motivation to a large extent. The ultimate effect is that the organism's "*milieu interne*" as far as the nutrient and energy intake is concerned is regulated to a large extent according to a hedonistic homeostatic regulation principle.

[1] The chapters by Beauchamp and Cowart, Rolls and Drewnowski (9, 11 and 12) deal with this issue more extensively.

Studies investigating the validity of Cabanac's theory have resulted in conflicting evidence. Fantino (1984), in discussing the theory's present status, notes that a number of experiments in this field were inadequately designed. For example, they were carried out at the wrong time of the day, subjects used were not on their set-point weights, and in some "loading studies" the glucose load was too low in caloric content. As particularly damaging to the theory an experiment of Wooley et al. (1972) is often cited. These investigators found that a cyclamate "loading" was as effective as a glucose "loading" in reducing the post-ingestive sweetness pleasantness of a number of sucrose solutions. Rolls and Rolls and co-workers have interpreted this and other findings as an indication that post-ingestive decrease in the pleasantness of sweetness is not a function of biological usefulness per se (e.g. Rolls et al. 1982, 1984). The taste and odour (flavour) of a food itself is a factor that contributes to affective post-meal changes. Repetitious presentation of the same flavoured food is sufficient for what has been named "sensory specific satiety". From these studies it has become clear that "negative alliesthesia" and "sensory specific satiety" are different mechanisms.

Attitudes Towards Sweetness and Sweet Foods

Sensory affective experiences of taste and odour are important for the regulation of eating and drinking in the limited sense, since these factors govern approach/avoidance behaviour (start/stop). However, before actual intake takes place, foods have to be selected from the environment. For food selection attitudes, cognitions, learned food preferences and aversions are important, in addition to social norms about food, food use and eating behaviour.

Sweetness as sensorily perceived and the pleasantness of sweetness as sensorily experienced can be coded into cognitions in the short-term (working) memory of the observer. In fact this is exactly what happens in a psychophysical experiment where the subject rates the sweetness intensity of a particular stimulus. The rating is a cognitive entity and no longer a sensory one: that is, it is a cognitively coded sensation. When processed further into the long-term memory, sweetness can become a psychological attribute ascribed to or associated with the stimulus that generated the sensory percept. If this is the case, sweetness has become a part of the psychological representation of the food and at the same time it is a part of the mental attitude of the observer towards the food concerned. That many foods are "seen" as sweet has been demonstrated by Meiselman (1977), who found that 40% of the 50 foods most preferred by enlisted men out of 378 food items were classified by them as being sweet.

It is surprising that despite the importance of sweetness from an attitudinal point of view only a few empirical studies have been published on this issue. One reason for this may be the difficulty of designing such studies, since it is difficult to decide whether sweetness in itself can be distinguished from sweet foods, sugar-containing foods, health orientation, fear of body weight or caries etc. However, in principle the specific contribution of sweetness as a cognitive attribute of a particular product type could be studied by applying Fishbein's multi-attribute model (Fishbein and Ajzen 1975) developed to predict choice of alternatives. This has yet to be done. The same can be noted about the attitude to normal sugars, or foods containing them, compared with alternative sweeteners.

When designing a study on sweet foods account should be taken of the fact that the cognitive structure associated with a particular type of food can be very specific for that product. This was shown by Prättälä and Keinonen (1984), who studied the constellation of cognitions associated with ice cream, sweetened yoghurt and soft drinks. They selected 22 psychological attributes and obtained ratings from their respondents on five-point Likert scales for these attributes. The average ratings are given for each product in Fig. 5.1. Of these 17 attributes, four basic dimensions were extracted for each of the three products. The dimensions for ice cream and sweetened yoghurt appeared to be identical. These were labelled as Personal Use, Healthfulness, Nutritiveness, and Psychosocial Meaning. For soft drinks the dimensions were labelled as Emotional Meaning, Social Meaning, Healthfulness, and Personal Use. In addition they showed that there was a relationship between frequency of use and meaning structure. The positive emotional, social and psychosocial meanings were more important for daily users of these food items than for moderate or occasional users.

A more specific study on sweetness was carried out by Tuorila-Ollikainen and Mahlamäki-Kultanen (1985). They developed a reasonably reliable nine-item Likert-type scale for assessment of attitude to sugar (Table 5.2) (although it is not clear in the present author's view whether the scale taps only the attitude towards sugar, or to a host of different psychological entities such as sugar, sweetness, sweet-

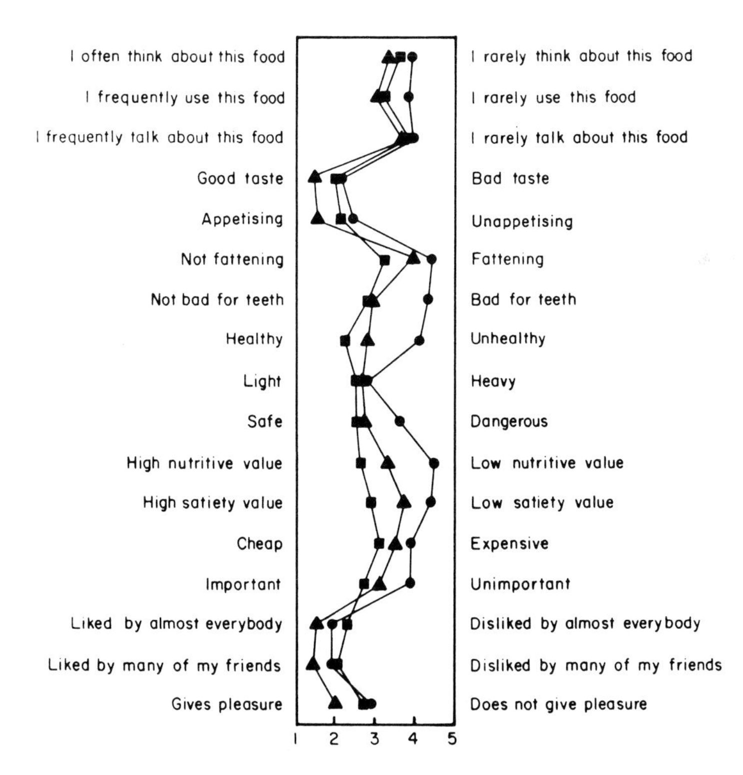

Fig. 5.1. The average ratings (1–5) on 17 psychological attributes of ice cream (▲), sweetened yoghurt (■) and soft drink (●) (n = 515). (Prättälä and Keinonen 1984)

Table 5.2. Sugar Attitude Scale (Tuorila-Ollikainen and Mahlamäki-Kultanen 1985)

Item	Item—total correlation
Positive statements	
To object to the use of sugar is superfluous faddism	0.45
Use of sugar is not as harmful as health educators say	0.37
A sweet bite a couple of times a day does not harm a healthy person	0.32
Negative statements	
Every effort should be made towards restriction of the use of sugar	0.59
Restraining of the use of sugar is one of the most important tasks of health education	0.45
Eating pastry is a part of celebration, not of everyday life	0.43
To sell and to eat sweets is unnecessary	0.40
The serving of sweet desserts in canteens should be restricted to provide a good example	0.38
The price of sugar should be increased if this would help to cut down consumption	0.38

tasting foods, sugar-containing foods, use of sweet foods etc). They also collected data on the reported frequency of use and liking for nine sweet-tasting foods. Factor analysis on these data showed that the foods could be grouped into four clusters: soft drinks, candies, delicacies and pastry. This categorisation suggests that foods within the same grouping are each other's substitute from a food-use point of view and it may well be that they have similar cognitive representations. Another finding in this study was the association between (a) the attitude score of the respondents and (b) their verbally reported liking for and use of soft drinks. Especially women with an unfavourable attitude towards "sugar" reported a lower liking for and a less frequent use of soft drinks. It was also found that women have a more unfavourable attitude towards "sugar" than men. This observation may be related to the effect they attribute to sugar and sugar-containing foods on body weight and its development. Van Strien et al. (1986) have found that women who report a high degree of dietary restraint also report that they do not like sweetness. To what extent this attitude influenced actual intake of sweet foods was not assessed. Other indications that this is an attitudinal effect and not a sensory one comes from two other studies. Frijters (1984) found no relationship between degree of self-reported restrained eating in women and the maximally preferred sweetness intensity as sensorily assessed. Esses and Herman (1984) found that restrained eaters rated the sweetness of 20% and 40% (w/v) sucrose solutions as less pleasant than did unrestrained eaters. In both studies "restrained eaters" were defined by a score on a scale, not by actual food intake in a period of time.

Despite the enormous publicity given in the past decade to the claimed and suggested potential effects on health of the use of sugar or of some alternative sweeteners, no empirical studies have been reported in the literature on the effect of such claims on the attitudes and cognitive structure related to sweetness, sugar, alternative sweeteners and sugar-containing or sweet foods. Also it has not been investigated what role these factors play in the "personal theory" individuals have about body weight regulations and health management, and how these factors affect food choice and food selection. The public debate about the potential "catastrophogenic" effects of the use of sugar and of artificial sweeteners still continues (for example, Clark et al. 1985; Shell 1985). To assess the effects of this debate and of the cognitive dissonance often resulting from these discussions on personal well-being and on food choice and intake behaviour, further studies need to be carried out.

References

Amerine MA, Pangborn RM, Roessler EB (1965) Principles of sensory evaluation of food. Academic Press, New York

Bartoshuk LM (1979) Preference changes: sensory vs hedonic explanations. In: Kroeze JHA (ed) Preference behaviour and chemoreception. I.R.L., London, pp 39–47

Beauchamp GK, Cowart BJ (1985) Congenital and experiental factors in the development of human flavor preferences. Appetite 6: 357–372

Beauchamp GK, Moran M (1984) Acceptance of sweet and salty tastes in 2-year-old children. Appetite 5: 291–305

Booth DA, Thompson A, Shahedian B (1983) A robust, brief measure of an individual's most preferred level of salt in an ordinary foodstuff. Appetite: 4: 301–312

Cabanac M (1971) Physiological role of pleasure. Science 173: 1103–1107

Cabanac M (1979a) Sensory pleasure. Q Rev Biol 54: 1–29

Cabanac M (1979b) Gustatory pleasure and body needs. In: Kroeze JHA (ed) Preference behaviour and chemoreception. I.R.L., London, pp 275–289

Cabanac M, Duclaux R (1970) Specificity of internal signals in producing satiety for taste stimuli. Nature 227: 966–967

Cabanac M, Fantino M (1977) Origin of olfacto-gustatory alliesthesia: intestinal sensitivity to carbohydrate concentration. Physiol Behav 18: 1039–1045

Cabanac M, Rabe E (1976) Influence of monotonous food on body weight regulation in humans. Physiol Behav 17: 675–678

Clark M, Gosnell M, Katz S, Hager M (1985) The perils of a sweet tooth. Newsweek no. 34

Coombs CH (1979) Models and methods for the study of chemoreception-hedonics. In: Kroeze JHA (ed) Preference behaviour and chemoreception. I.R.L., London, pp 149–169

Coombs CH, Avrunin GS (1977) Single-peaked functions and the theory of preference. Psychol Rev 84: 216–230

Crook CK (1978) Taste perception in newborn infant. Infant Behav Dev 1: 52–69

Desor JA, Maller O, Turner RE (1973) Taste in acceptance of sugars by human infants. J Comp Physiol Psychol 84: 496–501

Desor JA, Maller O, Greene LS (1977) Preference for sweets in humans: infants, children and adults. In: Weiffenbach JM (ed) Taste and development: the genesis of sweet preference. U.S. Government Printing Office, Washington D.C., pp 161–172

Esses VM, Herman CP (1984) Palatability of sucrose before and after glucose ingestion in dieters and nondieters. Physiol Behav 32: 711–715

Fantino M (1984) Role of sensory input in the control of food intake. J Auton Nerv Syst 10: 347–358

Fantino M, Baigts F, Cabanac M, Apfelbaum M (1983) Effects of an overfeeding regimen on the affective component of the sweet sensation. Appetite 4: 155–164

Fishbein M, Ajzen I (1975) Belief, attitude, intention and behaviour. Addison-Wesley, Reading, Mass

Frijters JER (1983) Sensory qualities, palatability of food and overweight. In: Williams AA, Atkin RK (eds) Sensory qualities in foods and beverages: definition, measurement and control. Ellis Horwoods, Chichester, pp 431–447

Frijters JER (1984) Sweetness intensity perception and sweetness pleasantness in women varying in reported restrained of eating. Appetite 5: 103–108
Frijters JER, Rasmussen-Conrad EL (1982) Sensory discrimination, intensity perception, and affective judgment of sucrose-sweetness in the overweight. J Gen Psychol 107: 233–247
Lipsitt LP (1977) Taste in human neonates: its effect on sucking and heart rate. In: Weiffenbach JM (ed) Taste and development: the genesis of sweet preference. U.S. Government Printing Office, Washington D.C., pp 51–69
Mackay DAM (1985) Factors associated with the acceptance of sugar and sugar substitutes by the public. Int Dent J 35: 201–209
Maller O, Desor JA (1973) Effects of taste on ingestion by human newborns. In: Bosma JF (ed) Fourth symposium on oral sensation and perception: development of the fetus and infant. U.S. Government Printing Office, Washington D.C., pp 279–291
McBride RL (1982) Range bias in sensory evaluation. J Food Technol 17: 405–410
McBride RL (1983a) A JND-scale/category-scale convergence in taste. Percept Psychophys 34: 77–83
McBride RL (1983b) Psychophysics: Could Fechner's assumption be correct? Aust J Psychol 35: 85–88
Meiselman HL (1972) Human taste perception. CRC Crit Rev Food Technol 3: 89–119
Meiselman HL (1977) The role of sweetness in the food preference of young adults. In: Weiffenbach JM (ed) Taste and development: the genesis of sweet preference. U.S. Government Printing Office, Washington D.C., pp 269–279
Moskowitz HR (1970) Ratio scales of sugar sweetness. Percept Psychophys 7: 315–320
Moskowitz HR (1971) The sweetness and pleasantness of sugars. Am J Psychol 84: 387–405
Moskowitz HR, Kluter RA, Westerling J, Jacobs HL (1974) Sugar sweetness and pleasantness: evidence for different psychological laws. Science 184: 583–585
Mower GD, Mair RG, Engen T (1977) Influence of internal factors on the perceived intensity and pleasantness of gustatory and olfactory stimuli. In: Kare MR, Maller O (eds) The chemical senses and nutrition. Academic Press, New York, pp 104–118
Nowlis GH, Kessen W (1976) Human newborns differentiate differing concentrations of sucrose and glucose. Science 191: 865–866
Pangborn RM (1970) Individual variations in affective responses to taste stimuli. Psychon Sci 21: 125–128
Peryam DR, Girardot NF (1952) Advanced test methods. Food Eng 24: 58–61
Pfaffmann C (1980) Wundt's schema of sensory affect in the light of research on gustatory preferences. Psychol Res 42: 165–174
Prättälä R, Keinonen M (1984) The use and the attributions of some sweet foods. Appetite 5: 199–207
Riskey DR, Parducci A, Beauchamp GK (1979) Effects of context in judgments of sweetness and pleasantness. Percept Psychophys 26: 171–176
Rolls BJ, Rowe EA, Rolls ET (1982) How sensory properties of foods affect human feeding behavior. Physiol Behav 29: 409–417
Rolls BJ, van Duijvenvoorde PM, Rolls ET (1984) Pleasantness changes and food intake in a varied four-course meal. Appetite 5: 337–348
Shell ER (1985) Sweetness and health. The Atlantic 256: 14–20
Steiner JE (1973) The human gustofacial response. In: Bosma JF (ed) Fourth symposium on oral sensation and perceptions: development in the fetus and infant. U.S. Government Printing Office, Washington D.C., pp 254–278
Steiner JE (1979) Oral and facial innate motor responses to gustatory and to some olfactory stimuli. In: Kroeze JHA (ed) Preference behaviour and chemoreception. I.R.L., London, pp 247–258
Stevens SS (1969) Sensory scales of taste intensity. Percept Psychophys 6: 302–308
Stevens SS (1975) Psychophysics. John Wiley, New York
Thompson DA, Moskowitz HR, Campbell RG (1976) Effects of body weight and food intake on pleasantness ratings of a sweet stimulus. J Appl Physiol 41: 77–82
Tuorila-Ollikainen H, Mahlamäki-Kultanen S (1985) The relationship of attitudes and experiences of Finnish youths to their hedonic responses to sweetness in soft drinks. Appetite 6: 115–124
Van Strien T, Frijters JER, Roosen RGFM, Knuiman-Hijl WJH, Defares PB (1986) Eating behaviour, personality traits and body mass in women. Addict Behav 10: 333–343
Wagenaar WA (1975) Stevens vs Fechner: a plea for dismissal of the case. Acta Psychol 39: 225–235
Weiffenbach JM (ed) (1977) Taste and development: the genesis of sweet preference. U.S. Government Printing Office, Washington D.C.
Wooley OW, Wooley SC, Dunham RB (1972) Calories and sweet taste: effects on sucrose preferences in the obese and nonobese. Physiol Behav 9: 765–768
Young PT (1977) The role of hedonic processes in the development of sweet taste preference. U.S. Government Printing Office, Washington D.C., pp 399–417

Commentary

Booth: Responding relative to one's ideal point on an ascribed dimension is not in itself a "scale". It is using the category label of maximum acceptance to assign the stimulus a position on that dimension towards another category label like "no sweetness" or "too little sweetness to be acceptable". As Frijters mentions, ideal relative rating was the response format that Booth et al. (1983) found useful to obtain robust and precise measurements of individuals' perceptive performance. However, this improvement in psychophysical scaling depended on the stimulus presentation procedure, not on any assumptions about measurement properties in the response format. Also we should be clear that this rating method, as distinct from our objectivist use of it, is not a recent development. Rating relative to ideal is well known in the food industry (see Chap. 17). Indeed, Moskowitz (1972) used ideal-relative sweetness rating to measure sweetener optima. He also expounded some of the advantages, including individualisation, but he never pursued those merits against "magnitude estimation" doctrine and that paper has almost never been cited by him or anyone else.

References

Booth DA, Thompson AL, Shahedian B (1983) A robust, brief measure of an individual's most preferred level of salt in an ordinary foodstuff. Appetite 4: 301–312
Moskowitz HR (1972) Subjective ideals and sensory optimisation in evaluating perceptual dimensions in food. J Appl Psychol 56: 60–66

Beauchamp and Cowart: Frijters has presented a two-step procedure for estimating the pleasantness of sweetness rather than that of sugar concentration, as is more common, in order to separate sensory and sensory affective factors that might contribute to pleasantness judgements. This seems a reasonable approach, and it leads one to consider several other difficulties for subjects asked to make hedonic judgements of a sensory experience. As Frijters points out, sweetness-pleasantness functions often increase with increasing sweetness intensity up to an ideal point, after which pleasantness declines with increasing sweetness. There are substantial differences in the location of this ideal point, as illustrated in Table 5.1 of Frijters' paper. Most functions of this type have been generated with aqueous stimuli, and one wonders whether the downward side of the function might reflect factors other than the pleasantness of the sweetness, for example, increasing novelty and/or viscosity. More generally, when one asks for hedonic judgements it may be unclear to subjects what exactly is being asked. To what extent can naïve subjects separate "how much do I like this instantaneous sensory experience" from "how much would I like to drink/eat some of this" and/or "how much of this would I like to drink/eat"? It seems likely that these questions could lead to very different functions in at least some individuals.

Blass: Frijters' scaling procedures that reliably distinguish individuals according to peak pleasantness ratings for sucrose may provide a therapeutic entrée for classifying subpopulations among feeding disorders. Although Mower et al. (1976) did not obtain a relationship in normal populations between peak pleasantness and body weight, a relationship between peak sucrose pleasantness and aberrations in ingestive behaviour might have important clinical applications.

Reference

Mower GD, Mair RG, Engen T (1977) Influence of internal factors on the perceived intensity and pleasantness of gustatory and olfactory stimuli. In: Kare MR, Maller O (eds) The chemical senses and nutrition. Academic Press, New York, pp 104–118

Scott: It is good to see objective formulae relating intensity to hedonics and separating psychophysics from psychohedonics. Our electrophysiological and behavioural analyses of these factors imply that the external (sensory) and internal (motivational) characteristics, which are described here as paired factors determining ingestion, are actually aspects of the same process in the rat, while maintaining independence in the monkey and, presumably, the human.

Schutz: The discussion of sweetness preference after varying amounts of food deprivation is relevant to consumer panel testing of sweet products. Presumably one may find individual variation that is partially innate and partially due to varying food experiences immediately prior to the testing. Spurious preference groups and correlations may result unless this variable is accounted for in some way.

I also want to call attention to the fact that some work (presented in this series) has found malnutrition to depress sweetness preference (in infants), contrary to the general statement regarding underweight and malnutrition increasing sweetness pleasantness as presented by Dr. Frijters.

Section II

The Social Context of Sweetness

Chapter 6

Attitudes Towards Sugar and Sweetness in Historical and Social Perspective

Claude Fischler

Introduction

In the social psychology and the cultural anthropology of sweetness, one of the basic issues at stake is how individuals relate to pleasure and how their own society views that relationship. In most Western societies sweetness, in much the same way as sex, has raised the question of whether the pursuit of pleasure per se is socially and ethically acceptable. In recent decades the trend has been towards increasing tolerance of sexual pleasure, even when dissociated from reproduction. It may be seen as a paradox that, as far as pleasure from sweetness is concerned, there seems to have been comparatively less permissiveness.

The current situation contrasts rather sharply with that in the past. A long forgotten proverb, known in Italy (ca. 1475) and in France (Platine 1539, 1st edn. 1505). stated that "a little sugar never hurt a dish" ("*jamais sucre ne gâta viande*"). In ancient Rome, sugar was known only as a medicine (Pliny mentioned it as such in his *Natural History*: Pline 1949, French edn.). In fact from the time of its introduction in Europe, during the eleventh century (Charny 1965), to the seventeenth century, sugar was highly valued as a spice, as a medicine and as a form of conspicuous consumption (Wheaton 1983; Flandrin 1984; Laurioux 1985; Mintz 1985; Gillet 1985). Even in the eighteenth and nineteenth centuries, and well into the twentieth, after its use had become commonplace, it retained many of its supposed virtues. Today, however, in most Western industrial countries, the media, consumer organisations, also the dental and medical professions issue frequent warnings against sugar and to a certain extent its substitutes. Surveys in various countries indicate that consumers are often wary, and sometimes markedly hostile to sugar; most frequently, they are ambivalent. The old proverb is certainly no longer compatible with contemporary nutritional and culinary values.

This paper is concerned with shifts in attitudes and representations regarding sugar and sweetness. It will examine the process of representational change both in a long-term, historical perspective and in the shorter run. It will look at contemporary attitudes and try to relate them both to earlier trends and to the current social context. The first section is intended to expose general trends, mostly using recent historical literature. The second section is meant to document contemporary attitudes in various Western countries and hopefully to shed some light on smaller fluctuations in beliefs, opinions and knowledge over a shorter time span. Various types of information and data were used. First, results from consumer attitude surveys were obtained either from the literature or from market research organisations in various countries. Second, analysis of articles in the French press from 1975 to 1985 provided some information on the current public attitudes to sweetness and on some of the factors influencing it. Finally, I have examined some of the abundant contemporary literature on food, nutrition and popular dietetics, particularly through books specifically devoted to sugar (though no systematic research was carried out here).

Uses and Perceptions of Sweetness in Western European History: An Overview

The Early Status of Sugar: Spice and Medicine

There are two striking characteristics of the uses of sugar in early European history. Firstly, sugar "was for a very long time associated with medicinal preparations, entering medieval cookery hesitantly (with the exception of England). . . . It was used primarily in dishes for the sick in fourteenth century French manuscripts" (Laurioux 1985). All authors confirm this medical dimension (Tannahill 1974; Flandrin 1982; Mintz 1985 etc.). Wheaton (1983), following Wiegelmann (1967), describes the introduction of sugar as a three-step phenomenon: first it was an expensive medicine; then it became an item in diets for invalids; lastly, as is confirmed by Laurioux, after the fourteenth century sugar assumed a growing importance in general cookery.

The second feature is related to a well known characteristic of medieval cookery. It is now common knowledge that extensive use of spices was a dominant feature in medieval cuisines. Historians agree that, in the culture of the time, sugar fell into the spice category (Mintz 1985; Laurioux 1985; Flandrin 1982).

The medical and culinary virtues of spices were closely intertwined. Literally all spices were believed to have some kind of medicinal significance. In fact, in the dominant medical system of the time, which was basically the humoral medical model derived from Hippocratic and Galenic medicine, all foods were specifically classified according to qualities of *heat, cold, wetness* and *dryness* associated with the four basic fluids in the body (sanguine, choleric, melancholic and phlegmatic humours) (Flandrin 1982; Anderson 1984). Sickness was basically believed to be related to some kind of imbalance among the four qualities. As a consequence, what would today be termed dietetics was a major part of medicine, and medical views played an important role in directing everyday life.

For five centuries both spices and medicines were sold by apothecaries, who can be regarded as the ancestors of both our pharmacists and our grocers (the French word for grocery is *épicerie*, from *épice*, spice). Among spices, it does appear that sugar was a *primus inter pares* of sorts in terms of medical properties. As Mintz (1985) reminds us, so useful was sugar in the medical practice of Europe from the thirteenth until the eighteenth century that the expression "like an apothecary without sugar" came to mean a state of utter desperation or helplessness.

There seems to have been few serious reservations, medical or other, about the use of sugar before the late seventeenth century. Before that its humoral qualities were consistently described in very similar terms, i.e. hot and wet. Albertus Magnus, for instance, circa 1250, wrote: "It is by nature moist and warm, as proved by its sweetness, and becomes dryer with age. Sugar is soothing and solving, it soothes hoarseness and pains in the breast, causes thirst (but less than honey) and sometimes vomiting, but on the whole it is good for the stomach if it is in good condition and free of bile" (*De Vegetabilibus*, quoted in Mintz 1985). In the sixteenth century, in a French version of an earlier Italian book, sugar was also described as hot and wet, good for the stomach, the chest and lungs, able to cure colds and cough, and good for the bladder and kidneys (Platine 1539).

All sweets were not equally praised by the doctors. Honey was contantly described as different from, and by and large inferior to sugar (in its humoral qualities). To a certain extent it seems to have been viewed as an inferior, primitive version of sugar, as if at the lower end of a continuum ranging from primeval, coarse, crude forms of sweets to purified, attenuated, more "civilised" ones. A constant feature of opinions inspired by humoral medicine about sugar was that perfect whiteness was regarded as highly desirable. Tobias Venner confirmed this in the seventeenth century: "Sugar by how much the whiter it is, by so much the purer and wholesomer it is, which is evident by the making and refining of it" (quoted in Mintz 1985).

In the time from the introduction of sugar to the late seventeenth or early eighteenth century one finds no hint of a moral reservation of any kind against sugar, and particularly not from the church. Sugar, provided it is purified and white, is literally immaculate. In fact Mintz quotes from Thomas Aquinas discussing the delicate issue of whether one could eat sweets during fast without breaking religious rules. The answer is very clearly in favour of sugar, thanks to its medical value: "Though they are nutritious themselves, sugared spices are nonetheless not eaten with the end in mind of nourishment, but rather for ease in digestion; accordingly, they do not break the fast any more than taking of any other medicine." All in all, the implicit assumption that the association of sin with sweetness originated in Judgements from the Catholic church is not substantiated by evidence. In fact objections to sweetness and sugar seem to have emerged in Protestant territory and in medical literature.

The Emergence of Reservations About Sugar

Even when early medical reservations were expressed concerning the use of sugar, they did not at first question its traditionally acknowledged qualities, keeping instead well within the logic of the humoral view. It was first *inappropriate* consumption (i.e. by people whose humoral temperament was incompatible with sugar's qualities), then *excessive* consumption, that was advised against by physicians. The shift from qualitative to quantitative judgements seems to have been a gradual one, but once the notion of excess had settled, moral judgements came into play: "excessive" bordered on "immoderate"; immoderate was improper and close to immoral. James Hart, in his 1633 *Klinike or the Diet of Diseases*, wrote: "Sugar is good for abstersion in diseases of the brest and lungs. That which were commonly call Sugarcandie, being well refined by boiling, is for this purpose in most frequent request, and although Sugar in it Selfe be opening and cleansing, yet being much used produceth dangerous effects in the body: as namely, the immoderate uses thereof, as also of sweet confections, and Sugar-plummes, heateth the blood, ingendreth the landise obstructions, cachexias, consumptions, rotteth the teeth, making them look blacke, and withall, causeth many time a loathsome stinking-breath. And therefore let young people especially, beware how they to meddle with it." (quoted in Mintz 1985; Dufty 1975).

In the seventeenth century a definitely anti-sugar trend seems to have developed among physicians. Garencières, a French physician who had emigrated to England, attributed *Tabes Anglica*, or pulmonary phtisis, to excessive consumption of sugar. Dr. Thomas Willis, commonly considered the discoverer of glycosuria (1674), expressed the opinion that sugar was responsible for scurvy. A heated debate seems to have developed in the following decades over Willis' theses. In England, Willis and his followers were opposed by an enthusiastic pro-sugar militant, Slare, who denied any detrimental effect of sugar and went as far as to give a recipe for a sugar-based dentifrice (Moseley 1800).

In France, at the onset of the eighteenth century, the medical community did not seem to have become as upset about sugar as some of the English physicians of the time. Louis Lémery acknowledged that "several authors regard sugar as a most pernicious food", but was sceptical of such views. Sugar remained in his view basically good, particularly for colds and coughs (it is interesting to note in this regard that to this day most cough medicines in France and some other countries come in the form of syrups). Honey, in Lémery's opinion, was much more dangerous than sugar (Lémery 1755, 1st edn. 1702).

Later in the eighteenth century, British reservations about sugar occasionally took on a clearly ethical–political dimension, for instance with the creation in 1792 of the Anti-Saccharite Society, which opposed slavery and boycotted sugar. The same concern was apparent in the French Enlightenment, but it did not in the least spoil the remarkable enthusiasm for sugar and sweets in Diderot's *Encyclopédie*. Bonnet (1976) reports that, throughout the articles about food, pâtisserie and cuisine, sweetness epitomises apparently mutually exclusive values. On the one hand, sweets are associated with qualities of nature: they are infused with fragrances of exotism, with idyllic views of lush forests and good savages; they evoke images of peace and love, of simple though generous ways of life. On the other hand, they also signify modernity and progress: Jaucourt, the author of most of these articles, was highly enthusiastic about the technology of sugar, confections and pastry, which he described and illustrated in endless detail (Bonnet 1976). As to Jean-Jacques Rousseau, he strongly recommended a sweet, vegetarian diet. Evidence that meat is rude, unnatural food, he contended, was to be found in the fact that children spontaneously prefer "dairy foods, pastries, fruits" while Britons love it (*Emile*, quoted in Wheaton 1983; Bonnet 1975).

Spices, which had played such a prominent role, tended to wane after the sixteenth century. Flandrin and his associates (Flandrin et al. 1983) suggest that the typical French incompatibility between sweet and salty tastes gradually began to develop after the decline of spices. Sugar was thus not excluded from food, it was simply kept more and more within well defined boundaries. While in medieval cuisine sugar was liberally added to almost any dish, from the seventeenth century onward French classic cookbooks, like L.S.R.'s *L'Art de bien traiter* (where sweet sauces were sneered at and ridiculed), gradually established rigid rules, a good number of which have lasted to this day. While sweetness was increasingly excluded from most courses, it firmly settled in others.

"Sweet Revolution" in the Nineteenth Century: the "Vulgarisation" of Sugar

According to Johnston (1977), the consumption of sugar (and that of tea) doubled in Britain during the second half of the nineteenth century. According to Mintz (1985), "there is no doubt that the sucrose consumption of the poorer classes in the United Kingdom came to exceed that of the wealthier classes after 1850". Whatever the exact extent of the phenomenon, attitudes and representations related to sugar were probably bound to change in a dramatic way. In the wider population sugar came to be regarded no longer as a medicine, but rather as a gratifying complement to the daily bread. It is reasonable to assume that increasingly widespread consumption probably accelerated a loss of status for sugar in the higher strata of English society.

Although by no means as marked in England, similar phenomena occurred on the continent. The price of sweetness fell sharply in France and consumption also increased there, particularly after the development of production of sugar from beet. Sugar became an indispensable commodity, as is indicated by the fact that, until recently, critical political or economic situations immediately triggered massive hoarding by consumers. But this probably occurred no earlier than well into the twentieth century.

Thus at all points in the history of sugar it seems that increased availability and consumption have brought about changes in its social perception and usage. Such was the case when the first medical anti-sugar theories were brought forward. Such seems to have been the case, again, when sugar became highly consumed by all strata in society. In the process known as "social distinction" (Bourdieu 1982), the upper class (or certain groups in society) adopts or rejects certain practices (i.e. consumption of certain goods, use of socially codified rules of behaviour etc.) according to how effective they are in distinguishing it from other strata, i.e. preserving its identity and status. Obviously, it is tempting to speculate that similar mechanisms could have played a role in the evolution of attitudes to sugar. Widespread use of a commodity in the lower classes can make that commodity lose much of its former aura in the ruling class, or it can determine adoption of new forms of usage. It can also produce favourable ground for new ideological biases to develop.

Contemporary Attitudes

Some modern attitudes to sweetness seem to be based on a complete reversal of the early views described in the first section of this paper. To a certain extent an uncanny form of continuity between past and present can be found in such an inversion. It is comparable to that between a photographic negative and the final print. It is particularly conspicuous in some of the modern forms of "saccharophobia".

Modern "Saccharophobia"

Modern saccharophobia seems to have achieved, to a large extent, an improbable synthesis between two apparently incompatible movements: Nature-worshipping Rousseauist vegetarianism and the late seventeenth century English medical discourse against sugar. There is a possibility that the transition could in part have been effected through nineteenth century English and American vegetarian-oriented movements (S. Kaplan, personal communication; Zafar, unpublished manuscript). Modern vegetarianism has objections to sucrose (Ossipow 1985) while militant anti-sugar literature can be infused with vegetarian themes.

In recent decades a number of books have been published in North America or Britain which, though different in many ways, were all devoted to one topic: the evils of sugar (Abrahamson and Pezet 1971; Fredericks and Goodman 1969; Cleave et al. 1969; Yudkin 1972; Du Ruisseau 1973; Dufty 1975; Starenkyj 1981). One of the most striking — and popular — among them is Dufty's *Sugar Blues* (1975). The book is a compendium of the ills and evils attributed to sugar. The author in fact brought together almost every charge or suspicion ever made against sugar. Hardly any hierarchy or distinction is made in the book between scientific and pre-scientific sources of criticism. The result is a rather apocalyptic picture. Listed in Table 6.1 are some of the problems for which Dufty holds sugar directly or indirectly responsible (the list is not exhaustive).

Comparing the major themes in Dufty's views on sugar with those of the sixteenth or seventeenth century provides a good illustration of the "reversal" principle. The colour of sugar, for instance, has acquired a meaning which is completely opposite to that in Tobias Venner or Platine (*supra*). Sucrose is seen as a "white powder", or "strange crystals", the purity of which is actually threatening: "So effective is the purification process which sugar cane and beet undergo in the refineries, that sugar ends up as chemically pure as the morphine or heroin a chemist has on his laboratory shelves." Thus whiteness and purity have become the mark of doom: sugar is "pure, white and deadly" (Yudkin). Conversely, honey is now consistently regarded as more "natural" and less devoid of nutritious contents. This reversal is no minor

Table 6.1. Some problems allegedly related to sugar consumption (as found in Dufty 1975)

Acne	Hypoglycaemia
Addiction	Immunological malfunction
Alcoholism	Impotency
Allergies	Inability to concentrate
Blood composition problems	Inability to handle alcohol
Brain malfunction	Insect bites
Bubonic plague	Insulin addiction
Cachexia	Irritability
Cancer	Liver overload
Caries	Loss of calcium, minerals
Catatonia	Loss of memory
Childhood diseases	Loss of taste
Circulatory malfunction	Low blood pressure
Coronary thrombosis	Lung cancer (sugar-cured tobacco)
Craving for sweets	Menstrual colic
Criminal behaviour	Migraine
Depression	Negativism
Diabetes	Nervousness
Divorce rate (Starenkyj)	Obstinate resentment of discipline
Drug addiction	Overweight
Fatigue	Saccharine disease (Cleave et al.)
Fevers (mostly in children)	Schizophrenia
Freckles	Scurvy
Gum decay	Sleepiness
Hair loss	Streptococcal throat
Hebephrenia	Suicide rate
Haemorrhoids	Sunburns
High cholesterol	Tuberculosis
Highway accidents (functional hyperinsulinism)	Ulcer (duodenal or gastric)
Hyperactivity	Varicose veins
Hypoadrenocorticism	Vitamin deficiencies

phenomenon. It implies a vision of the world in which Nature is no longer viewed as untamed and menacing, but rather threatened by technology. It may seem to be closer to the Rousseauist view of the worship of Nature. But the *Encyclopédie* had great trust in technology and Rousseau's warnings were against *society,* not technology.

When Dufty compares white sugar to morphine or heroin, it is in his mind no mere metaphor. "The difference between sugar addiction and narcotic addiction is largely one of degree", he writes. Sugar is also presented as akin to alcohol in more than one way. Firstly, the author contends that it is a substitute for alcohol, as he believes is evidenced by the fact that consumption of sugar "zoomed" during Prohibition. Secondly, it turns into alcohol anyway: it "produces alcohol in tummies instead of bathtubs."[1] Addiction has become a major theme in contemporary discourse, one that is by no means restricted to extreme "militant" views such as Dufty's or Starenkyj's (1981). The addiction model provides a modernised expression of the conflict between ethics and pleasure. In *Sugar Blues* there is a pervasive sense of impending sin and punishment. The author reports on the early history of his addiction: his "first sins" (it was "deep purple grape pop" that started him "on the road to perdition"). The road to perdition happily turned out to lead to Damascus: he saw the light, became converted, and set out to pay for his sins and, in pursuit of salvation, became a world missionary.

Ambivalent Attitudes and the Social Management of Pleasure

Obviously the views expressed in this type of literature cannot be considered perfectly representative of general attitudes. Yet studies show that some of the themes found in such works are also present in global representations. In fact, the one most striking feature of current attitudes is ambivalence.

In most qualitative attitudinal studies respondents do associate sweetness to a certain degree with some form of pleasure. Sweetness is perceived as gratifying. It also somehow makes for emotional security. Through offering and sharing, sweets symbolise and help bring about feelings of togetherness, social bonding, festive activity. This appears with equal clarity, for instance, in a British study of the sixties (*Motivation study on consumers' attitudes toward sugar,* conducted by David de Boinot Associated Ltd, for the British Sugar Bureau, February 1966), in a French research report of 1984 (*Vous avez dit sucre? Rapport de synthèse réalisé pour le CEDUS par le CCA,* Paris, December 1984), and in a Belgian "focus group" of the same year.

Yet at the same time, as studies reveal, sweetness is also associated with an obscure sense of danger. It is widely recognised by respondents that sugar does indeed have some drawbacks, although they are not always explicitly described. Disapproving moral overtones are almost always present and they are reminiscent of the guilt-ridden views in saccharophobic literature. Sugar appears as an often unnecessary gratification, one in which one should not indiscriminately indulge.

Is there a salient dimension in the way sweets are used that could make it possible to predict when and under what conditions consumption can be socially and morally

---

1. In a recent publication, a French army physician was apparently more concerned about draftees drinking soft drinks than alcohol. Sweetened beverages, it was contended, "must be forbidden in military restaurants, in which drinks containing less carbohydrates or even wine and beer must be offered" (Fromantin 1985).

tolerable? The answer seems obvious: socialised and ritualised consumption is more acceptable than individual, solitary use. In very much the same way as alcohol or, to a certain extent, food in general, sweets require a clear social context and meaning in order to be perfectly proper: they must be shared, exchanged, given, or consumed in a commensal way. Thus he or she who eats sweets alone runs the risk of incurring implicit reprobation for indulging in solitary pleasure, one that is not put to any useful, legitimate social use.

In the late eighteenth and nineteenth centuries masturbation was believed to be the cause of devastating consequences in both individuals and societies for analogous reasons: semen could not be wasted just for pleasure. Its emission had to be legitimised by reproduction. In fact, the analogy may well be worth closer examination. Biezunski (1983) has drawn a parallel between a text by Dr. Tissot (a Swiss physician who spread the same theories in France which had been made public in England by the unknown author of the famous *Onania, or the Heinous Sin of Self Pollution* — Anonymous 1710; Tissot 1760) on the appalling consequences he attributed to masturbation, and an excerpt from a recent book (Starenkyj 1981) on the dangers of sugar, particularly hypoglycaemia, which are almost equally appalling. In both cases, Biezunski emphasises, the very definition of the symptoms is constructed almost exclusively in *social* terms. The sufferers are described as unable to work, to relate to others, as devoid of any sense of challenge or accomplishment, etc. The victims are seen as responsible for their condition; the alleged symptoms cover almost the whole range of human pathology. In fact Dufty himself confirms the striking similarity between his views on sugar and eighteenth century medical beliefs on masturbation. But the reason, he contends, is that physicians of the time would not admit that the ills they attributed to masturbation were actually caused by sugar.

Another source of ambivalent attitudes resides in the close association of sweetness with childhood. Sweetness is often seen as having a dark, dangerous, addictive side. Sweets are compared to drugs ("*La drogue douce*", — the sweet/soft drug — according to the Paris weekly *Le Point*, 24–30 July 1978) and children to "junkies" (in *Le Matin de Paris*, 5 July 1977). The energy benefit often attributed to sucrose can turn into a danger: hyperactivity. In the words of a Santa Monica woman interviewed in one American study (*The Shopping Crisis, a Research Report from Needham, Harper and Steers* , 1975): "Sugar is really an energy but it is just such a false energy. I mean your energy level goes up and it won't sustain it. And with hyperactive children, they traced a lot of it back to really high sugar content in their diet". The same pattern applies to calories in sugar, which are seen as "false", or empty, i.e. devoid of any *really* nutritious substance.

The question again arises of what the critical element is, if any, that can predict the shift from positive to negative in parental attitudes. In the case of France, an analysis of interviews with a sample of mothers (Fischler 1986a,b) indicates that the answer is parental *control* over what is eaten by children. When sweets acquired by children on their own were mentioned they were contantly referred to as "junk" and described as "filling but not nutritious". Sweets, as well as nibbling and snacking, were seen by parents as a threat to their authority and control over the children's eating and general behaviour. Conversely, sweets seem to be tolerated when "administered" under strict parental control and in meaningful context, for instance as a gratification. Or, in the words of a housewife quoted in the American study mentioned above: "They don't need the candy. But kids do deserve a little treat now and then. The time they do influence me is if they're extra special good. Then I want them to have something."

Fluctuations and/or Trends in Attitudes and Usage

Examining the data or documents made available for the purpose of this paper, mostly from private applied research organisations, it appears that the evolution of attitudes since World War II, particularly in the United States, has been marked by a fairly constant trend with shorter term fluctuations. The basic trend is one of generally increasing diffidence towards sugar. The shorter term fluctuations seem to be related to specific "scares", probably triggered by specialised medical controversies being reported or amplified by the media.

In 1967 a quantitative study of use and attitudes was conducted for the industry by Elmo Roper and Associates on a nationwide cross-section of 2000 men and women 18 years and over. Some of the results allowed comparison with those from similar research in 1945 by the same organisation. The 1967 report reveals changes in responses to the same questions asked in 1945 and 1966. While sugar was believed with equal assertiveness to cause tooth decay, significantly fewer people accepted in 1967 "Too much sugar causes diabetes" and "Sugar is more fattening than most other foods" as true statements. In both studies, the statement that "Sugar is the best source of quick energy" received high approval (70% in 1945, 74% in 1966). The authors of the study assessed as "positive" the audience's response to an advertising campaign in which energy-from-sugar themes had been stressed. Results also showed that only a minority (11%) accepted the statement that "People would be better off in every way if they never ate any sugar". 79% rejected the statement as false.

However, data also showed that a significant part of the sample reported practically never eating sweets (29%), desserts (19%) or sugar (12%) "because they were watching their weight". (In 1982, according to a study sponsored by *Psychology Today*, one-third of the sample reported having cut out sugar). Thirty-nine per cent drank their coffee with no sweetener at all. A significant portion of the sample were using artificial sweeteners (11% regularly, 11% occasionally), and almost one-third were using low calorie soft drinks "a lot". In 1975, a qualitative research based on open-ended interviews (*The Shopping Crisis, a Research Report from Needham, Harper and Steers*, 1975) showed a less sympathetic image of sugar among the general public than did the 1967 study. The report mentioned concern among women about the effects of sugar on children. Not only was tooth decay a common problem, but respondents also tended to believe that sugar, instead of "quick energy", gave "an induced sense of energy, a high of sorts, particularly harmful to children". Weight gain appeared as a major preoccupation of women. Finally sugar was reported as being strongly associated with additives, artificial flavouring and colouring, and processing, and it was considered antithetical to good nutrition. According to a typical verbatim comment from an Atlanta housewife, "they must take out all the nutrition when they put in all the sugar".

At approximately the same time (1974), National Analysts Inc. made a study of attitudes of professionals and opinion leaders towards sugar. Results were interpreted as showing that those in the dental care professions were most consistently negative in their attitudes, with dieticians and opinion leaders following closely. The medical profession was rated as "more temperate" in their evaluations ("over-indulgence" rather than sugar itself constituted a potential danger to health in their opinion). A large majority of the sample thought current sugar consumption was too high in the United States. Yet most professionals did not "subscribe to the need for normal, healthy adults" to restrict sugar intake.

Almost a decade later, the Food Marketing Institute (1983, 1985) introduced in its yearly survey *Trends* new questions to consumers about how concerned they were with the nutritional content of what they ate. Sugar content was a source of concern for 21%. In 1985, the same source reported that "concerns about nutritional content of foods appear to have declined modestly" since 1983, particularly in male shoppers. The *nature* of the concern also had been changing: mentions of the vitamin and mineral content (from 24% to 17%), preservative concerns (22%–13%), and "natural food" concerns (12%–5%) had all declined. In contrast, mention of fat and cholesterol content had clearly increased (9%–13% and 5%–10% respectively). However, concerns about sugar content had remained stable and, because other concerns had declined, "they have become the first and second most often mentioned by shoppers".

In a 1984 survey of a panel of consumers (Better Homes and Gardens Consumer Panel, $n=426$) three-quarters of the respondents reported that their family consumed less sugar than 2 years earlier, and the top two reasons given were health concerns and weight control. Yet 71.1% of the total respondents reported adding sugar to a food product consumed at home by their family. Beverages alone reflected a decrease in sugar usage. At the time of the study, almost 80% of the respondents had tried substitutes and approximately 68% were currently buying them. However, concern was apparent in the panel about suspected health effects of artificial sweeteners.

An intriguing example of rapid short-term fluctuation is provided by Australia. In 1982 overall levels of sugar consumption in Australia had begun to show a declining trend. A quantitative study of consumer attitudes towards (and usage of) sugar conducted by the Dangar Research Group found heightened consumer consciousness of foods and nutrition and showed that the image of sugar was being seriously damaged by publicity about alleged adverse effects of sucrose on general health (apart from "traditional weight and teeth problems"). Two years later another quantitative study was carried out in order to "track any changes in key attitudes" and also to assess the effects of ad hoc advertising campaigns.

The survey, conducted in Melbourne and Sydney ($n=800$), found no evidence of further increase in concern for dietary change since 1982. However, 59% of the population were found to have effected some kind of change in their diet. Sugar appeared to be one of the easiest targets for diet modification and was, indeed, the most common subject of change. As a consequence, the proportion of interviewees reporting they had cut down on sugar had gone up to one-third of the population from a quarter in 1982. However, in spite of this phenomenon "the proportion of people with positive feelings about sugar had increased" in the 16 months between the two surveys "by 14% to 38%", while those with negative feelings had decreased by 14%.

The researchers were somewhat at a loss to explain the positive trend in attitudes (and they did not expand on the negative one in reported behaviour). They admitted it could not be attributed solely to advertising and speculated that the public could be tired "of health scares which are often later contradicted or withdrawn", and could consequently adopt "a more moderate attitude" and accept that "a little of everything makes for a balanced and relatively 'safe' diet". Finally, they hypothesised that "the improved economy" might have "lifted a cloud of negative thinking" (the "state of the economy" is a factor which tends to be resorted to rather liberally in the absence of other, more satisfactory, explanations).

In fact, the remarkable feature is the apparent contradiction between attitudes and behaviour. Reported behaviour seems to be evolving rather consistently in the same direction. Throughout the studies one finds an increase in the proportion of the population reporting they have cut down on sugar or dropped it altogether. Attitudes are less constant, with the two notable exceptions of concerns about tooth decay and fatness (actually we have seen that they date back at least to the sixteenth and eighteenth centuries respectively). What is seen by the market research studies as "ups" in the attitudinal trends might in fact reflect the end of a particularly negative period and a return to more moderate, basically ambivalent attitudes.

But if there are no real "ups", how can we explain the "downs"? An obvious hypothesis is that they are "shock waves" caused by public controversies involving the scientific community or a part of it, consumer organisations, government agencies, and the various economic interests involved, all this being echoed and amplified by the media. The hypothesis seems to be supported by the fact that, in many of the studies reviewed, clear increases in negative attitudes are often associated with specific new themes, for instance hyperactivity in 1974–75 studies (see above).

Sweets and Sugar in the French Press: a Content Analysis

In order to test some of these assumptions and to acquire a better understanding of the relationship between the media, sweetness-related concerns and other factors possibly involved in fluctuations of attitudes, articles published from 1975 to 1985 in 72 French publications in all sectors of the press were analysed. Negative comments on sugar and sweets were found at least once in 268 articles and 42 publications (Fig. 6.1). For each occurrence the nature of the themes discussed was coded, as well as the presence or absence in the story of scientific references (i.e. reports on conferences, interviews with or quotes from scientists or physicians, etc.), the size of the story and whether it was addressing sugar and sweets as a specific topic or not (for instance en passant, in a story on nutrition or other topics), etc.

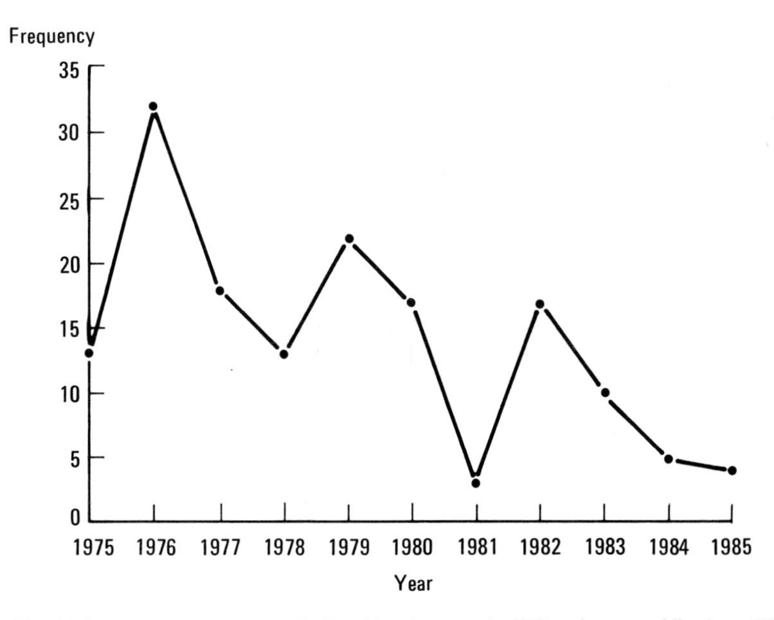

Fig. 6.1. Frequency of stories specifically addressing sugar in 42 French press publications, 1975–1985.

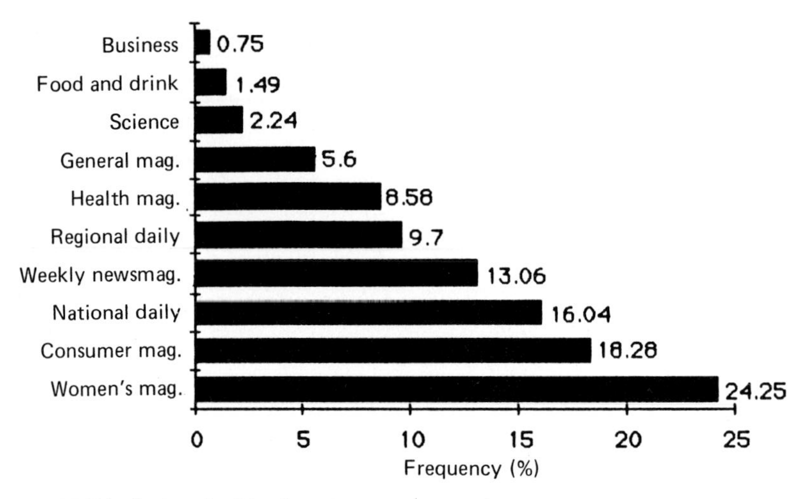

Fig. 6.2. Distribution of articles discussing sugar by type of publication.

Figure 6.2 shows how the relevant stories in the sample were distributed across types of publication. Women and family magazines, publications from the consumerist press, and large circulation dailies and weeklies logically formed the bulk.

Variation in frequency failed to show any clear trend over time. However, the number of occurrences significantly peaked in 1976 (13.43% of all occurrences), in 1978–79 (10.07% and 15.67% respectively) and again in 1982 (10.82%).

The various issues addressed in the stories were unevenly distributed. As Fig. 6.3 shows, overconsumption ("excess"), dental decay ("caries") and obesity were among the most frequently evoked topics. Behavioural effects of sugar, in contrast, is a theme that never really became an issue in France, apparently in contrast to the United States (see above) and probably also Great Britain, where stories on sugar

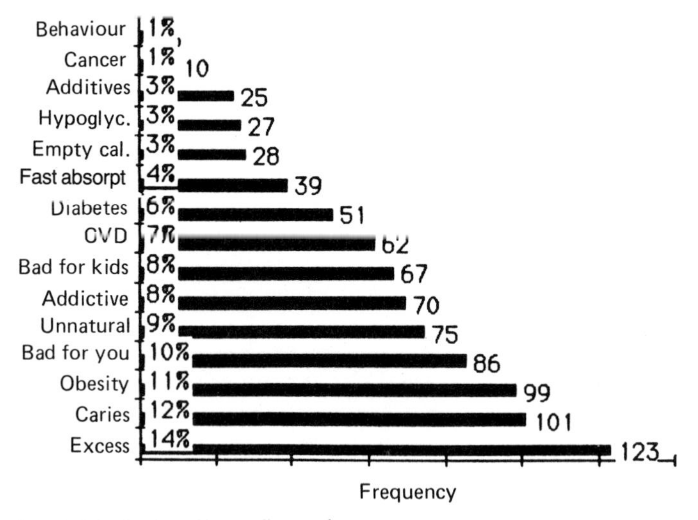

Fig. 6.3. Distribution of issues discussed.

and child behaviour or sugar and crime appeared in the press (see for instance: "Sugar's Bitter Harvest", *The Times,* 1 August, 1983).

Most of the themes were directly or indirectly health oriented. Some were highly specific (caries, diabetes, behaviour); others were totally non-specific (sugar is "bad for you"; we are eating "just too much sugar"). Some distinctly carried strong moral overtones. Overconsumption ("excess"), for instance, was often commented upon in terms of "self-indulgence" or "loss of control". Non-specific categories such as "Bad for kids" and "Bad for you" also frequently bore the imprint of strong judgemental implications.

While, as we have seen, the overall distribution of the stories over time showed rapid, apparently unpredictable ups and downs, certain themes appeared significantly more frequently at given times. Contingency tables clearly show that, indeed, some of them emerged in a transient way. For instance, reference to chemical additives (sugar was criticised for being associated with colouring agents and other additives in sweets or other products) occurred mostly in the late 1970s, with a peak in 1976 (χ^2=24.666, P<0.006). In fact additives were a major issue in France in the early 1970s, when consumer organisations began to acquire influence. Similarly, the "fast absorption sugar" (χ2=29.382, P<0.001) and "hypoglycaemia" (χ^2=20.035, P<0.02) issues emerged around 1977, increased in 1982 and culminated in 1983 (they were mentioned, respectively, in 43.5% and 26.1% of the stories discussing sugar that year). Some of the issues with less specific implications, for instance "Bad for kids", also occasionally went through high activity periods (1975–76). Other themes were constantly present and showed no significant fluctuations over time. Such was the case with obesity, tooth decay and cardiovascular disease. They can indeed be regarded as basic in media concerns regarding sugar and sweetness.

In recent years a significant trend seems to have been in the direction of a decrease in the frequency of specific issues. Since 1982 the stories specifically dealing with sugar have become infrequent (60.7% of the stories were specifically centered on sugar in 1982; 17.4% in 1985) (χ2=63.599, P<0.00001). In many cases no mention of any specific issue was made. Sugar was only alluded to in general, taken-for-granted terms ("The drawbacks of sugar are well known" was a typical statement). It should be noted that this trend seems to have coincided with publication in the press of a small number of stories, based on recent scientific literature, actually partly "rehabilitating" sugar[2] (in previous years few if any pro-sugar stories were published to my knowledge in the French press, except in the form of recipes for preserves or jams). This can probably be interpreted as an indication that 1985 was one of the relatively low-key periods discussed earlier in connection with recent Australian situation. Typically, nevertheless, despite the fact that specific controversy was not currently taking place, underlying general reservations against sugar were still present.

Another question addressed in the study was whether the influence of the scientific, medical community was associated with the emergence in the press of specific issues related to sweetness. Again, simple statistics (cross-tabulation of themes, with the variable "presence of medical reference", χ^2 test) showed that such was indeed the case on a number of occasions. The "fast sugars" issue, for instance (the view according to which sucrose, as a fast-absorption sugar, is inferior to "slow", complex carbohydrates), arose significantly more frequently in stories quoting from inter-

2. "Le sucre réhabilité!", *Le Parisien Libéré,* 2 May 1985, following the publication of a report on *Carbohydrates and dietary balance* (Debry 1985). In the United States literature see: Shell 1985, also on a "rehabilitation" of sugar.

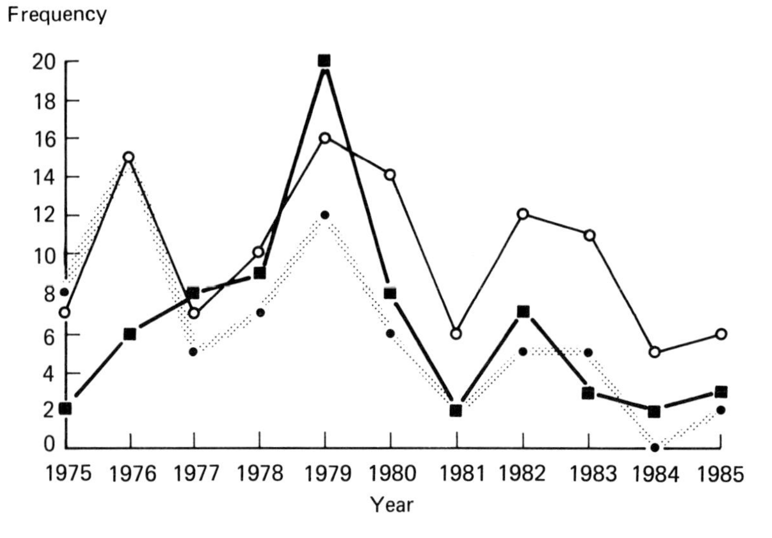

Fig. 6.4. Presence of medical references in press stories (○—○) compared with the presence of two themes: "sugar is addictive" (■—■) and "sugar is bad for children" (●- - -●).

views with physicians or including reports on medical conferences (61.5% of the cases, $P<0.003$). The same holds for "hypoglycaemia" (66.7%, $P<0.003$), while medical science was called upon to testify on "obesity" in less than half the cases (48.5%, $P<0.004$).

The influential role of physicians, in particular, is not limited to specialised, technical issues. It extends well into matters with moral and social implications. For example, the "bad for kids" theme, although rather non-specific and loaded with moral judgements, was indeed very significantly associated with medical events and references (58.2%, $P<0.0001$) (Fig. 6.4).

A closer look at the data shows that, quite logically, it is the paediatricians who have often taken up the issue of sugar and children. But what these rather predictable data draw attention to is that certain members of the medical profession have often given highly judgemental advice of an educational and moral nature, with considerable echo in the media. One medical round table, for instance, at the 1979 *Entretiens de Bichat* (a well-attended and well-covered yearly series of updating sessions aimed at general practitioners) gave rise to no less than nine full stories in various magazines and dailies. The convenor was quoted as stating that "sweets can be addictive to children" and headlines read: "Sweets: Kindergarten Narcotics" (*Ici-Paris,* Nov. 1979) and "Sweets: a Habit-forming Drug" (*Le Figaro, La Croix,* Oct. 1979).

Medicine, Social Ideologies and Individual Behaviour

Two additional remarks can be made from the above results. First, media coverage has been selective. It has favoured certain themes, according to criteria the nature of which was not always demonstrably predictable, but did include *perceived* scientific legitimacy. Second, members of the medical profession have taken public stands which implied educational and behavioural advice. Such stands are indeed reminiscent of those taken frequently in the past, at least since the seventeenth century. The medical profession appears to have played an important role in the construction and evolution of social norms of behaviour. Although the history and sociology of science have as yet not thoroughly examined this role for matters to do with sweetness, one might hypothesise that medicine, in this respect, has been influenced by ideological trends in society as much as it was influencing them.

An important issue raised by this phenomenon should probably be further studied in the future. It is to do with the relationship between science, public health policy and individual behaviour. Karl Popper gave an often-quoted definition according to which, in order to be of a true scientific nature, a theory has to be both demonstrable and *falsifiable* at all times. If one accepts this definition, problems follow when it comes to drawing on scientific knowledge to form or reform individual behaviour, directly or indirectly. If the theory is "falsifiable" at all times, it is at the very least difficult to use it to establish advisable, reasonably durable rules of behaviour. If it is not, then it cannot claim to be scientific, and scientists cannot legitimately use the social prestige of science to impose any form of behaviour (or moral values, for that matter) upon their fellow citizens.

Acknowledgements. I am indebted to Jean-Louis Flandrin, who provided invaluable historical information and made useful comments on the paper, and to Philip and Mary Hyman, who were also extremely helpful in both respects and in addition helped me with my English. The collecting of the data used in the second part of the paper simply could not have taken place without Marie-France Moyal's help.

References

Abrahamson EM, Pezet AW (1971) Body, mind, and sugar. Pyramid, New York
Anderson EN (1984) 'Heating' and 'cooling' foods re-examined. Soc Sci Inform 23: 755–773
Anonymous (1606) Le trésor de Santé. Paris
Anonymous (1710) Onania, or the heinous sin of self-pollution. London (quoted in Biezunski 1983)
Biezunski (1983) Réflexions sur la rationalisation médicale de la saccharophobie. Cedus, Paris
Bonnet J-C (1975) Le système du repas et de la cuisine chez Rousseau. Poétique 6 (22): 244–267
Bonnet J-C (1976) Le réseau culinaire dans l'Encyclopédie. Annales, Economies, Sociétés, Civilisations 31: 891–914
Bourdieu P (1982) La distinction — critique sociale du jugement. Editions de Minuit, Paris
Charny F (1965) Le sucre. Presses Universitaires de France, Paris
Cleave TL, Campbell GD, Painters NS, (1969) Diabetes, coronary thrombosis and the saccharine disease. John Wright, Bristol
Debry G (1985) Part des glucides dans l'équilibre alimentaire. UCC-Communications Economiques et Sociales, Paris
Dufty W (1975) Sugar Blues. Warner Books, New York
Du Ruisseau J-P (1973) La mort lente par le sucre. Editions du Jour, Montréal
Fischler C (1986a) Diététiques savantes et diététiques spontanées: la "bonne alimentation" enfantine selon les mères de famille. Culture Technique 16
Fischler C (1986b) Learned versus "spontaneous" dietetics: French mothers' view of what children should eat. Soc Sci Inform 25(4)
Flandrin J-L (1982) Médecine et habitudes alimentaires anciennes. In: Margolin JC, Sauzet R (eds) Pratiques et discours alimentaires à la Renaissance. Actes du colloque de Tours 1979, 85–95. Maisonneuve et Larose, Paris

Flandrin J-L (1984) Internationalisme, nationalisme et régionalisme dans la cuisine des XIV° et XV° siècles: le témoignage des livres de cuisine. Actes du colloque de Nice (15–17 octobre 1982) Tome 2. "Cuisine, manières de table, régimes alimentaires", 75–91. Publications de la faculté des lettres et sc. humaines de Nice. No. 28. 1st series
Flandrin J-L, Hyman M, Hyman P (1983) Le cuisinier François. Montalba, Paris
Food Marketing Institute (1985) Trends: consumer attitudes and the supermarket. Washington DC
Fredericks C, Goodman H (1969) Low blood sugar and you. Constellation International, New York
Fromantin M (1985) L'éducation nutritionnelle lors du service national. Médecine et armées 13 (8): 809–816
Gillet P (1985) Par mets et par vins. Voyages et gastronomie en Europe (16°–18° siècles). Payot, Paris
Johnston JP (1977) A hundred years eating. Food, drink and the daily diet in Britain since the late nineteenth century. McGill-Queens University Press, Montreal
Laurioux B (1985) Les épices dans l'alimentation médiévale, nouvelles approches. Food Foodways 1 (1): 43–75
Lémery L (1755) Traité des aliments, 3rd edn. Paris
Mintz SW (1985) Sweetness and power. The place of sugar in modern history. Viking, New York
Moseley B (1800) A treatise on sugar with miscellaneous medical observations. London
Ossipow L (1985) Les diététiques naturistes: pratiques et représentations en Suisse. Mémoire de D.E.A. sous la direction de Jean-Paul Aron. EHESS, Paris
Platine (1539) De l'honneste volupté. Paris (Italian edn. ca. 1475)
Pline (Pliny) (1949) Histoire naturelle (transl Ernout). Les Belles Lettres, Paris
Shell ER (1985) Sweetness and health. The Atlantic Monthly, August 1985, pp 14–20
Starenkyj D (1981) Le mal du sucre. Orion, Québec
Tannahill R (1974) Food in history. Stein and Day, New York
Tissot Dr. (1760, 1980) L'Onanisme. Le Sycomore, Paris
Wheaton BK (1983) Savoring the past. University of Pennsylvania Press, Philadelphia
Wiegelmann G (1967) Alltags — und Festspeisen: Wandel und gegenwärtige Stellung. Elwert, Marburg
Yudkin J (1972) Sweet and dangerous. Bantam Books, New York

Commentary

Bartoshuk: The position of sugar in medicinal preparations makes sense in the context of the physiology of the day. Both taste and smell were believed to be guardians of health. What tasted and smelled good was healthy and what tasted and smelled bad was unhealthy. This went beyond ingestion. It has been said that plague doctors wore a kind of mask with a nose cone that could be filled with good smelling substances. The good odours were believed to ward off the plague.

Booth: Does not criticism of presweetened breakfast cereals arise from ill-informed presuppositions? Even with further sugar added, they are still eaten with milk, and they are vitamin fortified too. As well as increasing the nutrient density, the milk buffers mouth acid. In any case the sugar is often washed off the teeth with a hot drink or later parts of the breakfast. Popular convenience breakfast calories prevent accidents and may well improve efficiency, reduce aggression and help prevent weight gain from non-nutritious snacks later in the day. Any healthily used product can be misused, of course, but in this case the problem is with habits of excessively frequent use of confections, not with its market positioning.

Chapter 7

Sweetness, Sensuality, Sin, Safety, and Socialization: Some Speculations

Paul Rozin

Sweetness, Sin, and Sensuality

Sugar is something special. From an agricultural perspective, in terms of calories per acre, it may be the most efficient food of humankind. Along with salt and water it is the only chemically pure substance that humans regularly consume. Sugar also represents purity in its whiteness. Its taste arouses positive responses innately in newborn infants and in almost all others of our species. It elicits delight and pleasure. The words "sweet" or "sugar" connote pleasure and goodness in many domains. At one time in England and other countries when availability of sugar was limited, it was an admired, upper-class food, used only to sweeten the noble's palate (Mintz 1985). The whiter, the purer, the better: a refined substance in more than one sense.

Sugar has been embraced by virtually every culture it has contacted. It stands as the dominant ingredient in some classes of food and is also often incorporated into main-course foods. Consequent upon its availability and reasonable price as a result of exploitation of the opportunities to grow it in the tropical Americas, the increase in sugar consumption in Western Europe was astounding. In England, for example, per capita consumption rose from 4 lbs (8.8 kg) per year at the beginning of the eighteenth century up to 18 lbs (39.6 kg) 100 years later (Mintz 1985). The subsequent falls in price in the nineteenth century that resulted from free trade later caused sugar to become a major foodstuff in English and other cuisines (Mintz 1985). In 1970, about 9% of the calories consumed in the world came from sugar (Mintz 1985). Sugar is a food that has had a major effect on history: on the colonizing of the Americas, the slave trade, the nutrition of the lower classes, and the rise of capitalism in England and other parts of Europe. As documented by Mintz (1985), the social consequences of sugar use were extensive. Over a period of a few hundred years, in England it moved from use primarily by upper classes, in a medicinal or spice-condiment context, to a basic foodstuff used primarily by lower classes.

Of possible relevance to the understanding of sugar acceptance is its relation to three of the most popular warm beverages in the world: coffee, tea, and chocolate. All three of these highly attractive substances contain xanthines (e.g., caffeine) and have attractive aromas and bitter tastes (only modestly so in the case of tea). Sugar played a powerful role in the acceptance of these beverages in the dietary. The sugar masked or counteracted the bitter taste. On account of its high solubility in water, a chemical property of no small significance in the history of the Western world, sugar could be added in large amounts to create a dominant sweet taste and to provide a caloric punch to go with the stimulant effect. In England it was primarily sweetened tea which became central to the nation's food habits and formed the first food incorporated into the work break (Mintz 1985). Unlike chocolate, coffee, and tea (or alcohol, often also combined with sugar), sugar does not produce any obvious pharmacological effects. In particular, it produces no symptoms that suggest illness, possession or loss of control. This early in its history sugar did not meet with the opposition generated by the pharmacologically active beverages, which were attacked as harmful to the body and soul (Mintz 1985).

Attitudes to this universally sensual substance, sweet and tasty, welcoming, a potent and rapid source of energy and calories, a favorite in foods of all sorts, have taken a turn for the worse in some quarters. In parts of American culture it has the ring of "sinful" and "unsafe." Where did this come from? I will deal with the "unsafe" side in the next section. I speculate that the sin may come from three sources. First is its association with sinful substances. It is the vehicle for acceptability of coffee, chocolate, and tea and some forms of sweetened alcohols, and the base for one major distilled beverage, rum. This may have had its effect. Just as sugar is used to make medicines palatable, it can also be used to make poisons or sinful substances palatable.

A second possible basis for the sinfulness of sugar or, in this case, sweetness, derives from an ascetic tendency that turns up in many cultures, often expressed in one form in the United States as Puritan values. It is the idea that anything that is extremely pleasurable, and that includes sex and sweetness, must be bad. There is, in this view, no justification for sweetness for its own sake (and unlike sex, it doesn't have the necessary function of making babies). Sugar-free, for some, has come to mean non-fattening, non-toxic, the moral high ground. This is reminiscent of a Freudian reaction formation. Only a substance once so dearly loved, that tastes so good, could be subject to such strong negative feelings. There is some parallel with the strong feelings of some people about eating meat. Meat, perhaps the main competitor of sugar as the favored food of mankind, and the most nutritionally complete of all foods, becomes an instantiation of both harmfulness and sinfulness. (see Chap. 6, in which Fischler argues that historically the harmfulness of sugar was emphasized much more than its sinfulness).

Thirdly, some people link sugar, but probably not sweetness, to obesity. Since obesity is believed by some people to be a moral failing, this may be another contributor to the sinfulness of sugar.

Why should eating sugar, or perhaps experiencing sweet, arouse such strong feelings? This may be partly because eating is an intimate act: it involves taking something into the body and is the dominant way in which we incorporate the outside world into the self (Rozin and Fallon 1981, 1986). Among traditional cultures there is often an explicitly stated belief that "you are what you eat" ("mann ist was mann isst") (Frazer 1959). Thus there is a common belief that one takes on the properties (personality, behavior) of what one ingests by virtue of incorporating it into one's body. Psychologically this seems like a reasonable inference. Eating lions should make one brave, eating antelopes should make one swift. Eating something offensive or sinful should make one offensive or sinful (Rozin and Fallon 1986). But are these beliefs true and do they influence behavior in modern Western society? We have recently uncovered evidence for an unacknowledged "belief" in "you are what you eat" in American college students (Numeroff and Rozin 1986). We asked them to read a one-page anthropological summary of a particular culture and then answer questions about the characteristics of the people in that culture. Half of our subjects read a summary in which it is mentioned that the people hunt boar only for the tusks, but hunt and eat turtles. The other half of the subjects read the same summary, except that the boar and turtle are reversed: boar is hunted and eaten, but turtle hunted only for its shell. Subjects rated the boar eating people as faster runners and give other indications in their results that suggest imparting of boar characteristics to the boar-eaters and turtle characteristics to the turtle-eaters. The results cannot be attributed to the requirements of hunting animals since both summaries state that both animals are hunted.

The intimacy of eating and some belief that you are what you eat may amplify the feelings about eating all kinds of things and account for some of the fervor of feelings about sugar. Eating sugar may be incorporating sin and becoming sinful; and the sin may be based on the sensuality of sweetness.

I have drawn a stark picture of the "good" and "evil" of sweetness, but at low intensities and for many people sweetness and sugar are unremarkable or just mild sources of pleasure.

Sweetness and Safety

The innate preference for sweet tastes and avoidance of bitter tastes in many generalist species (e.g., omnivores) has a sound ecological basis. In nature, fruits are the most common items that produce the sweet sensation, which is, of course, associated with energy producing sugars. Similarly, a number of major classes of plant toxins taste bitter. This is not to say that in nature what tastes good *is* good (nutritionally) and what tastes bad *is* bad (harmful). But it is to say that there is some truth in the view that good taste goes with good nutrition, and that bad taste goes with bad nutrition. Of course, culture-based advances in agricultural and food technology have led to production of supersweet foods and artificial sweeteners, energy-rich, bland-tasting starches and flours, etc. As a result, the validity of the "tastes good, *is* good" rule of thumb may be further compromised, but it still has some validity in most cultural contexts.

What do children think about sweetness and safety? We know that sugar is a primary determinant of preference in young American children (Birch 1979). We also have evidence that young American children believe that if something is bad for you it will taste bad (Fallon et al. 1984). I know of no data that directly support the "tastes good *is* good" relationship. However, there are indications that 4- to 5-year-old American children tend to think that *anything* people eat will help them grow and be good for them (Rozin et al. 1986a). In short, young children tend to confound the sensory qualities of a food with its consequences. Among adults, or even among children over about 7 years of age, sensory properties and consequences become quite independent attributes of foods. In fact the idea that things taste too good to be good for you (either on moral or health grounds) seems not uncommon among adults in American culture. Sugar is a natural target of this belief.

Such beliefs are reinforced by reports (of varying degrees of reliability) of potential health risks of high levels of sugar intake. The aura of danger supported, perhaps, by some scientific evidence, and the "good is bad" view, are further amplified by a second traditional belief that comes from the same source as "you are what you eat." This belief goes under the name of the law of contagion, one of two laws of sympathetic magic put forward by Frazer (1959) and Mauss (1972) to account for a set of beliefs common in traditional societies. The law of contagion says "once in contact, always in contact." In other words, when two objects touch they may permanently transmit something of their "essence" to one another. Thus a food touched by an offensive entity, such as a person of a lower caste in India or a cockroach retains the properties of these offensive entities, perhaps by a physical but invisible trace. Critically a trace of a contaminant such as a cockroach can impart the properties of that contaminant to the food it touches. We have recently demonstrated that the behavior that gave rise to these attitudes (avoiding objects which have a history of contact with negative, offensive entities) can be commonly observed in members of our culture (Rozin et al. 1986b). For example, people normally reject foods that have briefly contacted offensive or disgusting items like cockroaches or unsavory persons. They also reject laundered clothing that has been worn by disliked or unsavory people. Insofar as contagion is an operative principle in our culture, it may account for a particular aspect of the reaction of some Americans to sugar. Some people treat sugar as a toxin. They seem to feel that any contact with sugar is a source of danger. For example, 11 of 63 parents of children in a paediatric practice in Philadelphia responded "false" to the statement "small amounts of sugar or salt added to foods are perfectly safe" (Casey and Rozin, unpublished data; see also Chap. 6). Given their acceptance of the view that at some levels of intake sugars are toxic, the contagion principle carries this into the micro-level: a trace of sugar transmitted by contact conveys the properties of sugar into the contact substance. This justifies the search for "sugar-free" foods and encourages the labeling of foods in this way.

One might think the positive aspects of sugar might also be transmitted by contagion. This may be true, but contagion seems much more effective in the transmission of negative effects (Rozin et al. 1986b).

People also have particular difficulty in dealing with the idea that something may be toxic at one level but beneficial at another (usually lower) level even though this is the case for many modern medicines. Hence in the minds of some adults it follows from belief in the harmfulness of high levels of sugar that *any* sugar is harmful. Inability to understand this "dose-effect reversal" is more common in children (Rozin et al. 1986a). In our studies with American children we found that most older children (9–12 years old) recognized that something (e.g., potato crisps, sugar) may be acceptable nutritionally in modest amounts but not at high levels, whereas younger children typically thought that if something was not harmful at low levels it would also not be harmful at high levels. Whether we subscribe to either the magical or the inverse dose-effect account, we must confront the fact that what was once a delicious, precious, and highly valued food and taste becomes, for some moderns, a substance that is believed to have potent toxic effects. Fischler (see Chap. 6) has noted some of the same conflicts about sugar that are discussed here and aptly describes the attitude to sugar as ambivalence.

Some Semantics: What's in a Name?

We have invoked two traditional beliefs, "you are what you eat" and the law of contagion, to help to explain the sinful and safety aspects of the psychology of sugar and sweetness. In this section we will begin with a third traditional belief and apply it to reactions to the word "sugar." The belief in question is the second law of sympathetic magic: similarity. According to the law of similarity, the "image equals the object". That is, if two objects have a superficial resemblance (e.g., in appearance) then they share a common essence and have a deep similarity (Frazer 1959; Mauss 1972). In traditional cultures one manifestation of this law, in sorcery practices, is that harm is believed to be caused to a person if replicas of that person, such as a wax likeness, is harmed. One example of the operation of the law of similarity in American culture is that most people are disinclined to eat what they know is a perfectly edible item if it is shaped to appear like an offensive substance, such as a piece of fudge shaped to look like dog feces (Rozin et al. 1986b). Another example is a disinclination to damage photographs of loved or admired persons (Rozin et al. 1986b).

Frazer and Mauss referred to a special instance of the law of similarity in which the name of an object, as one attribute of that object, is treated as if it has all the properties of the object. Thus in traditional cultures there are often prohibitions on saying a person's name under cetain unfavourable conditions. Insofar as this belief is operative in Western cultures, and many of us can sense it in our reluctance to tear up a piece of paper on which the name of a loved one is written, it may affect food selection. Labels with words like "poison," "natural" or "sugar" may alter reactions to the contents of the containers they are attached to, independent of what is *known* to be the nature of the contents. Our demonstration of this phenomenon, ironically, uses sugar as the "safe" or "control" name. We studied the effects of a label, "sodium cyanide," on a bottle that was known to contain sugar (Rozin et al. 1986b). Subjects were presented with two empty and clean bottles. In full view of the subject, sugar (sucrose) from a commercial package bought at a local supermarket, was poured into both bottles. The bottles were closed. The subjects from a university community, were then given two printed labels. One said "sugar (sucrose)" and the other said "sodium cyanide." Subjects were asked to choose one bottle and attach whichever label they selected to that bottle and then to put the other label on the other bottle. We then opened each bottle, and mixed two glasses of sugar water, each with a spoonful from one of the bottles. Subjects then indicated on a rating scale how much they would like to drink sugar water from each glass. Ratings for the sugar water that was made from sugar coming from the cyanide-labeled bottle were significantly lower than the ratings for the water from the sugar-labeled bottle. For some subjects the cyanide label, even though it was known to be placed on a bottle containing sugar, deterred ingestion. It is possible that for some people the word "sugar" might also have a weak magical effect: perhaps positive for some, negative for others. It is also possible that the current practice of indicating that foods do *NOT* contain "harmful" substances, e.g., contains "no caffeine," "no artificial substances," "no sugar," might backfire. The power of the word may not respect the negation. We (Rozin and Ross, unpublished observations) have evidence that the label "not sodium cyanide" has a significant deterrent effect. So much, perhaps, for "no sugar" labels.

There is another aspect of the semantics of sugar. People like to think of categories as distinct and well defined: sugars are "bad for you". In the view of many adults, sugars and starches are totally different entities. Though often chemically equivalent after digestion, they are perceived as fundamentally different entities with respect to health and the body. Sugars are bad, starches are good. Recent evidence that suggests that some starches, (e.g., from potato) are absorbed as rapidly as sugars (Crapo et al. 1977; Bantle et al. 1983) threatens this dichotomy, but it will be hard for the public to digest or assimilate this information. So diabetics have been warned to stay away from the category sugars more than from starch because of the supposed more rapid absorption. We don't always cut nature, semantically, the way the body does.

There is a curious piece of the history of ulcer treatment that illustrates the same point. There was some evidence that the irritant spices, chili pepper and black pepper, might exacerbate existing ulcers, probably by inducing gastric secretion of acid. Since these substances belonged to the generic category, spices, this knowledge was translated as advice to ulcer patients to eat bland foods. Generations of ulcer patients were denied access to tasty foods, as they were advised to avoid foods seasoned with cinnamon, nutmeg, cumin, garlic, and many other spices for which there is no evidence supporting ulcer exacerbation. These items simply fell under the common language term that covered the few irritant spices.

There is another side to this. In the absence of the sugar or sweet "label" or designation, and in the absence of a salient sweet taste, the presence of sugar may be tolerated and enjoyed by those who claim to be averse to it. Thus when sugar is integrated into a cuisine and used more as a spice, it may go unnoticed. Sugar phobics love Chinese cuisine although sugar is present, in modest amounts, in most Chinese dishes. And catsup, a basic American flavoring, contains much sugar but is not perceived or labeled as a sweet substance. The case of honey is most perplexing. In some quarters, despite its sweet taste and high sugar content, it is considered highly preferable to "sugar", probably because it is viewed as "natural".

Sweetness and Socialization

We will take socialization to mean the acquisition of the rules, values, and attitudes of one's culture. At first glance this seems irrelevant to sugar and sweetness because one does not have to *learn* to like or value sweetness. However, there are three ways in which sweetness and socialization interact (see also Chaps. 9 and 15). First, sugar or sweetness play roles as socializing agents. Second, in some cases sugar or sweetness is used in such a way as to oppose some aspects of socialization. Third, we must learn the location of sweetness in our environment (e.g., what foods contain it), its appropriate culinary context, and its meaning within the culture. Before discussing these processes, we must cover some more general ground. We consider first the critical contrast between intrinsic and extrinsic value, and then a framework for describing the different psychological categories of potential edibles. These discussions will clarify fundamental issues in adult food selection which form the targets of socialization.

Intrinsic and Extrinsic Motivation

There are two very different ways through which one can come to conform with cultural attitudes. One is to internalize the attitudes or values in question so that they become one's own. For example, when one comes to like what one's culture values and dislike what it rejects, one has internalized these cultural values. Under these conditions we say one is motivated intrinsically: one seeks the cultural values for their own sakes (Lepper 1983). Alternatively one may modify one's behavior to conform with social values that one has not internalized. Here, the behavior in question may be described as compliance (Kelman 1958). It is concern about the outcome of failing to behave in the prescribed matter that maintains the behavior. Under these conditions we say that the motivation is extrinsic, and the value in question (e.g., stopping one's car at red lights) is not internalized. In the domain of foods, if one likes a food (that is, responds positively to its sensory properties) the motivation is intrinsic; whereas if a food is consumed primarily because we believe it is healthy or polite to do so, the motivation is extrinsic. The intrinsic–extrinsic distinction is critical because intrinsically motivated behavior is self-maintaining and continues in the absence of socializing agents. This distinction is connected with the taxonomy of food acceptance and rejection that we will now present.

A Psychological Taxonomy of Food Acceptance and Rejection

On the basis of interviews and questionnaire studies with American adults, we have described three motivations that explain rejection or acceptance of potential foods; sensory-affective factors (liking or disliking the taste or smell), anticipated consequences (e.g., negative or positive physiological or social events), and ideational factors (knowledge of the nature, origin, or significance of a food) (Rozin and Fallon 1980, 1981). Combinations of these attributes generate four psychological categories of rejection. *Distaste* is a category that applies to substances that are rejected primarily because of dislike of the taste, small, texture, and/or appearance of a food (e.g., lima beans or hot peppers, for most people who do not like them). The *danger* category applies to substances that are rejected primarily because of the anticipated harmful consequences of ingestion. The consequences may be short-term physical or social discomfort or long-term harm (e.g., cancer caused by carcinogens in foods). *Inappropriate* is the category that applies to substances that are rejected primarily on ideational grounds. The substances in question are simply not considered food (e.g., grass, sand or paper). *Disgust* is the category that applies to substances that are rejected on both ideational grounds and because they are thought to be distasteful. The thought of consuming these substances produces nausea. Furthermore, contact of an acceptable food with a disgusting substance tends to make the acceptable food inedible (psychological contamination or contagion). Worms, insects, and feces are examples or substances in the disgust category in American culture.

On the positive side we propose corresponding categories. The category of *good taste* applies to foods that are accepted primarily because of positive sensory properties (e.g., sweets). *Beneficial* is a category that applies to items that are eaten primarily because of their anticipated positive consequences (e.g., medicines, health foods, or socially desirable foods). Of course, many items (e.g., meat or fruit for most Americans) fall between these categories, so that there are two salient motivations for acceptance. Ideational factors in food acceptance are much less salient than in rejection.

These categories represent idealizations and refer to primary as opposed to exclusive motivations. Furthermore a food may fall into different categories for different members of the same culture. Thus for most reasonably affluent people, sugar is a good taste, with secondary anticipated consequences: beneficial (energy generating) for some, dangerous (e.g., energy generating, for dieters) for others. For a few it is a rejected, dangerous substance, and for a very few, a distaste.

Sensory-affective motivation matches up rather closely with intrinsic values. We say we like or dislike foods in this category. We prefer them because we like them. In contrast a dieter who prefers (chooses) cottage cheese in preference to ice cream is probably demonstrating extrinsic motivation and probably likes the ice cream better. The interesting problem in the acquisition of food preferences has to do with intrinsic value; that is, the acquisition of good and bad tastes. It is not puzzling that people come to eat foods that they are told are nutritious, and to avoid foods that are claimed to be harmful. What is puzzling is that some foods, whatever their physiological consequences, come to be liked. They are acquired tastes. How and why does this happen (see Rozin 1984, for a review)?

Sweetness as a Socializing Agent

We do not know how good tastes (or other intrinsic values) are created. There are many suggestions, and some evidence, but definitive studies have yet to be done. The situation is somewhat better for the acquisition of distastes, where nausea has been specifically implicated as a potent cause (Pelchat and Rozin 1982). I will review mechanisms that have been implicated in acquired likes, and possible roles for sweetness in each. There are many counter examples for each of these mechanisms, which indicates that there are surely multiple ways to acquire likes. Furthermore, cultural contexts may create constraints on the operation of specific mechanisms.

The acquisition of likes usually occurs in the course of several to many exposures to a food. Hence exposure is essentially a prerequisite for acquired likes (see Chap. 9 for a discussion of the role of exposure in preference for sweets). Some claim that exposure is a sufficient condition to produce liking (Zajonc 1968). In any event, sweetness may play a role here, as an inducer to acceptance. By sweetening a potential food one may induce temporary acceptance, and hence exposure.

Two explicitly associative (Pavlovian) processes have been proposed to account for acquired likes. In one, the food in question is paired with rapid satiety. In the experimental work of Booth and his collaborators (Booth 1982; Booth et al. 1982), flavors associated with highly satiating foods are liked more than flavors associated with less satiating foods. Since sugars have satiety value, as major components of a food they could contribute to liking in this manner. However, there are many highly satiating foods that are not widely liked. As we indicated above, this suggests the operation of other mechanisms and/or some sort of cultural constraints.

A second Pavlovian process is pairing of a neutral or disliked food (conditioned stimulus, CS) with an already liked food (unconditional stimulus, US). There is no question that within certain constraints one can enhance palatability by adding a desirable food to a less desirable one. The question is whether a change in liking for the CS will result, in the *absence* of the US. This process, termed evaluative conditioning (Martin and Levey 1978), has been demonstrated in a number of non-food domains. The only experimental demonstration with foods for humans used sugar as the US (Zellner et al. 1983). Flavor A was served to young adults in a sweet (palatable) beverage, while flavor B was served equally often in an unsweetened, and hence less palatable, form. There was an enhanced liking for flavor A, even when both flavors were served in the unsweetened form. As with the satiety US, the good taste US may only work in constrained situations. For example, adding sugar to a food that is not considered an appropriate context for sweetness (e.g., some meats) may be counterproductive for adults in American culture.

On the other hand, pairing with sugar may be an important component of the acquisition of preferences for bitter beverages. Coffee, tea, and chocolate are typically served with high amounts of sugar. Those who come to drink and like these beverages without sugar may have a history of initial exposure with sugar. This is, in some sense, a real world replication of the sugar-pairing experiment we just described. But there is one difference. In the sugar-pairing experiment, subjects sampled sweet and non-sweet flavored drinks on a minute-by-minute basis. Therefore, they had no way of learning that one of the flavors was associated with (sugar-induced) satiety: it had to be the good taste of sugar. But when one sips a cup of sweet tea or coffee one often consumes the same item for some period, so that the satiating effect of the sugar is confounded with its sweetening effect. In fact, in the case of consumption of the xanthine-containing beverages, the drinker is experiencing a pleasant aroma, a sweet taste, satiating effects of the sugar (and milk or cream, if added), and the positive pharmacological effects produced by the xanthine. This might amount to quite a positive "jolt" and could contribute to the popularity of these beverages. However, there is no evidence that positive physiological effects other than satiety lead to good tastes. Foods paired with such effects usually become classified as beneficial (Pliner et al. 1985).

Perhaps the most potent mechanism for socialization is, appropriately, social. Basically the perception that people who are admired (for children: parents, appropriate other adults, admired peers, or older children) like to eat a particular food seems to induce liking. This has been demonstrated in the laboratory on a number of occasions, most convincingly by Birch and her collaborators (Duncker 1938; Marinho 1942; Birch 1980; Birch et al. 1980; reviewed in Birch, 1986). Preference for a particular food by people regarded as significant or the use of a food as a reward by such people can enhance liking for that food. The mechanism through which this social process operates has not been fully described. In some instances it could be a case of Pavlovian conditioning in which the positive emotion expressed by the "significant" person induced a similar emotion in the subject by empathic processes. This emotion in the self becomes a US, attached to the food in question (CS). Whatever the mechanism, social factors seem to be powerful influences, but they are far from infallible. Many a parent has failed to induce liking by feigning (or legitimately displaying) pleasure at consuming a particular target food. The role of sugar in socialization by example of "significant" people is indirect. It is simply that sugar is a major factor in inducing positive expressions and likes among adults, and so would indirectly contribute to the socialization experience.

Many parents may recognize the potentially powerful role of social factors in their dealings with sugar, since the majority of American parents in one study claimed that they avoided employing sweets in a positive social context with their children (Ritchey and Olsen 1983). However, the same study reports no significant correlations between three measures of parental behavior or attitudes to sweets, frequency of use of sweets, or reported parental behaviors towards their children with respect to sweets, and the children's preference for sweet foods (Ritchey and Olson 1983).

Sweetness as Reward: Reversing or Preventing Liking

The importance of the perception of social value in acquired liking is highlighted by the conditions under which the converse effect, the disappearance of intrinsic value, occurs. According to self-perception theory (Bem 1967), people infer their attitudes from their own behavior. If a person eats a food without clear extrinsic motivation, there is a tendency to justify the behavior by increasing the value of the food. However, if there is a clear extrinsic factor (e.g., being forced to eat it, the absence of any alternative), people view their behavior as externally constrained, and the object in question declines in value. This "overjustification" effect has been shown to operate in a number of domains (Lepper 1983). In the case of foods these principles suggest that a food will become less liked if there is a well-defined reward for consuming it. Perhaps, as well, a neutral (not yet intrinsically valued food) will not acquire such value if its ingestion is explicitly rewarded. Again, the intuition here is that the person (usually a child in the studies on this subject) concludes that if a reward is needed to induce ingestion the food in question is not desirable in itself. Studies by Birch, Lepper, and their colleagues support these ideas in the food domain (Birch et al. 1982; Lepper et al. 1982). Rewarding a child for consuming a particular food may temporarily increase rate of ingestion, but after the reward is discontinued, liking for the food drops and remains below initial levels. The role for sweetness as the prototypical innate good taste, is to serve as the reward in the form of candy or other sweets.

In summary, results indicate that when a food is used as a reward (indicating high valuation by a significant other), its intrinsic value rises. If ingestion of a food is rewarded with an already positive item (e.g., a sweet), its intrinsic value ultimately drops. Sugar and sweetness can apparently work on either side of the socialization process.

Acquiring Contexts and Meanings

Sugar is a pure white substance, sweetness is a taste quality. The taste quality is distributed in many food items within any cuisine, with no visual indication of the presence of sugar. One thing every child must learn is where the sweetness is: fruits have it, meats don't, etc. In some cuisines the use of sweets is clearly marked in terms of location in the meal (e.g., dessert), in the daily cycle (e.g., sweet snacks), and on special occasions. In Chinese cuisine the sweetness may be spread more evenly throughout the foods available than in the cuisines of Western Europe and North America. Some rules are explicit, others implicit. In the United States it would be considered peculiar to sprinkle sugar on one's hamburger, but catsup, a very sweet substance, is the standard garnishing for hamburgers. In different cultures, at different times, sugar has different meanings. Historically, when it was scarce and expensive, consumption of sweets was a sign of social elevation. In Western countries, under current conditions, the social implications are not very salient.

We know very little about how children acquire the contextual rules of their cuisine for sugar or any other substance. A recent study reports that 3- to 4-year-old American children already know that some foods are appropriate for breakfast, others for other meals (Birch et al. 1984). Beauchamp and Moran (1984) report that children only a few years old distinguish sugar in different contexts, finding it much less pleasant in water than in fruit drinks. But there is much to be learned, and preschool children show notable differences from adults. One is an appreciation of combination rules. Preschool children tend to follow the simple and appealing rule that if they like A and they like B, then they will like A+B (Fallon et al. 1984; Rozin et al. 1986a). Hence these young children typically like chocolate with hot dogs, or cookies with catsup. Apparently, it takes a while to learn appropriate contexts, and that includes where sugar or sweetness is appropriate.

Summary

We started with the statement that sugar and sweetness are something special. The salience and sensuality of sugar and sweetness make them subject to special attention and strong feelings. For some, sweetness stands for sensuality and sin and a threat to safety. It is employed as a socializing agent and, inadvertently, to oppose internalization. It has created and destroyed social class differences. It may provide more pleasure to humankind than any other single substance, and may have had more influence on Western history, in the last 500 years, than any other food. And it all starts with the sweet taste of innocuous-looking white substance.

References

Bantle JP, Laine DC, Castle GW, Thomas JW, Hoogwerf BJ, Goetz FC (1983) Postprandial glucose and insulin responses to meals containing different carbohydrates in normal and diabetic subjects. N Eng. J Med 309: 7-12

Beauchamp GK, Moran M (1984) Acceptance of sweet and salty tastes in 2-year-old children. Appetite 5: 291–305

Bem D (1967) Self-perception: an alternative interpretation of cognitive dissonance phenomena. Psychol Rev 74: 183–200

Birch LL (1979) Dimensions of preschool children's food preferences. J Nutr Educ 11: 77–80

Birch LL (1980) Effects of peer models' food choices and eating behaviors on preschooler's food preferences. Child Dev 51: 489–496

Birch LL (1986) The acquisition of food acceptance patterns in children. In: Boakes R, Popplewell D, Burton M (eds) Eating habits. Wiley, Chichester

Birch LL, Zimmerman SI, Hind H (1980) The influence of social-affective context on the formation of children's food preferences. Child Dev 51: 856–861

Birch LL, Birch D, Marlin DW, Kramer L (1982) Effects of instrumental consumption on children's food preference. Appetite 3: 125–134

Birch LL, Billman J, Richards S (1984) Time of day influences food acceptability. Appetite 5: 109–112

Booth DA (1982) Normal control of omnivore intake by taste and smell. In: Steiner J, Ganchrow J (eds) The determination of behavior by chemical stimuli. ECRO Symposium Information Retrieval, London, pp 233–243

Booth DA, Mather P, Fuller J (1982) Starch content of ordinary foods associatively conditions human appetite and satiation, indexed by intake and eating pleasantness of starch-paired flavors. Appetite 3: 163–184

Crapo PA, Reavan G, Olefsky J (1977). Postprandial plasma-glucose and -insulin responses to different complex carbohydrates. Diabetes 26: 1178–1183

Duncker K (1938) Experimental modifications of children's food preferences through social suggestion. J Abnorm Soc Psychol 33: 489–507

Fallon AE, Rozin P, Pliner P (1984) The child's conception of food: the development of food rejections with special reference to disgust and contamination sensitivity. Child Dev 55: 566–575

Frazer JG (1959) The new golden bough: a study in magic and religion (abridged). Macmillan, New York (edited by Gaster TH, 1922; original work published 1890)

Kelman HC (1958) Compliance, identification, and internalization: three processes of opinion change. J Conflict Resol 2: 51–60

Lepper M (1983) Social-control processes and the internalization of social values: an attributional perspective. In: Higgins ET, Ruble DN, Hartup WW (eds) Social cognition and social development. Cambridge University Press, New York, pp 294–330

Lepper M, Sagotsky G, Dafoe JL, Greene D (1982) Consequences of superfluous social constraints: effects on young children's social inferences and subsequent intrinsic interest. J Pers Soc Psychol 42: 51–65
Marinho H (1942) Social influence in the formation of enduring preferences. J Abnorm Soc Psychology 37: 448–468
Martin I, Levey AB (1978) Evaluative conditioning. Adv Behav Res Ther 1: 57–102
Mauss M (1972) A general theory of magic. (Brain R, Transl). W.W. Norton, New York (original work published 1902)
Mintz S (1985) Sweetness and power. The place of sugar in modern history. Viking, New York
Nemeroff C, Rozin P (1986) Evidence for "belief" in the magical maxim, "you are what you eat" in American culture. (submitted for publication)
Pelchat ML, Rozin P (1982) The special role of nausea in the acquisition of food dislikes by humans. Appetite 3: 341–351
Pliner P, Rozin P, Cooper M, Woody G (1985) Role of medicinal context and specific postingestional effects in the acquistion of liking for tastes. Appetite 6; 243–252
Ritchey N, Olson C (1983) Relationship between family variables and children's preference for and consumption of sweet foods. Ecol Food Nutr 13: 257–266
Rozin P (1984) The acquisition of food habits and preferences. In: Matarazzo JD, Weiss SM, Herd JA, Miller NE, Weiss SM (eds) Behavioral health. A handbook of health enhancement and disease prevention. Wiley, New York, pp 590–607
Rozin P, Fallon AE (1980) The psychological categorization of foods and non-foods: a preliminary taxonomy of food rejections. Appetite 1: 193–201
Rozin P, Fallon AE (1981) The acquisition of likes and dislikes for foods. In: Solms J, Hall RL (eds) Criteria of food acceptance: how man chooses what he eats. A symposium. Forster, Zurich, pp 35–48
Rozin P, Fallon AE (1986) A perspective on disgust. Psychol Rev
Rozin P, Fallon AE, Augustoni-Ziskand M (1986a). The child's conception of food: the development of categories of acceptable and rejected substances. J Nutr Educ
Rozin P, Millman L, Nemeroff C (1986b) Operation of the laws of sympathetic magic in disgust and other domains. J Pers Soc Psychol: 50: 703–712
Zajonc RB (1968) Attitudinal effects of mere exposure. J Pers Soc Psychol 9 (2): 1–27
Zellner DA, Rozin P, Aron M, Kulish C (1983) Conditioned enhancement of human's liking for flavors by pairing with sweetness. Learning and Motivation: 14: 338–350

Commentary

Booth: A difference in terminology between fields should be registered because it is unfortunately liable to confuse thinking about food attitudes unless the difference is made very clear. Social psychology has for decades made the distinction between intrinsic and extrinsic motivation described by Rozin. It is the common sense (and philosophical) distinction between goals valued in themselves and means valued only for their ends. However, consumer psychology also used these words to make the distinction between characteristics of the product itself (intrinsic ascriptions) and its attributes as marketed (extrinsic ascriptions).

How can it be said that "ideational factors are much less salient in food acceptance than in food rejection"? The breakfast food classification is very powerful and very early acquired (Birch et al. 1984). I (Booth 1977) have suggested that poor satiability of milk, fruit juices, and wine arose from their "culinary promiscuity." Is it not possible that there is powerful "eugust" for chili, coffee, and the like, as I have suggested for cigarettes in smokers?

References

Birch LL, Billman J, Richards S (1984) Time of day influences food acceptability. Appetite 5: 109–112
Booth DA (1977) Appetite and satiety as metabolic expectancies. In: Katsuki Y, Sato M, Takagi SF, Oomura Y (eds) Chemical senses and food intake. University of Tokyo Press, Tokyo, pp 317–330

Beauchamp and Cowart: Rozin makes the interesting point that while it is widely agreed that sugars taste good, there is a tendency to view them as dangerous and to be avoided. However, it is important to distinguish sweetening agents from sweetness. Is there any evidence that the sensory experience of sweetness, when divorced from the calories of sugars or the presumed evils of refined white sugar, is considered dangerous? The huge success of non-caloric sweeteners, as well as acceptance of "natural" sweeteners (e.g., honey, brown sugar) by many who reject both refined sugar and artificial sweeteners, would argue against a psychological proscription against sweetness itself.

The recent papers by Beauchamp and Moran (see Chap. 9 by Beauchamp and Cowart) may well be relevant to Rozin's discussion (last section) of the development of the child's conception of the appropriateness of sugars in various food contexts.

Blass: Rozin's identification of transitions during childhood may serve as a prospective diagnostic device for the development of feeding disorders. Aberrant feeding patterns will not generally be attributed to altered sensory systems. More likely candidates are hedonic mechanisms, differing belief systems, and conflict resolutions. Undoubtedly more refined classifactory systems will be able to identify retrospectively (and ultimately prospectively) different etiologies and their appropriate clinical interventions.

Section III

Inborn and Acquired Aspects of Sweetness

Chapter 8

Opioids, Sweets and a Mechanism for Positive Affect: Broad Motivational Implications

Elliott M. Blass

Introduction

Responsiveness to sugars and sweetness is phylogenetically ancient, being manifest as chemotaxis even in motile bacteria such as *E.coli* (see Adler 1971 for review). Electrophysiological and behavioral characteristics of sugar receptors and feeding have been identified and analyzed in certain invertebrates, most notably the blowfly (see Dethier 1978 for review), and taste receptors and electrophysiological responsiveness to sugar solutions have been identified in frogs and fishes (Bardach and Atema 1971; Sato 1969; Sato et al. 1969).

In certain birds, e.g., the nectar-feeders and their phylogenetic predecessors (linnets, orioles), and in mammals the sugar taste exerts profound alterations in behavior. Sugar responsiveness, the apparent perception of sweet, and the positive affective state associated with sugar characterize the mammalian system at birth (Steiner 1979; Blass et al. 1984), and indeed prenatally (Mistretta and Bradley 1986; Hall and Terry 1985, cited in Hall 1985). Thus sugar responsivity is a phylogenetically ancient, ontogenetically viable sensory, perceptual (motivational) quality (Pfaffmann 1961; Pfaffmann et al. 1977; Sheffield and Roby 1950).

Sugar solutions, as best exemplified by sucrose, have a unique characteristic in rats. The more concentrated the sucrose solution, the greater the affective state it engenders. This is a robust phenomenon that has been reliably captured by different measurements of affect. Thus, in rats for example, the more concentrated sugar solution is always initially chosen over the less concentrated one (of the same sugar) in brief exposure tests and this holds for all pairwise comparisons (Shuford 1959; Young 1957, 1966; Young and Trafton 1964). Similarly, as sucrose concentration increases, so does the frequency of ingestive responses during brief (1-min) exposures (Schwartz and Grill 1984, 1985). When postabsorptive factors are bypassed by esophagal fistula, sucrose intake reflects concentration. In his classic study, Mook (1963) demonstrated that rats sham drinking through an esophageal fistula–gastric cannula preparation, drank increasingly more fluid of increased sucrose concentration. Moreover rats that received water intragas-

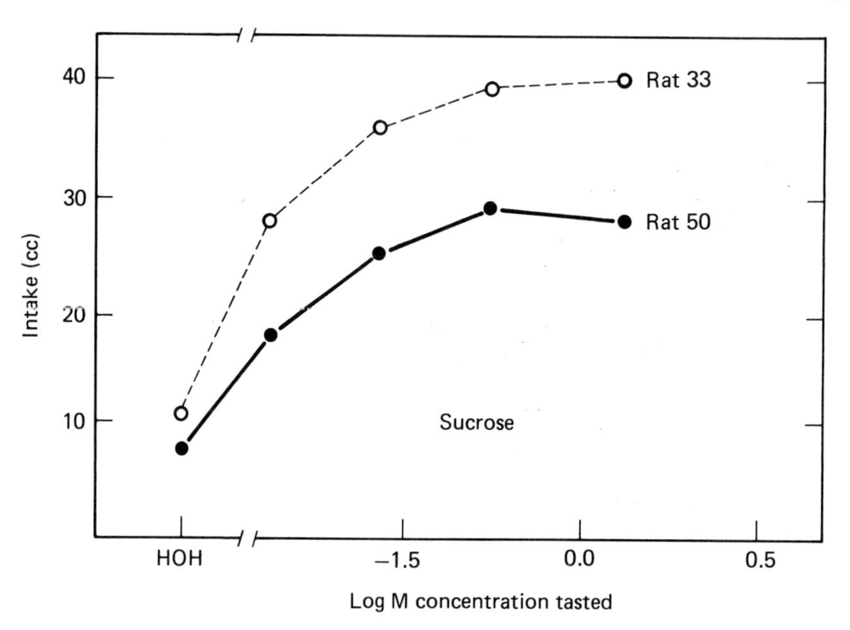

Fig. 8.1. Sucrose intake in rats with an esophageal fistula–gastric cannula. Ingested fluid escaped via the fistula. It was not replaced in rats with exaggerated intake. It was replaced with water, delivered through the cannula during drinking. (Adapted from Mook 1963)

trically at the rate at which drinking occurred, ingested more with increased sucrose concentration. This latter finding is shown in Fig. 8.1, which demonstrates a monotonically increasing rate of intake with increasing concentration. Guttman's (1953) study also addressed the relationship between motivation and sucrose. Guttman demonstrated that, in addition to enhanced preference and volume intake, work output on a fixed interval 1-min schedule accelerated with sugar concentration. Thus rats expanded operant rates with apparent increased sweetness.

The sweet system in rats is unique. Increased sucrose concentration has several effects: (a) it enhances neurophysiological responses at all levels of the sensory neuraxis studied (Scott and Perrotto 1980); (b) in normal rats it increases the frequency of consummatory responses in short-term tests that are not affected by postabsorptive factors; (c) it enhances intake in sham-drinking situations that reflect hydrational need (Blass et al. 1976) but especially taste properties of a solution (McCleary 1953); (d) it causes enhanced preference; and (e) it increases work output. No apparent sensory limit has been established. This is clearly not the case for any other class of tastes or for other modalities where increased stimulus intensity at some point leads to rejection, aversion, or downright pain.

The properties of the sweet system appear to be present at birth. A number of studies have demonstrated that newborn human and rat infants prefer the taste of sugar upon its first presentation (Desor et al. 1973; Steiner 1979; Beauchamp and Moran 1982; Crook 1978) even when the sugar is the very first food that the infants have ever tasted (Blass et al. 1984). According to Hall and Bryan (1980), who studied 3- and 6-day-old albino rats, sweet taste produces an affective state as judged by behavioral arousal and excitement, that increases with increased concentration. Hall and Terry (cited in Hall 1985) have extended this to rat fetuses. In human newborns this sensitivity becomes translated into a state of positive affect that has been captured photographically by Steiner (1974, 1977), who presented a smile elicited by sucrose in a newborn human. More objective studies in humans by Desor et al. (1973) confirm the preference effects seen in rats.

The taste of sucrose has an additional motivational quality in human infants: it reinforces conditioning. Blass et al. (1984), studying newborn human infants in their first 2–48 h of age, classically conditioned forehead stroking that predicted the delivery of a 20% sucrose solution intraorally. Sucrose immediately following stroking allowed the previously neutral stroking to gain control over the suckling system. The major components of suckling were elicited by forehead stroking. This is demonstrated in Fig. 8.2, which plots the orient response (normally elicited in hungry infants by cheek or lip contact with the breast, nipple, or finger) during the 12 10-s epochs of training. Clearly sucrose was reinforcing. Stroking on the forehead midline that predicted sucrose elicited many more orient responses during stroking presentation than did stroking that did not predict stimulation. The acquisition of this temporal relationship is not confined to the conditional stimulus of stroking, but can also be obtained through auditory conditioning. Clicks from castanets also very effectively elicit the orient response when predicting sucrose delivery (Blass and Hoffmeyer 1986, unpublished work).

What are the mechanisms that underlie the positive affect associated with

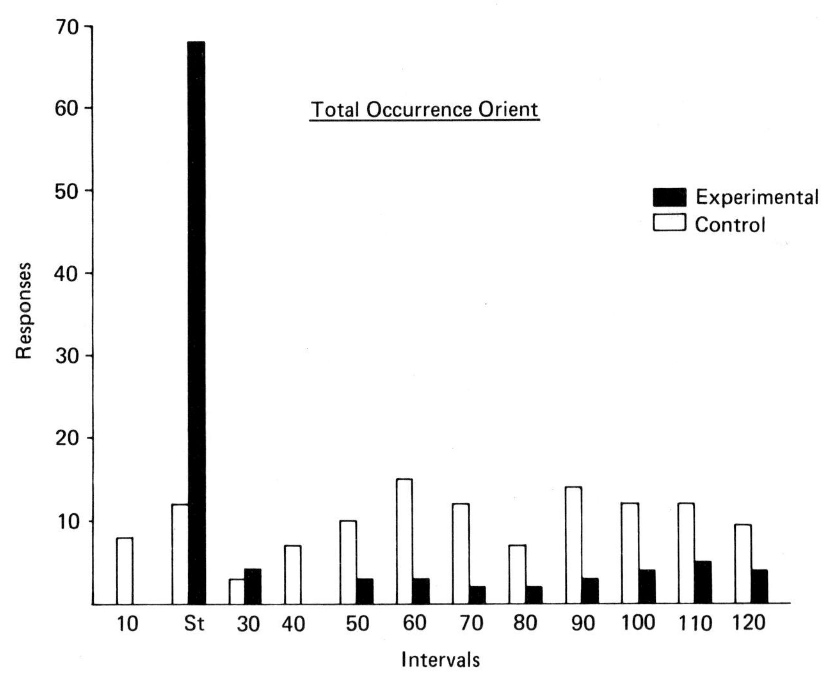

Fig. 8.2. Cumulative number of orient responses during each of the 12 10-s intervals that constituted a conditioning trial. There were eight infants in each condition and each infant received 18 trials. *Solid histogram*: infants for whom forehead stroking predicted immediate sucrose delivery. *Open histogram*: infants for whom sucrose followed stroking by delays of 10, 20, or 30 s. (Blass et al. 1984)

sugar? Through what mechanism might increased electrophysiological activity become transduced into sweet taste, positive affect, and enhanced motor output? Endogenous central opioids are good candidates. They may control food intake in general (Yim and Lowy 1984; Morley et al. 1983), but they especially contribute to sugar ingestion (Siviy et al. 1982).

Opioid antagonists, such as naloxone or naltrexone, markedly reduced the intake of sweet solutions. For example, Ostrowski et al. (1980) demonstrated that 2 mg naloxone/kg body weight reduced intake of a 10% (w/v) sucrose solution by 36%–46% in both non-deprived and 23-h deprived rats. In fact opioid antagonists reduced saccharin intake at all concentrations tested and flattened the preference aversion function, especially for the more concentrated solutions of saccharin (Le Magnen et al. 1980; Siviy and Reid 1983). This also holds true for the sugars. Le Magnen et al. (1980) showed convincingly that daily administration of the opioid antagonist naltrexone made animals indifferent to the taste of either glucose or saccharin solutions. The normal preference was immediately re-expressed with the cessation of the naltrexone regimen. Thus sweetness affect as expressed in choice and volume drunk depends upon the availability of receptors for endogenous opioids.

These findings are supported by morphine administration. According to Siviy et al. (1982) morphine in a dose of 2 mg/kg enhanced the preference of saccharin solutions, especially the more highly concentrated (0.15%–1.0%) ones that are not necessarily preferred to water. Rats simultaneously choosing between a saccharin solution and water drank considerably more saccharin than water following morphine injection. Saccharin choice was not very much affected, however, when the saccharin was diluted. It is of interest in this regard that animals feeding under distress preferentially select sweet substances (Bertiere et al. 1984). This stress-induced preference can be blocked by opioid antagonists. Yet blocking receptor availability does not reduce intake below baseline water levels, suggesting a modest role for opioids in control of fluid to satisfy hydrational need.

Failure to reduce intake below baseline provides a very interesting parallel with how naloxone affects the rate at which rats work for electrical stimulation of the brain. Administering morphine to rats that are pressing a lever to deliver current to the lateral posterior hypothalamic areas enhances the rate at which self-stimulation occurs (Belluzi et al. 1976; Broekkamp et al. 1976). The enhanced rate can be reversed by naloxone administration (Belluzi et al. 1976). Yet naltrexone does *not* reduce baseline levels of self-stimulation (van der Kooy et al. 1977). A parallel phenomenon of non-opioid baseline behavior that can be opioid enhanced can also be established for pain-withdrawal baseline thresholds. Rats taken out of their home cages and tested during opioid antagonist treatment perform essentially like control rats (Lieblich et al. 1983; North 1978; Fanselow 1985; Goldstein et al. 1976). Yet stress (Fanselow 1985) or opioids (Chance 1980) elevate pain thresholds and these elevations *are* naloxone reversible (very low doses are effective, Amir et al. 1979). To the point of the present communication, drinking in response to dehydration or deprivation is not blocked by naloxone treatment, or only marginally so, but its expansion by sweet tastes is almost totally blocked (see Reid 1985 for review).

Sucrose might be perceived as hedonically positive because the sucrose taste itself might cause the release of endogenous opioids. Support for this idea comes from enhanced preference and intake following morphine administration and preference blockade and fluid reduction following antagonist injections. Support for this idea also comes from an unlikely source: feeding under stress. Stress-induced feeding is a well established and clinically very interesting phenomenon. Stress in rats, as elicited by tail-pinch for example, is a potent elicitor of ingestive behavior in rats and this effect is naloxone reversible (Bertiere et al. 1984). Moreover stress-enhanced pain thresholds are also normalized by opioid antagonists (Fanselow 1985; Lester and Fanselow 1985).

Experimental Analyses

Because sucrose appears to cause positive affect through opioid release, and because stress-induced feeding is apparently effective through the opioids released by stress, it became of considerable interest to determine whether and how sucrose ingestion would influence pain thresholds and response to stress.

Our initial understanding of the relationship between sweet taste and pain was provided by Lieblich et al. (1983), who studied rats bred for high levels of self-stimulation (Lieblich et al. 1978) and a marked preference for high concentrations of sweet solutions (Ganchrow et al. 1981). Paw lift latencies were considerably elevated after these rats had drunk a saccharin–sucrose cocktail solution for 28 consecutive days. These elevated latencies were reversed by naloxone treatment at testing, thereby implicating opioid mediation. Accordingly, when chronically drinking this sweet solution, these selected rats may have sustained elevated levels of circulating opioids, and/or did not habituate to elevated normal opioid levels.

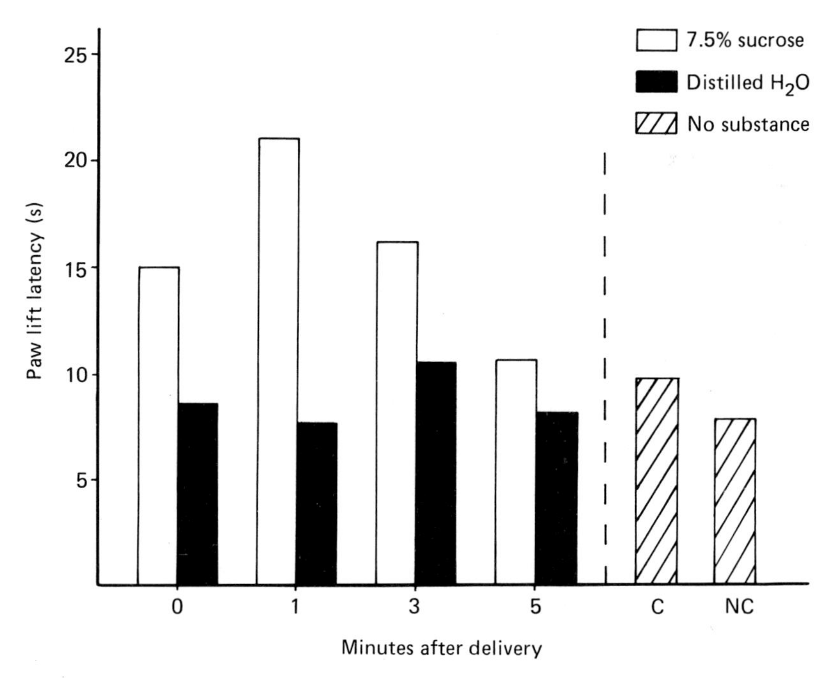

Fig. 8.3. Paw lift latencies in 10-day old rats following intraoral water or sucrose infusions (*C*, control — cannula; *NC*, control — no cannula). (Blass et al. 1986, unpublished work)

More direct evidence on this point comes from our studies on interactions among intraoral sucrose infusions, pain thresholds, and isolation distress in developing rats (Blass et al. 1986, unpublished work). Ten-day-old rats that were group-housed away from their mother received infusions of various concentrations of sucrose and were then tested for paw lift latencies on a 48°C hot plate (Kehoe and Blass, to be published, a,b). As demonstrated in Fig. 8.3, intraoral sucrose infusions markedly elevated paw lift latencies. These elevations were naloxone-reversible. Intraoral sucrose infusion therefore apparently causes the release of endogenous opiates that become available to pain systems (Blass et al. 1986, unpublished work).

These endogenous opioids released by sucrose taste also appear to be available to systems that mediate social distress. In a companion experiment (Blass et al. 1986), 10-day-old rats were individually isolated from their mother for 8 min during which time ultrasonic vocalizations, which are thought to reflect distress (Noirot 1972; Hofer and Shair 1978), were recorded. Figure 8.4 compares the rate of distress vocalizations in animals receiving sucrose with that of various control siblings. Clearly vocalization rate was markedly decreased by sucrose infusion, a decrease that was protracted well beyond termination of the infusion. Moreover this decrease was reversed by a low dose of opioid antagonist pretreatment.

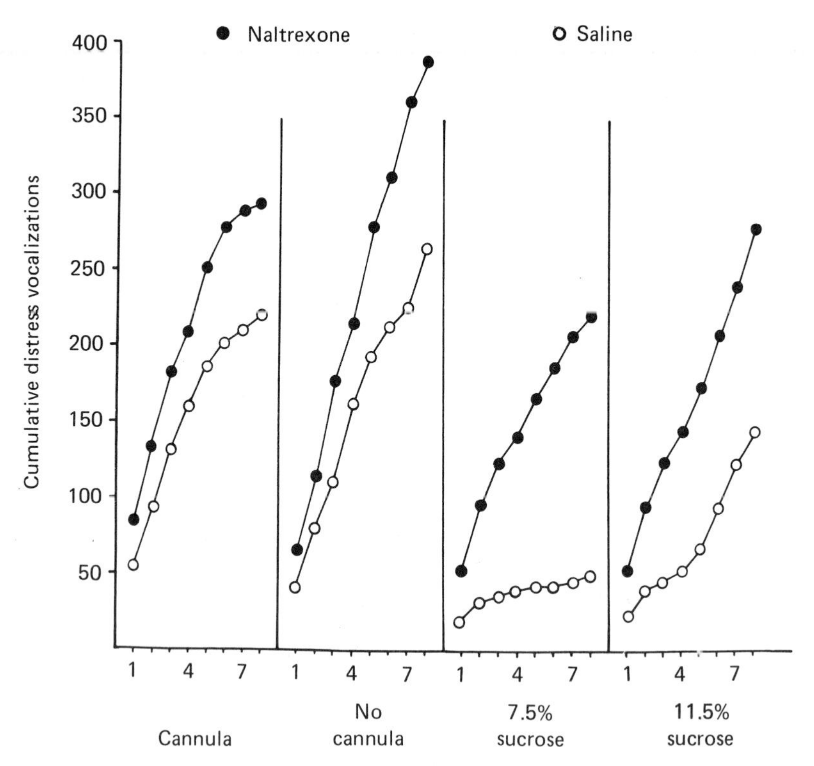

Fig. 8.4. Rates of distress vocalizations during 8 min of isolation in 10-day-old rats during and after intraoral infusions of water or sucrose in rats systemically injected with naltrexone 0.5 mg/kg or isotonic saline vehicle. (Blass et al. 1986, unpublished work)

Again, this supports the notion that sucrose ingestion caused the release of endogenous opioids that become available to negative affective systems (Blass et al. 1986).

General Discussion

These data imply that it is opioid release that confers the positive affective state elicited by sugar solutions. To the extent that there is less opioid effectively available under conditions of receptor blockade, it follows that smaller volumes of sugar solutions will be ingested. Furthermore it makes the specific prediction that animals administered the antagonist will no longer prefer a sweet solution (Le Magnen et al. 1980). One infers, based on human reports, that the affective quality of sugar is no longer the same under naloxone treatment (Fantino et al., to be published).

Electrophysiological evidence of chorda tympani mediation of sweet (Pfaffmann 1961) and behavioral evidence of diminished orofacial responses to sucrose following trigeminal (somatosensory) deafferentation (Berridge and Fentress 1985) suggest somatosensory and gustatory synergism in sweet perception and its concomitant affect. These classes of evidence predict that rats tasting sucrose following somatosensory or gustatory surgical insults would not present either increased pain thresholds or stress alleviation.

The opioid peptides therefore, among other functions, provide a neurochemical complex that may be available to both positive and negative affective systems (see also Le Magnen et al. 1980). A possible role for this system is to expand an animal's capacity for sustained environmental interactions. In the cases of sweet tastes and certain types of olfactory stimuli, endogenous opioids might sustain behaviors that allow non-deprived animals to seek and ingest nutritionally rich foods, thereby building available energy stores, possibly in response to seasonal fluctuations in food availability.

In the case of negative affect the opioids, as presently conceived, may reduce discomfort, allowing the animal to continue functioning to reach shelter, food, etc. A role for the opioids in combatting distress is suggested by the calming effects of both morphine and sucrose on distress vocalizations. It is not difficult to envisage a phylogenetic advantage provided by a system that enables individuals either to continue performing through pain or behavioral distress or to seek the nutritionally rich foods of sweet carbohydrates.

This conceptualization provides structure for classes of data that are not readily handled by thinking about separate and functionally independent positive and negative affective systems. The phenomenon of stress-induced feeding serves as an example. Why should the response to stress, be it that of tail-pinch in rats or of isolation in people, cause marked overeating? Why should these two systems interact? How does feeding palliate the circumstances that gave rise to the perceptions of stress? Why are sweet tasting substances, sugars and chocolates in particular, especially effective?

The converse also holds: infant rats that emit stress vocalizations during maternal isolation reduce their callings and exhibit an increased pain threshold when sucrose is infused intraorally. Yet the isolation that gave rise to distress remains. Moreover decreased vocalization is normalized by naloxone. It is inconceivable that these systems would ever interact in a functionally adaptive fashion. The isolated animal is no less isolated for having obtained sucrose intraorally, yet vocalizations are reduced. If we conceptualize individual motivational systems as functionally independent in their particularities and certainly in their valance, then these findings should not obtain. If on the other hand we conceptualize an additional unitary affective system that has evolved to provide animals and humans with flexibility beyond that provided by primitive drive systems, then these findings can be accommodated. Moreover, specific predictions can be made to direct future experimental efforts concerning sweets, other palatable food (no studies to my knowledge have addressed this issue), opioids, and the relief of stress.

In summary, the sweet taste has been placed in a larger biological and ecological context concerning the delicate balance and interactions among pleasurable and aversive events. Endogenous opiate mechanisms are seen as becoming available to both systems. This interpretation provides a parsimonious and unique orientation for certain types of experimental data and makes explicit predictions for certain types of behavioral, neurological, and neurophysiological interactions.

Acknowledgments. Research in the author's laboratory was supported by grants in aid of research AM 18560 and HD 19278 from the National Institute of Arthritis, Metabolism, and Digestive Diseases and the National Institute of Child Health and Human Development, respectively. Support was also provided by Research Scientist Award MH 00524 from the National Institute of Mental Health to the author.

I am especially grateful to Dr. Charlotte Mistretta, whose time, wisdom, and grace have enhanced this chapter.

References

Adler J (1971) The sensing of chemicals by bacteria. Sci Am 234: 47–76

Amir S, Brown ZW, Amit Z (1979) The role of endorphins in stress: evidence and speculations. Neurosci Biobehav Rev 4: 77–86

Bardach JE, Atema J (1971) The sense of taste in fishes. Handbook of Sensory Physiology 4: 293

Beauchamp GK, Moran M (1982) Dietary experience and sweet taste preference in human infants. Appetite: J Intake Res 3: 139–152

Belluzzi JD, Grant N, Garsky V, Sarantakis D, Wise CD, Stein L (1976) Analgesia induced in vivo by central administration of enkephalin in rat. Nature 260: 625–626

Berridge KC, Fentress JC (1985) Trigeminal-taste interaction in palatability processing. Science 228: 747–750

Bertier MC, Sy TM, Baigts F, Mandenoff A, Apfelbaum M (1984) Stress and sucrose hyperphagia: role of endogenous opiates. Pharmacol Biochem Behav 20: 675–679

Blass EM, Jobaris R, Hall WG (1976) Oropharyngeal controls of drinking in rats. J Physiol Psychol 90: 909–916

Blass EM, Ganchrow JR, Steiner JE (1984) Classical conditioning in newborn humans 2–48 hours of age. Infant Behav Dev 7: 223–235

Broekkamp CL, van den Bogaard J, Heijnen HJ, Rops RH, Cools AR, van Rossum JM (1976) Separation of inhibiting and stimulating effects of morphine on self-stimulation behaviour by intracerebral microinjections. Eur J Pharmacol 36: 443–446

Chance WT (1980) Autoanalgesia: opiate and non-opiate mechanisms. Neurosci Biobehav Rev 4: 55–67

Crook CK (1978) Taste perception in the newborn infant. Infant Behav Dev 1: 52–69
Desor JA, Maller O, Turner RE (1973) Taste in acceptance of sugars by human infants. J Comp Physiol Psychol 84: 496–501
Dethier VG (1978) Other tastes, other worlds. Science 201: 224–228
Fanselow MS (1985) Odors released by stressed rats produce opioid analgesia in unstressed rats. Behav Neurosci 99: 589–592
Fantino M, Hosotte J, Apfelbaum M (to be published) An opioid antagonist, naltrexone, reduces the preference for sucrose in man. Am J Physiol
Ganchrow J, Lieblich I, Cohen E (1981) Consummatory responses to taste stimuli in rats selected for high and low rates of self-stimulation. Physiol Behav 27: 971
Goldstein A, Pryor GT, Otis LS, Larsen F (1976) On the role of endogenous opioid peptides: failure of naloxone to influence shock escape threshold in the rat. Life Sci 18: 599–604
Guttman N (1953) Operant conditioning, extinction, and periodic reinforcement in relation to concentration of sucrose used as reinforcing agent. J Exp Psychol 46: 213–224
Hall WG (1985) What we know and don't know about the development of independent ingestion in rats. Appetite 6: 333–356
Hall WG, Bryan TE (1980) The ontogeny of feeding in rats. II. Independent ingestive behavior. J Comp Physiol Psychol 94: 746–756
Hofer MA, Shair H (1978) Ultrasonic vocalization during social interaction and isolation in 2-week-old rats. Dev Psychobiol 11: 495–504
Kehoe P, Blass EM (to be published, a) Behaviorally functional opioid systems in infant rats: I. Evidence for olfactory and gustatory classical conditioning. Behav Neurosci
Kehoe P, Blass EM (to be published, b) Behaviorally functional opioid systems in infant rats: II. Evidence for pharmacological, physiological and psychological mediation of pain and stress. Behav Neurosci
Le Magnen J, Marfaing-Jallat P, Miceli D, Devos M (1980) Pain modulating and reward systems: a single brain mechanism? Pharmacol Biochem Behav 12: 729–733
Lester LS, Fanselow MS (1985) Exposure to a cat produces opioid analgesia in rats. Behav Neurosci 99: 756–759
Lieblich I, Cohen E, Beiles A (1978) Selection for high and for low rates of self-stimulation in rats. Physiol Behav 21: 843
Lieblich I, Cohen E, Ganchrow JR, Blass EM, Bergmann F (1983) Morphine tolerance in genetically selected rats induced by chronically elevated saccharine intake. Science 221: 871–873
McCleary RA (1953) Taste and postingestion factors in specific-hunger behavior. J Comp Physiol Psychol 46: 411–421
Mistretta CM, Bradley RM (1986) Development of the sense of taste. In: Blass EM (ed) Handbook of behavioral neurobiology 8. Plenum, New York, pp 205–236
Mook DG (1963) Oral and postingestional determinants of the intake of various solutions in rats with esophageal fistulas. J Comp Physiol Psychol 56: 645–659
Morley JE, Levine AS, Yim GKW, Lowy MT (1983) Opiate modulation of appetite. Neurosci Biobehav Rev 7: 281–305
Noirot E (1972) Ultrasounds and maternal behaviour in small rodents. Dev Psychobiol 5: 371–387
North MA (1978) Naloxone reversal of morphine analgesia but failure to alter reactivity to pain in the formalin test. Life Sci 22: 295–302
Ostrowski NL, Foley TL, Lind MD, Reid LD (1980) Naloxone reduces fluid intake: effects of water and food deprivation. Pharmacol Biochem Behav 12: 431–435
Pfaffmann C (1961) Sensory and motivating properties of the sense of taste. In: Jones MR (ed) Nebraska symposium on motivation. University of Nebraska Press, Lincoln, pp 71–108
Pfaffmann C, Norgren R, Grill HJ (1977) Sensory affect and nutrition. Ann NY Acad Sci 18–33
Reid LD (1985) Endogenous opioid peptides and regulation of drinking and feeding. Am J Clin Nutr 42: 1099–1132
Sato T (1969) The response of frog taste cells. Experientia 25: 709–710
Sato M, Yamashita S, Ogawa H (1969) Afferent specificity in taste. In: Pfaffman C (ed) Olfaction and taste III. Rockefeller Press, New York, pp 470–487
Schwartz GJ, Grill HJ (1984) Relationships between taste reactivity and intake in the neurologically intact rat. Chem Senses 9: 249–272
Schwartz GJ, Grill HJ (1985) Comparing taste-elicited behaviors in adult and neonatal rats. Appetite 6: 373–386
Scott TR, Perrotto RS (1980) Intensity coding in pontine taste area: gustatory information is processed similarly throughout rat's brain stem. J Neurophysiol 44: 739–750

Sheffield FD, Roby TB (1950) Reward value of a non-nutritive sweet taste. J Comp Physiol Psychol 43: 471–481
Shuford EH Jr (1959) Palatability and osmotic pressure of glucose and sucrose solutions as determinants of intake. J Comp Physiol Psychol 52: 150–153
Siviy SM, Reid LD (1983) Endorphinergic modulation of acceptability of putative reinforcers. Appetite 4: 249–257
Siviy SM, Calcagnetti DJ, Reid LD (1982) Opioids and palatability. In: Hoebel BG, Novin D (eds) The neural basis of feeding and reward. Haer Institute for Electrophysiological Research, Maine, pp 517–524
Steiner JE (1974) Innate discriminative human facial expressions to taste and smell stimulation (discussion paper). Ann NY Acad Sci 237: 229–233
Steiner JE (1977) Facial expressions of the neonate infant indicating the hedonics of food-related chemical stimuli. In: Weiffenbach JM (ed) Taste and development: the genesis of sweet preference. U.S. Department of Health, Education, & Welfare, Washington DC, pp 173–189
Steiner JE (1979) Human facial expression in response to taste and smell stimulation. Adv Child Dev 13: 257–295
van der Kooy D, LePiane FG, Phillips AG (1977) Apparent independence of opiate reinforcement and electrical self-stimulation systems in rat brain. Life Sci 20: 981–986
Yim GKW, Lowy MT (1984) Opioids, feeding, and anorexias. Fed Proc 43: 2893–2897
Young PT (1957) Psychological factors regulating the feeding process. Am J Clin Nutr 5: 154–161
Young PT (1966) Hedonic organization and regulation of behavior. Psychol Rev 72: 59–86
Young PT, Trafton CL (1964) Activity contour maps as related to preference in four gustatory stimulus areas of the rat. J Comp Physiol Psychol 58: 68–75

Commentary

Rodin: It is interesting to see the body of data suggesting that animals feeding under distress preferentially select sweet substances. This is something that any of us who does clinical work with humans believes, and yet it is impressively difficult to find it experimentally in human research, either with normal populations, with obese, or with bulimic patients. Perhaps we can learn from the animal experiments about how best to provoke these responses in humans in order to measure them.

Blass: It is premature confidently to provide an animal model for human stress-induced feeding. Stress-induced feeding in animals has been elicited primarily by exerting pressure on the base of the tail by clamping it with a padded haemostat. Too much pressure disrupts ongoing behavior; too little pressure is apparently insufficient. It is of interest that the stress in humans that appears to elicit feeding is moderate and somewhat poorly defined. For at least certain bulimics in whom binge eating may be stress induced, treatment with opioid antagonists may provide symptomatic reduction.

Rodin: The intriguing notion that opioids mediate our preference for sweet is well argued in the present chapter. One wonders whether the data have been pushed sufficiently to allow us to determine whether opioids especially contribute to sugar ingestion. A study by Siviy and co-workers in 1982 compared sweet to water, which is not an especially useful comparison for this question. More work needs to be done to determine whether sweet, as compared to other flavors, is selectively influenced by opioid administration or opioid blocking.

Blass: Rodin identifies a shortcoming in the opioid–sweet literature. Opioid antagonists reduce the intake of sweet substances in single bottle tests and flatten sweet preferences against water. There are no data to my knowledge, however, that evaluate the influence of opioid blockade on sweet preferences against substances other than water.

Scott: It is quite exciting to see the proposal that sweetness causes an opioid release which serves to mediate its reinforcing properties. We have studied some of the parameters associated with this hypothesis in both behavioral and electrophysiological experiments. Twelve years ago we infused sucrose over the tongue through an intraoral catheter while rats were self-stimulating (LH electrodes). We observed an increase in s/s rate which was not caused by any other taste solution, by general activation through other sensory input, or by thermal reinforcement. Presently we are recording taste-evoked activity in the rat's nucleus tractus solitarius during intravenous infusions of β-endorphin to determine whether the sensory response to sugars might be enhanced. Evidence from Dr. Blass' manuscript fits very nicely with our expectations.

Bartoshuk: The observation that sucrose fails to produce aversion at high concentrations is very interesting. However, the fact that other modalities often do produce aversion or pain at high intensities may be misleading. Pain is produced by specific neurons that are excited by high intensity stimulation, not by overstimulation of other sensory modalities. For example the pain produced by strong acid on the tip of the tongue is produced by stimulation of trigeminal neurons (chemical pain neurons) rather than by overstimulation of sour taste neurons. Since pain usually results because tissue damage is about to occur, sucrose may never produce pain because even high concentrations of sugar are not damaging to tissue.

Hirsch: The discussion of sucrose concentration and affect ignores the fact that in humans the relationship is quite different. In addition, the statement that "the more concentrated the sucrose solution the greater the affective state it ingenders" is not a completely accurate description of the examples that follow, which are all behavioral observations in rodents.

There are two bodies of data that may relate to the opioid model outlined in this chapter and could be integrated with it. Firstly, there is considerable data showing that sucrose consumption leads to sympathetic activation. For example, Lipsitt (1977) showed that sucrose sucking increases heart rate in infants tested from birth to 4 days of age. More recently Landsberg and Young (1981) have shown that sucrose leads to an increase in norepinephrine turnover. Secondly,

Grunberg et al.'s work on nicotine and sweet food consumption may also relate to the opioid model.

References

Grunberg NE, Bowen DJ, Morse DE (1984) Effects of nicotine on body weight and food consumption in rats. Psychopharmacology 83: 93–98
Grunberg NE, Bowen DJ, Maycock VA, Nespor SM (1985) The importance of sweet taste and caloric content in the effects of nicotine on specific food consumption. Psychopharmacology 87: 198–203
Landsberg L, Young JB (1981) Diet-induced changes in sympathoadrenal activity: implications for thermogenesis. Life Sci 28: 1801–1819
Lipsitt LP (1977) Taste in human neonates: its effects on sucking and heart rate. In Weiffenbach JM (ed) Taste and development. The genesis of sweet preference. U.S. Government Printing Office, Washington DC, pp 125–142

Drewnowski: There are people for whom stress may be associated with under-eating. For example, stress and depression are thought to play a role in the etiology of eating disorders. Young women with anorexia nervosa deliberately deprive themselves of sweet and starchy foods, yet our data show that their responsiveness to sweet taste may in fact be elevated (Drewnowski et al. 1985). Here we have a case where taste responsiveness is clearly dissociated from food consumption.

Reference

Drewnowski A, Duberstein P, Gibbs J, Halmi KA, Pierce B, Smith GP (1985) Taste and eating disorders: hedonic responsiveness in anorexia nervosa and bulimia. Paper presented at Association for Chemoreceptor Sciences Annual Meeting, Sarasota, Fla

Chapter 9

Development of Sweet Taste

Gary K. Beauchamp and Beverly J. Cowart

Introduction

The sense of taste in conjunction with other senses plays a crucial role in decisions about whether a potential food will be accepted and swallowed or rejected, and it is intricately involved in ensuring that an organism consumes sufficient nutrients. In humans the chemosensory response to some sugars includes the perception of sweetness, which is almost universally described as a pleasant sensation. Thus for humans, and many other omnivorous and herbivorous species, sweet sugars are highly acceptable, stimulating ingestion (see Foman et al. 1983) and often preparatory reflexes for digestion as well (for review see Brand et al. 1982).

It is the contention here that the widespread existence of chemosensory responsiveness to carbohydrate sugars reflects evolutionary pressure to ensure detection and recognition of food sources likely to be high in calories (cf. McBurney and Gent 1979). If this is so, the hedonic value of sweet stimuli should be evident early in life and persist throughout the life span. Such is generally the case in humans. It is also suggested that acceptability of sweet-tasting substances, and their perceived pleasantness, might therefore be affected by nutritional state. By analogy with salt appetite, which in animal models has been shown to vary markedly with sodium balance, acceptability of, preference for, and pleasantness of sweet tastes might be expected to vary as a function of energy state or its correlates (e.g., blood glucose or insulin levels). Some experiments provide support for this view (see Cabanac 1979), although this issue is controversial (see Chaps. 2, 10–12).

This general perspective, then, suggests that the perception of sweetness and the pleasantness associated with that perception is innate and may be regulated, in part at least, by nutritional factors. In the present essay the literature on development of sweet perception in humans will be discussed within this framework. The majority of the available work has assessed acceptance (i.e., amount consumed when a single sample is available), preference (choice), or pleasantness (usually a verbal hedonic response), and that work will be emphasized here. Only a few developmental studies have attempted to evaluate changes in perception of intensity as a function of changes in concentration and virtually none have evaluated threshold (see Osepian 1958, for one attempt). Our purpose is to highlight specific issues rather than to provide an exhaustive review.

Newborn Infants

Investigations of newborn infants' taste responses have uniformly indicated that sweet sugars are highly accepted (for reviews see Blass, this volume; Beauchamp 1981; Cowart 1981). It is less clear whether they elicit the same sensation of pleasantness in infants as they do in adults, although pleasantness has been inferred from several kinds of measures, including facial expressions. It has been reported, for example, that drops of concentrated sucrose solution elicit expressions resembling smiles in neonates (Steiner 1973; Ganchrow et al. 1983), although this finding has been disputed (Rosenstein and Oster, to be published). Rozin (1979) has pointed out that acceptance of, or preference for a substance cannot be taken as unequivocal evidence that the substance is pleasant, since preference can be influenced by such factors as perceived health value, convenience, and economic factors (Rozin and Vollmecke 1986). None of these factors, however, is likely to influence relative acceptance in newborn or very young infants, so it seems probable that not only are sucrose solutions well accepted but that they are also perceived as pleasant from birth.

Even given an innate basis for the acceptability and pleasantness of sweet-tasting substances, a variety of factors could impact on, and thereby alter the expression of, responses to these substances. Among these are the following: (a) postnatal maturation of the structures involved in recognizing and processing the chemosensory information could occur; (b) environmental factors (e.g., exposure or lack of exposure to sensory stimulation) could alter avidity for sweet substances; (c) age-related changes in such factors as caloric need or hormonal status could chronically alter perception and/or hedonic response.

Maturation and Sensitivity

Little information is available concerning possible changes in responsiveness to sweetness arising from postnatal, maturational changes in the sensory channel itself. There is no strong evidence of substantial changes in sensitivity (threshold; perceived intensity) during childhood or between childhood and young adulthood. As critically reviewed and discussed by Cowart (1981), some studies report that thresholds for sucrose or saccharin (detection and recognition) are moderately higher (lower sensitivity) in young children compared with older children or adults (e.g., DiMattei 1901; Richter and Campbell 1940; Yasaki et al. 1976). However, these studies are difficult to interpret and are often confounded by possible non-sensory difficulties that young children may have in performing the tasks.

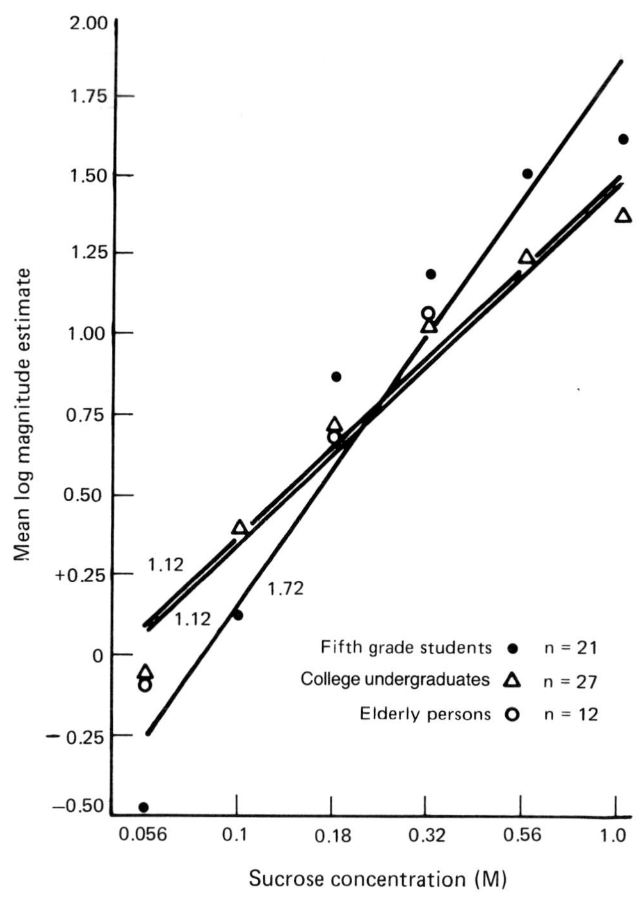

Fig. 9.1. Mean logarithmic magnitude estimates of sucrose intensity by fifth grade students, college undergraduates, and elderly persons (Enns et al. 1979).

Similarly, very few studies have evaluated age differences in suprathreshold scaling of sweet taste intensity. Using magnitude estimation techniques, Enns et al. (1979) reported that children (fifth grade) produced a steeper slope when scaling sucrose solutions than did college undergraduates or elderly persons (Fig. 9.1). They concluded that there were two possible explanations for their results: (a) children may have had a greater tendency to use extreme scores; or (b) capacity to discriminate between the sweet solutions was enhanced in childhood. A third possibility is that the slope of the psychophysical function may reflect factors other than number usage or capacity to discriminate. Cowart (1982) used cross-modal matching to evaluate taste across a wide age range (6–77 years). She found that younger subjects (6–22 years) rated higher concentrations of sucrose as tasting sweeter than did older subjects (40–58 and particularly 77 year olds). There were, however, no differences between subjects of 6, 11, and 22 years of age. Thus these data are not entirely consistent with the Enns et al. study and the issue has not yet been settled.

Dietary Experience and Sweet Acceptability

The role of dietary experience in creating, altering, and/or maintaining the preference for sweets has elicited much speculation (e.g., Jerome 1977; LeMagnen 1977; Mistretta and Bradley 1977) but remarkably little experimental analysis and very few relevant human data are available. Animal studies have yielded conflicting results. Wurtman and Wurtman (1979) fed rats diets of 0%, 12%, or 48% sucrose between days 16 and 30 of postnatal life. When given a simultaneous choice of these three diets between 30 and 63 days of age, no group differences were noted. Similarly, Bernstein et al. (1985) reported no differences in adulthood in response to sucrose solutions among groups of rats exposed to varying levels of sucrose stimulation during the pre- and postweaning periods. In contrast, Marlin (1983) found that rats pre- and postnatally exposed to high sucrose diets did show a greater preference for a high sucrose diet relative to animals exposed to a high glucose diet, presumed by the authors to be less sweet, during the same period. The reason for these different results is not known, although there were many methodological differences among the studies.

If the degree of sweet acceptability in humans were molded by dietary experience, one might expect children's preferences to be related to those of their siblings and parents. Ritchey and Olson (1983), however, found no relationship between children's frequency of consumption of sweet foods and their parents' frequency of consumption. This result is generally consistent with data on food choice, which show a very modest relationship, at best, between parents' and their children's food preferences and choices (see Rozin and Vollmecke 1986). Similarly, Greene et al. (1975) found very low correlations between twins, both identical and fraternal, in the concentration of sucrose solution maximally preferred.

Brown and Grunfeld (1980) varied the amount of sugar which babies consumed during their first 3 months on solid food by instructing an experimental group of mothers on how to avoid sweet foods while permitting the control group mothers to use sweetened baby foods. Following this presumed differential exposure to sweets, infants were tested by providing the parents, on alternate weeks, sweetened or unsweetened jars of fruits with which to feed the infants. The experimental group did not consume more of the sweetened food compared with the control group, thereby providing no evidence for an effect of exposure on acceptability of sweetened foods. Curiously, the authors suggested that infants of this age may not have a sweet preference at all because equal amounts of unsweetened and sweetened fruit were consumed by both groups. In the face of all the other data indicating a strong preference for sweets in infants and young children (e.g., Desor et al. 1977), this seems unlikely. The amounts consumed under this type of procedure are probably more a function of the ability of the mother to induce consumption than of the infant's preference (cf. Filer 1978).

We have conducted a longitudinal study aimed in part at assessing the possibility that different kinds and levels of dietary exposure to sweets may be related to subsequent individual differences in sweet acceptance (Beauchamp and Moran 1982). Relative intake measures were employed to assess sweet preference in a group of children at birth, at 6 months of age, and again at 2 years of age. At all three ages, intake of water was compared with intake of 0.2 M and 0.6 M sucrose solutions.

As expected, newborn infants ingested more of both sucrose solutions than of water and more 0.6 M than 0.2 M sucrose. At 6 months of age, infants were divided into two groups determined by 7-day dietary histories: subjects who were regularly fed sweetened water by their mothers and those who were not (Beauchamp and Moran 1982). Infants regularly fed sweetened water consumed more sucrose solution, but not more water, than did infants not given sweetened water by their mothers. However, these two groups of infants had not differed at birth in their relative acceptance of sucrose solutions. Generally the relative (to water) acceptability of sucrose solutions declined between birth and 6 months of age in those subjects not fed sweetened water while it remained the same for those subjects who were being fed sweetened water.

Based on retrospective data gathered from the mothers, these same children were divided into three groups when tested at 2 years of age: those never fed sweetened water (NF), those fed sweetened water for 6 months or less (Fed ≤ 6 mos) and those fed sweetened water for more than 6 months (Fed > 6 mos) (Beauchamp and Moran 1984). While again all three groups had an equal preference for sucrose at birth, a significant ingestive preference for sucrose solutions at 6 months and 2 years of age was observed only in the two groups that had received postnatal experience with sweetened water (Table 9.1). Interestingly there were no differences between the ingestive responses of the group Fed ≤ 6 mos and the group Fed > 6 mos. Early exposure to sweetened water, even though it had been discontinued, appeared to have lasting effects on the acceptance of 0.2 and 0.6 M sucrose in water. None of the other dietary variables measured (e.g., frequency of consumption of sweet foods) was related to ingestive responses during the taste tests. Finally, the 2-year-olds in this study were also tested for relative acceptance of 0.6 M sucrose in Kool-Aid. When sucrose was presented in this context there were no differences between the NF, Fed ≤ 6 mos, and Fed > 6 mos groups: all children ingested significantly more of the sweetened than of the unsweetened Kool-Aid.

One further notable finding in these studies was that sucrose acceptabilities at 6 months and 2 years were related. Specifically, the amount of 0.6 M sucrose consumed during the taste tests at 2 years of age was significantly correlated with the amount of 0.2 M and 0.6 M sucrose consumed at 6 months of age. Furthermore the relative intakes of sucrose to water were correlated at these two ages (see

Table 9.1. Volume (mean ± SEM) of water and sucrose solutions consumed by 63 black children at birth, at 6 months, and at 2 years of age

Group[a]	Birth			6 months			2 years		
	Water	0.2 M	0.6 M	Water	0.2 M	0.6 M	Water	0.2 M	0.6 M
Never Fed (n = 16)	8.3 ±1.2	24.7 ±4.0	32.1 ±5.1	13.3 ±3.0	21.3 ±5.5	15.4 ±4.3	62.8 ±6.9	66.9 ±7.8	57.8 ±7.2
Fed ≤ 6 mos (n = 18)	13.1 ±2.0	27.7 ±3.7	32.9 ±4.4	13.4 ±2.3	28.9 ±5.6	38.1 ±6.6	55.7 ±8.0	101.3 ±20.1	116.4 ±22.0
Fed > 6 mos (n = 29)	11.3 ±1.1	28.1 ±3.1	31.4 ±3.4	15.9 ±2.5	28.2 ±4.8	39.3 ±6.5	58.5 6.8	84.9 ±11.8	105.6 ±17.3

[a] Never Fed = mothers reported they never fed infants sugar water
Fed ≤ 6 mos = mothers reported they fed infants sugar water for 6 mos or less ($\bar{x}$ = 2.8 months)
Fed > 6 mos = mothers reported they fed infants sugar water for more than 6 months ($\bar{x}$ = 12.1 months)

Beauchamp and Moran 1984, for further discussion). Whether these correlations are specific to sucrose or represent a more general consistency in willingness to ingest flavored liquids or foods remains to be determined. Chiva (1979) reported that consistent individual differences in facial responses to tastes became evident at these early ages.

These correlational studies (Beauchamp and Moran 1982, 1984) suggest that ingestive expression of the innate preference for sweet tastes may be subject to modification quite early in life. The effects of experience on this response are apparently specific to the context in which sweet is experienced. In the absence of postnatal exposure to sweetened water, relative acceptance of sugar water diminishes. At the same time ingestive response to sweet tastes in other contexts appears unaffected. A sense of what should and should not be sweet, rather than generalized hedonic responsiveness to sweetness itself, may be shaped through dietary experience. It is reasonable to hypothesize that neophobia may underlie this contextual effect. Given a familiar food or beverage that has only been experienced without added sweetness, addition of even this highly preferred taste may not enhance the substance's acceptability, and may even render it less acceptable, because the familiarity of its taste is thereby reduced. However, given a novel food or beverage with and without added sugar, the inherent pleasantness of sweet may lead to ingestive preference for the sweetened version.

Age and Sweet Acceptability

Children love sweets. This common belief implies that sweet preference in children is greater (i.e., greater amounts consumed; higher concentrations maximally preferred) than in adulthood, and there is considerable evidence to support it. Before confronting this evidence, however, it should be noted that there are apparently no studies of preference across the years from birth to early adolescence. This may be due in part to the difficulties in measuring preference or acceptability of various concentrations of sweet stimuli in infants and very young children. Particularly between birth and 3–4 years of age, so many developmental changes occur that devising comparable measures of taste function for newborn infants and 2-year-old children, for example, is very difficult. Relative intake measures (as described in the previous section) may be the only way to address this issue.

Table 9.2 presents brief summaries of six studies that have experimentally investigated possible differences in preference for sweets (generally sucrose) in children and adults. In general the concentrations of sweetener judged most pleasant by the different age groups were the parameters of interest. Four of the six reported significantly lower preferences in older subjects and a fifth (Cowart 1982) reported a trend in this direction. Even in the one paper reporting no age differences (Enns et al. 1979), the published figure suggests a similar trend. Finally, using a questionnaire format, Hill (1980) evaluated changes in liking for foods among college students. She found that liking for sweet tastes decreased more than it increased but the report was so brief as to make it difficult to interpret.

Table 9.2. Studies on changes in sweet preference with age

Study	Age range (yrs)	Taste substance	Finding
Desor et al. (1975)	9–15 vs 18–64 (cross-sectional)	Sucrose and lactose solutions (0.075–0.60 M)[a]	Most preferred concentrations lower in older group; trend for males to prefer higher concentrations
Catalanotto et al. (cited in Cowart 1981)	7–15 (cross-sectional)	Sucrose solutions (0.15–2.40 M)[a]	Negative correlation between age and preferred solution
Grinker et al. (1976)	10–21; females (cross-sectional)	Sucaryl in Kool-Aid[b] (0–2 × manufacturers instructions)	Oldest subjects preferred sweeter solutions less
Enns et al. (1979)	11 vs 19 vs 71 (cross-sectional)	Sucrose solutions[b,c] (0.056–1.0 M)	No age difference; males greater preference than females in youngest and oldest groups only
Cowart (1982)	6–77 (cross-sectional)	Sucrose solutions[d] (0.056–1.8 M)	No significant age differences at any concentration; trend for youngest group to judge higher concentrations most pleasant
Desor and Beauchamp (1986)	11–15 vs 19–25 (longitudinal)	Sucrose solutions (0.075–0.60 M)[a]	Most preferred concentrations decreased in subjects tested at older age

[a] Subjects asked to pick most preferred solution
[b] Used hedonic rating scale
[c] Used paired comparisons
[d] Used cross-modal matching

The first five studies listed in Table 9.2 were cross-sectional, i.e., individuals were drawn from different age groups to be tested. Group differences in this design cannot necessarily be attributed to aging itself. For example, the possibly important variable of differential prior exposure to sweet stimuli is not controlled. Thus the older individuals tested may, as children, have had considerably less experience with sweets or less experience with highly concentrated sweets than have younger groups. If early exposure plays a role in determining preferences, this could account for apparent age differences.

Some of this difficulty is avoided by using a longitudinal design. We (Desor and Beauchamp 1986, unpublished work) recently evaluated a sample of individuals ($n = 44$) who had first been tested approximately 9 years earlier when they were 11–15 years of age (Desor et al. 1975). A sucrose preference test identical to that originally given by Desor et al. was administered individually to each twin. Briefly, subjects were offered four cups containing different concentrations (0.075, 0.150, 0.300, 0.600 M) of sucrose in random order. The subject tasted each of the four samples without swallowing them and then ranked them in order of preference from the most to the least preferred. No data were collected on thresholds or suprathreshold intensity ratings.

Of the 44 individuals tested, 24 (56%) preferred a lower concentration when older, 14 (32%) showed no change, and 6 (14%) preferred higher concentrations (Fig. 9.2). These data provide further evidence that there is a decline between childhood and young adulthood in the concentration of sucrose maximally preferred, and they are consistent with consumption data (e.g., Page and Friend 1974). In this study too, there was evidence of long-term consistency in individual differences in preference. The test–retest correlation for the most preferred concentration ($\rho = 0.33$) was modest but significant ($P < 0.05$), indicating that in spite of a general shift downward in the most preferred sucrose concentration, relative rankings of individual preferences in a population are somewhat stable over a considerable period of years.

Why should there be a change in preference with age? As previously stated, it seems unlikely that these preference changes are a consequence of changes in the receptor apparatus. Such changes should alter sensitivity as well as preference. However, as also described above, children may even find increasing sucrose concentrations sweeter than do adults. A sensory difference in this direction would not account for preferences for higher levels of sweeteners in young individuals.

One hypothesis is that the change in preference reflects a change in caloric need. Children and young adolescents are in a phase of active growing whereas adults, on average, require relatively fewer calories. There are data suggesting that short-term alterations in caloric need affect avidity for sweets (e.g., Cabanac 1979). Johns and Keen (1985) have recently attributed an exceptionally high preference for glucose solutions among young adult Aymara Indians of Bolivia to

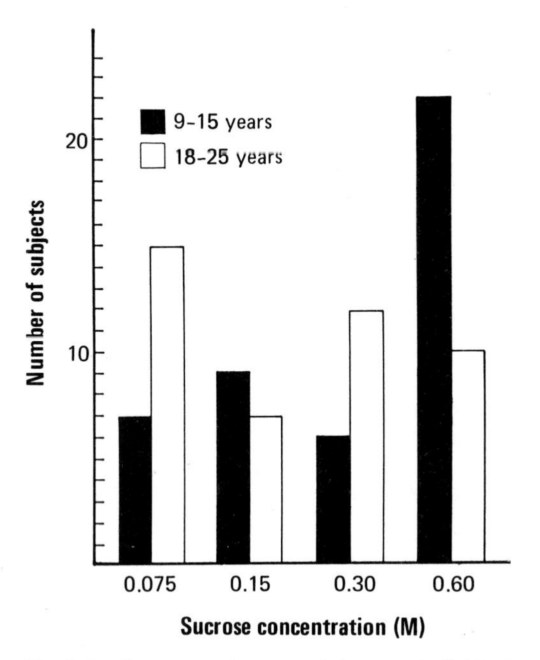

Fig. 9.2. Sucrose preferences of the same individuals at two different ages.

caloric need related to undernutrition and/or adaptation to high altitudes. In an attempt to test this issue in children directly, acceptability of sucrose solutions was evaluated in malnourished infants (2–24 months of age) and a control group of well-nourished infants (Vazquez et al. 1983). Malnourished infants did not exhibit a heightened response relative to well-nourished infants, but this may have been due to a ceiling effect since the well-nourished infants exhibited the normal high preference for the sucrose solutions. Interestingly, malnourished infants recently (≤ 5 days) admitted to a renutrition center actually exhibited a depressed acceptance of sucrose solutions relative to both malnourished infants tested more than 5 days after admission and to well-nourished infants. The reason for this depressed response is not known, although it is consistent with findings of anorexia in these infants and it could be due in part to nausea or other symptoms of illness. It is also consistent with pilot data suggesting a similar response in infants diagnosed as Failure to Thrive (Beauchamp and Moran 1986).

Other hypotheses might be advanced to account for the general downward shift in preference with age. In rats, sex steroids are known to modulate consumption of sweet solutions in complex ways, with females often showing elevated preferences (e.g., Wade and Zucker 1969). The data of Enns et al. (1979) and Cowart (1982) indicate the possibility of some shift upward in preference between childhood and adolescence in females only, despite average trends downward. Thus while steroids may also play a role in modulating human sweet preferences, there is no indication that their effects, if any, would contribute to an average decline in preference from childhood to adulthood. Again in rats, it has been shown that different diets alter expression of sucrose acceptability (e.g., Bertino and Wehmer 1981). Although a dietary shift in humans between childhood and adulthood could theoretically underlie the age change in sweet preference, no such change in dietary patterns has been documented, making this an unlikely explanation as well.

Inspection of Table 9.2 reveals that five of the six studies used aqueous sucrose solutions as the sweet stimuli. Studies with salt have underlined the importance of sensory context in perceived pleasantness and preference. For example, salt water solutions of increasing concentrations are generally judged increasingly unpleasant whereas increases in salt in many foods, such as soup, often increase their palatability (Bertino et al. 1983; Pangborn and Pecore 1982). This striking context-dependent effect has been observed in children as young as 36–60 months (Cowart and Beauchamp 1986) but not in children under 24 months of age (Beauchamp et al. 1986). Age-related changes in the relative importance of sensory context (i.e., changes in the acceptability of water as a vehicle for sweetness) could underlie or contribute to the apparent shift in sweet preference. Perhaps studies using a wider array of foods with varying levels of sweetness could address this issue.

Finally, such an age shift may not be specific to sweet perception but instead may reflect a generalized decrease with age in the preferred intensity of pleasurable stimuli. This is a difficult hypothesis to test but recent data on salt preference (Cowart and Beauchamp 1986; unpublished data) suggest there may be a substantial shift downward with age in the most preferred level of salt in soup. It is difficult to attribute this apparent shift in salt preference to any known nutritional alteration in sodium requirement. If the age changes in sweet and salt preference were to occur in parallel, this would support the hypothesis that the shifts may be due to other than nutritional factors.

Discussion

The majority of research into the development of responsiveness to sweetness in human infants and children has used sucrose as the sweet stimulus. Only a very few studies have assessed responses to other carbohydrate (glucose, lactose, fructose) or non-carbohydrate (saccharin, Sucaryl) sweeteners. Recently, evidence has been presented that there may be multiple peripheral mechanisms encoding the sweet taste (e.g., Lawless and Stevens 1983). Thus many of the generalizations made above about the development of responsiveness to sweetness and the factors influencing responsiveness must be qualified: the developmental course for sweetness perception could depend upon the sweetener of interest. For example, it is entirely possible that newborn infants may be less responsive to some of the non-carbohydrate sweeteners than are children and adults. There appears to be no experimental evidence relevant to this question.

With the above caveat in mind, the available data on the development of sweet perception in infants and children can be summarized as follows: (a) newborn and young infants and children exhibit an avid acceptance of sweet substances; (b) there is little evidence that degree of early exposure to sweets influences this avidity; (c) dietary exposure may determine the context in which sweetness is expected, thereby influencing the acceptability of specific sweetened foods; (d) the level of sweetness judged most pleasant tends to decline between childhood and adulthood.

Acknowledgements. Work described in this manuscript was supported by part by grant #HL 31736 from the National Institutes of Health.

References

Beauchamp GK (1981) The development of taste in infancy. In: Bard JT, Filer LJ Jr, Leveille GA, Thomson A, Weil WB (eds) Infant and child feeding. Academic Press, New York, pp 413–426

Beauchamp GK, Moran M (1982) Dietary experience and sweet taste preferences in human infants. Appetite 3: 139–452

Beauchamp GK, Moran M (1984) Acceptance of sweet and salty tastes in 2-year-old children. Appetite 5: 291–305

Beauchamp GK, Moran M (1986) Taste in young children. In Rivlin RS, Meiselman HL (eds) Clinical measurement of taste and smell. Macmillan, New York, pp 305–315

Beauchamp GK, Cowart BJ, Moran M (1986) Developmental changes in salt acceptance in human infants. Dev Psychobiol 19: 75–83

Bernstein IL, Fenner DP, Diaz J (1985) Effect of taste experience during the suckling period on adult taste preferences of rats. Chem Senses

Bertino M, Wehmer F (1981) Dietary influence on the development of sucrose acceptability in rats. Dev Psychobiol 14: 19–28

Bertino M, Beauchamp GK, Engelman K (1983) Long-term reduction in dietary sodium alters the taste of salt. Am J Clin Nutr 38: 1134–1144

Brand JG, Cagan RH, Naim M (1982) Chemical senses in the release of gastric and pancreatic secretions. Ann Rev Nutr 2: 249–276

Brown MS, Grunfeld CC (1980) Taste preference of infants for sweetened or unsweetened foods. Res Nurs Health 3: 11–17

Cabanac M (1979) Sensory pleasure. Q Rev Biol 54: 1–29

Chiva M (1979) Comment la personne se construit en mangeant. Communications 31: 107–118
Cowart BJ (1981) Development of taste perception in humans: sensitivity and preference throughout the life span. Psychol Bull 90: 43–73
Cowart BJ (1982) Age-related changes in taste perception: direct scaling of the intensity and pleasantness of basic tastes. Ph.D. dissertation, The George Washington University
Cowart BJ, Beauchamp GK (1986) The importance of sensory context in young children's acceptance of salty tastes. Child Dev 57
Desor JA, Greene LS, Maller O (1975) Preference for sweet and salty in 9- to 15-year-olds and adult humans. Science 190: 686–687
Desor JA, Maller O, Greene LS (1977) Preference for sweet in humans: infants, children and adults. In: Weiffenbach JM (ed) Taste and development: the genesis of sweet preference. US Government Printing Office, Washington DC, pp 161–172
DiMattei E (1901) La sensibilita' nei fanciulli in rapporto al sesso ed all'eta. Archives di Antropologia Criminale, Psichiatria e Medicine Legale 22: 207–228
Enns MP, Van Itallie TB, Grinker JA (1979) Contributions of age, sex and degree of fatness on preferences and magnitude estimations for sucrose in humans. Physiol Behav 22: 999–1003
Filer LJ Jr (1978) Studies of taste perception in infancy and childhood. Pediatr Basics 12: 5–9
Foman SJ, Ziegler EE, Nelson SE, Edwards BB (1983) Sweetness of diet and food consumption by infants. Proc Soc Exp Biol Med 173: 190–193
Ganchrow JR, Steiner JE, Daher M (1983) Neonatal facial expressions in response to different qualities and intensities of gustatory stimuli. Infant Behav Dev 6: 473–484
Greene LS, Desor JA, Maller O (1975) Heredity and experience: their relative importance in the development of taste preference in man. J Comp Physiol Psychol 89: 279–284
Grinker JA, Price JM, Greenwood MRC (1976) Studies of taste in childhood obesity. In: Novin D, Wyrwicka W, Brau GA (eds) Hunger. Basic mechanisms and clinical implications. Raven, New York, pp 441–457
Hill WF (1980) Changes in taste preferences as reported retrospectively by college students. J Genet Psychol 137: 149–150
Jerome N (1977) Taste experience and the development of dietary preferences for sweet in humans: ethnic and cultural variations in early taste experience. In: Weiffenbach JM (ed) Taste and development: the genesis of sweet preference. US Government Printing Office, Washington DC, pp 235–248
Johns T, Keen SL (1985) Determinants of taste perception and classification among the Aymara of Bolivia. Ecol Food Nutr 16: 253–271
Lawless HL, Steven DS (1983) Cross-adaptation of sucrose and intensive sweeteners. Chem Senses 7: 309–315
LeMagnen J (1977) Sweet preference and the sensory control of caloric intake. In: Weiffenbach JM (ed) Taste and development: the genesis of sweet preference. US Government Printing Office, Washington DC, pp 355–362
Marlin NA (1983) Early exposure to sugars influence the sugar preferences of adult rats. Physiol Behav 31: 619–623
McBurney DH, Gent JF (1979) On the nature of taste qualities. Psychol Bull 86: 151–167
Mistretta CM, Bradley RM (1977) Taste in utero: theoretical considerations. In: Weiffenbach JM (ed) Taste and development: the genesis of sweet preference. US Government Printing Office, Washington DC, pp 51–69
Osepian VA (1958) Development of the function of the taste analyzer in the first year of life. Pavlov J Higher Nerv Act 8: 766–772
Page L, Friend B (1974) Level and use of sugars in the United States. In: Sipple H, McNutt K (eds) Sugars in nutrition. Academic Press, New York, pp 93–107
Pangborn RM, Pecore SD (1982) Taste perception of sodium chloride in relation to dietary intake of salt. Am J Clin Nutr 35: 510–520
Richter CP, Campbell HK (1940) Sucrose taste thresholds of rats and humans. Am J Physiol 128: 291–297
Ritchey N, Olson C (1983) Relationships between family variables and children's preference for and consumption of sweet foods. Ecol Food Nutr 13: 257–266
Rosenstein D, Oster H (to be published) Taste-elicited facial expressions in newborns.
Rozin P (1979) Preference and affect in food selection. In: Kroeze JHA (ed) Preference behavior and chemoreception. Information Retrieval, London, pp 289–302
Rozin P, Vollmecke T (1986) Food likes and dislikes. Ann Rev Nutr 6
Steiner JE (1973) The gustofacial response: observations of normal and anencephalic newborn infants. In: Bosma JF (ed) Fourth symposium on oral sensation and perception: development in the fetus

and infant. US Government Printing Office, Washington DC, pp 254–278
Vazquez M, Pearson PB, Beauchamp GK (1983) Flavor preferences in malnourished Mexican infants. Physiol Behav 28: 513–519
Wade GN, Zucker I (1969) Hormonal and developmental influences on rat saccharin preferences. J Comp Physiol Psychol 69: 291–300
Wurtman JJ, Wurtman RJ (1979) Sucrose consumption early in life fails to modify the appetite of adult rats for sweet foods. Science 205: 321–322
Yasaki, T, Miyoshita N, Ahiko R, Hirano Y, Kamata M, Iizuka Y (1976) Study on sucrose taste thresholds in children and adults. Jpn J Dent Health 26: 20–25

Commentary

Booth: It seems to me that Table 9.2 and your new data could all be readily explained as older adolescents and adults being more successful at making some sense out of the very odd question how pleasing they find room-warm, unflavored sucrose solutions. Their experience of sweet drinks may now be sufficiently wide and prolonged for some generalization of an acquired "norm" in the region of 0.3 M sucrose and so there is some tendency to impose a descending limb on the innate monotone, even for sugar water.

Would you thus agree that conclusions c) and d) of your Discussion may well be the same phenomenon? Would you be prepared to strengthen this conclusion to the extent of asserting that sweetness is therefore much like any other salient sensory characteristic of a palatable food, once the experience-independent avidity has been overwhelmed by acquired appetite?

Beauchamp: I am not prepared to accept the view that the decline between childhood and adulthood in the level of sucrose solution (or in another study the level of Sucaryl in Kool Aid) most preferred can be attributed only to experience determining the appropriate or expected level. While this could well play a role as we noted in the chapter, the situation we cite with rats which also demonstrate a decline in sugar preference would argue that such cognitive explanations cannot solely account for the decline in maximally preferred levels. More generally I believe that sweetness is fundamentally different from most other salient sensory characteristics of food by virtue of its status as an innately preferred *taste* quality. In this view the sense of taste is a primitive system which has evolved in a rather simplistic way to deal with major nutritional problems (Beauchamp and Cowart, *Appetite*, 1985, 6: 357–372). As a consequence sweetness forms a fundamental building block on which food and flavor preference are constructed. This is not to say that cognitive factors play no role in sweetness preference since our chapter has, as one of its foci, the importance of context in the acceptability and preference for sweetness. However, I believe it would be a mistake not to note the salient difference between sweetness and, for example, lemon-ness (citral at a level it is tasteless). The degree of liking as well as the levels at which liking is greatest is, within limits, mainly determined by experience, expectation, and so forth in the latter case; I am not inclined to accept this to be the case for the former.

Booth: Accepting or preferring something *is* liking it and treating it as pleasing. "Please, Dr. Beauchamp" is asking you to do something, not playing with your feelings. The hedonic theory of reinforcement is either a tautology (we only do what pleases us) or false (we only do what gives us pleasure). Tastes (like caffeine, chili, and sugar), health, convenience, price, and genuinely stirring "gastronomic" delights (pleasures) all please us, increase our avidity, and influence hedonic ratings when the test design permits them to. I have yet to see anyone measure oral pleasures distinguishably from food values, sensory motivation, etc.

Beauchamp: We can induce certain 3-year-old children to sample unsweetened Kool Aid repeatedly with a high citric acid content (this being a very strong and, to most, an unpleasant stimulus) and, on another occasion, Kool Aid with a very high sucrose content without the acid (very sweet, but very acceptable to children). Although in both cases a child may accept the drink, other behavioral responses associated with the acceptance of the high acid drink, such as grimaces, delays in accepting subsequent presentations, verbal responses, etc., are in marked contrast to responses associated with the high sucrose drink. I submit that these two forms of accepting *are* fundamentally different. While I agree with Booth that the physiology and philosophy underlying the exact nature of these differences are not yet fully understood, an indication of the meaning of this distinction should be evident to anyone who has taken bitter medicine in juxtaposition to a good meal. In any case, the point of noting the distinction in our chapter was to emphasize the likelihood that such a distinction does not apply in very young, preverbal children.

Rodin: I have never been persuaded that the approach in studying whether degree of sweet acceptability in humans is molded by dietary experience should be to correlate children's preferences to those of their siblings and parents. As any of us who are parents know, children's dietary experience is altered at a very young age by many extrafamilial instances and I suspect that the significant dietary experiences may not be simply those that take place within the family. It is also the case that sweets may be much more vulnerable to this since it is often snack food that is shared more with peers than with family members.

Beauchamp: The extent to which one would expect a correlation between parent and child would certainly depend upon the age of the child. It might be expected that such a correlation would exist between very young children and parents (primarily the mother) since parental control over the child's food intake is greatest at this time. Even here, however, one might anticipate that parents would feed their children not what the parents like but what they think their children ought to eat. If this were the case there would be even less reason to expect a high parent–child correlation. None of these arguments, however, remove the puzzle of why the 9–15-year-old identical and fraternal twin pairs studied by Greene et al. (1975) were not more similar in the concentration of sucrose solution they chose as most preferred.

Blass: Beauchamp and Cowart's review of developmental influences on sweet perception and ingestion provides an empirical basis for intra- and intermodal ontogenetic comparisons. The development of sweet taste bears certain important similarities to that of sodium salt taste for example, because neither can be readily altered by dietary history (Mistretta and Bradley 1986). Efforts to change salt intake or preference by prenatal and (or) postnatal exposure to various concentrations of NaCl have not met with success. The same appears to hold for sugar preferences. The sweet taste may differ narrowly because extensive experience appears to maintain a specific preference (i.e., it does not decay over time). It is of interest that this maintenance did not extend to other sweet moieties. Comparable studies have not been undertaken for salt preference.

More complex dietary preferences are certainly induced by particular experiences. This induction represents an outstanding opportunity for students of development to identify the early events in the nest and at weaning that give rise to select cuisines.

Reference

Mistretta CM, Bradley RM (1986) Development of the sense of taste. In: Blass EM (ed) Handbook of behavioral neurobiology 8. Plenum, New York, pp 205–236

Rodin: Any of us concerned about the effects of sweetness on food intake in disordered populations as well as in a normal sample must try to understand with great interest issues regarding the development of sweet taste. I wondered on reading this whether one might speculate about a stronger correlation between sensation and hedonic evaluation in infants and young children than exists in adulthood, when other factors, such as those pointed out by Rozin as well as many others, influence hedonics as well as intake. What I am suggesting, then, is that the curves are more parallel in the early years and diverge with increasing age. One methodological approach to studying these kinds of questions exist in the armamentarium of our colleagues doing aging research, and that is the cross longitudinal design. In this design, cross sections are studied once and then again X number of years later. The cross sections are selected by virtue of the number of years later that the second measurement will take place, so that one may make both longitudinal and cross-sectional comparisons.

Section IV

Sweetness and Food Intake

Chapter 10

Sweetness and Food Selection: Measurement of Sweeteners' Effects on Acceptance

D.A. Booth, M.T. Conner and S. Marie

Objective Acceptability of Sweetness

There is no more obvious or mundane fact about human behaviour, feelings and thoughts than that we often like sweet foods and drinks. Nevertheless, practising food scientists are aware how complex are the roles of sweeteners in the perception and acceptance of food. The influence of sweetness on food selection is a mental process influencing an individual's behaviour. Therefore its scientific investigation cannot rely simply on polling of opinions, whether by aggregate market or dietary survey responses or by the more sophisticated statistical treatment of numbers produced by members of selected panels in answer to questions about food samples. Sensory testing coordinated with market research in the food industry as well as medical or dental research into disease-preventing behaviour would be advanced by objective measurement of influences on individuals' eating and drinking behaviour, using cognitive methods from experimental psychology (e.g. Poulton 1968; Fishbein and Ajzen 1976). However, such aspects of so-called hot cognition have yet to engage the attention of mainstream academic psychology (Booth 1986) and so this review can do little more than point to a way forward.

Perception of Sweetness

Sweetness is a sensation typically generated by sugar dissolved on the tongue or palate. In other words we can be conscious that something is sweet when sucrose, glucose, saccharin and similar-tasting substances stimulate chemoreceptors in the mouth.

However, no-one has yet adequately checked whether the effects of sweeteners on food selection are always, or even usually, mediated by an awareness of sweetness as such at the moment of acceptance or rejection (Booth 1979). Asking assessors to rate the sweetness as well as the pleasantness of a complex food could alter influences on selection that are normally pre-conscious (Farell 1984), and so

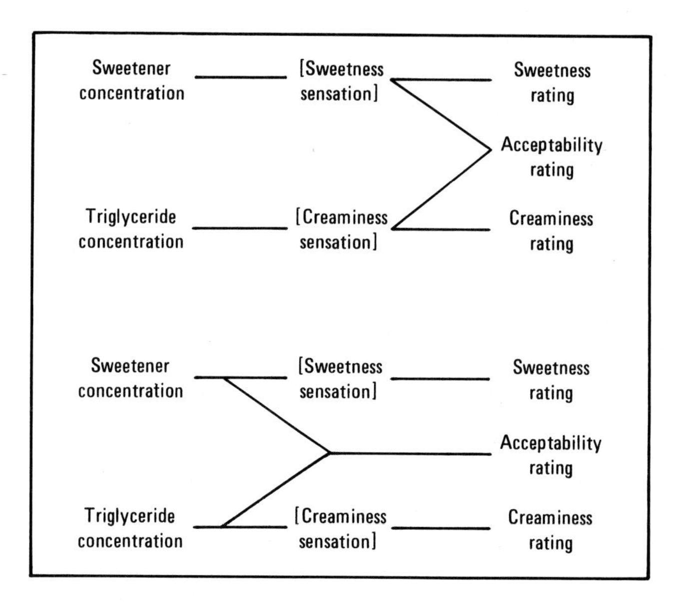

Fig. 10.1. Cognitive pathways from food constituents to ratings of acceptability and perceptual intensity. *Upper model*: classical assumptions. *Lower model*: the "straight-through" mechanism. All variables are observed except those marked [ ] — the sensations (contents of consciousness); structural modelling can provide evidence for the existence of such latent variables.

make self-fulfilling the assumption that sweetener preference is mediated via the "analytic" experience of a sensation of sweetness (Fig. 10.1).

This chapter will therefore be realistic and use a rather broader concept of sweetness: as a perceptual process, whether it is conscious or not on any particular occasion when a sweetener is demonstrably affecting the person's response.

Food Selection: Observed Datum or Hypothetical Process?

Strictly speaking, food selection is a single qualitative datum: which item was chosen from the available range. A quantitative measure of selection behaviour requires repeated testing of the individual in comparable circumstances, to give a frequency or probability. However, such measurements are an unduly costly way of investigating what matters for theory and practice: namely the perceptual–motivational disposition to ingest food and drink, i.e. momentary acceptability or appetite (the "orocline": Booth 1981a). This mechanism or state within the information-processing organisation of an individual's behaviour accounts for observed food selections but also for other aspects of food-oriented behaviour.

Behavioural research has used relative amounts eaten of two or more foods, even on different occasions, as a measure of dietary "selection". However, eating and drinking cause postingestional, sensory and learned cognitive changes (usually satiating) that are not present during the initial selection (Chap. 11; Booth, 1977, 1981b). The assumption that relative intakes reflect the processes determining instantaneous choice has proved erroneous, for example in infants and in animals faced with various concentrations of sugar solution.

The act of initial selection between foods and verbal choice, the latter being most relevant here, do not have this disadvantage.

Objective Influences of Sweetness on Selection

The measurement of perception or motivation is not by the verbal or behavioural response alone, but by its relation to the circumstances of its occurrence (Booth 1979, 1986). In itself a behaviour score is no more an objective measure of selection than is a verbal score. It is only that the verbal responses are symbolic, not concrete. All responses have the same problems of validation, whether psychophysical (as measures of actual food composition or bodily state, e.g. gastric wall tension) or psychometric (as predictors of the respondent's behaviour and mental processes during real-life food selection and intake).

Introspective language grammatically refers to subjective states, but that is no justification for assuming without empirical evidence that numerical, verbal or graphic ratings are estimates of the magnitudes of sensations or pleasures; or for throwing away the potential of the ratings as evidence of objective performance by inadequately designing the independent variables or the data analysis.

The impression in sensory evaluation of food is that acceptability data are more variable than "sensory" data. However, this is not so when psychological performance is measured scientifically. For example when sensory preference tests are designed and analysed relative to a determinate sensory optimum for each individual, the data can reveal rather precise relationships between food selection ratings and food composition parameters such as sweetener concentration (see below). This is largely because the individualisation allows for the "personal factor" that endangers panel averaging of intensity ratings, as well as rendering hedonic data almost useless. Indeed, when an acceptance reaction to a food does not require conscious perceptual analysis (Fig. 10.1), more precise measures of the sweetener contents of coffee drinks can be provided by suitable preference ratings than by sweetness intensity ratings (Marie 1986; Booth et al. 1986).

Such objective experimental measurement of preferred food composition has the further advantage of relating the design of the food directly to customer response (Booth 1979). The traditional effort to keep naïve or affective elements out of sensory data is at continual risk of developing an expert gastronomy, or just a humanly perceived food chemistry, that wrongly predicts food selection in the market.

Sweetness in the Selection of Familiar Foods

Development of Effects of Sweeteners on Food Selection

There is a congenital (probably innate) human liking for sweetness, i.e. an acceptance reflex to the taste of sweeteners (Chap. 8). Nevertheless, most and even perhaps all of the effect of sweetness on normal food selection is acquired.

The congenital preference increases monotonically with sweetener concentration. However, the acquisition of likings for foods and drinks containing sweeteners suppresses, competitively inhibits or modulates this reflex, so that sweetness greater than what is familiar in a particular eating or drinking context no longer elicits a stronger acceptance reaction. That is because the particular complex of sensory characteristics that an infant or older person notices in a beverage or solid food tends to become attactive to that individual. This automatic learning can go on throughout childhood and adult life.

So the liking for sweetness, like that for every other sensory characteristic, is for a particular level in a specific, familiar context. It only "generalises" to other sweetened formulations to the extent that they are similar to a familiar food. A taste that is relatively neutral at birth, such as salt in dilute solution, can become mildly aversive by 6 months of age because salt water differs from the solids or fluids in which the infant has become familiar with salt (Harris and Booth 1985). Similarly, 6-month-old infants who had been fed sugar water by their mothers took more of it in tests than those who had not been so exposed (Beauchamp and Moran 1982). In other words the acquired acceptance of particular foods appears as early as 6 months.

The induction of such specific habits and likings from early infancy is reviewed by Beauchamp and Cowart (Chap. 9). Several mechanisms ensure that salient sensory characteristics, including sweetness, have peak preferred levels in particular foods for each individual. Mere exposure to a food of course removes any dislike of novelty but it also seems to induce some genuine attraction to that particular substance (Birch and Marlin 1982; Pliner 1982). Also, various associations with eating a food determine increased liking for its particular characteristics, and even an increased appetite for it in the still broader complex which may include the familiar bodily or social circumstances of its consumption (Booth 1977, 1985a). These acceptance- and selection-inducing associations include prompt absorption of ready energy (Booth et al. 1974, 1982) or balanced amino acids (Booth 1974), interpersonal reward (Birch et al. 1980) or conformity (Birch 1980), and sweetener itself in the food (Holman 1975; Zellner et al. 1983).

Experience of unpleasant consequences of high sweetness or of no benefit at all from sweetness (Booth 1974) can reverse even the rat's original preference for sweetness, and for more rather than less of it. Similar direct experiences of overconcentrated sugar and less direct education into attitudes against sugar may do the same in people (Chap. 15), so that little or no sweetness becomes the preferred level, at least in a "danger" food (e.g. Witherley et al. 1980).

Preference for Sweetness Within the Food Complex

From the above it can be seen that mature uses of sweetening are drastic modifications of the innate reflex. No doubt, sweetener is added to bitter or sour flavourings because of the innate preference, although perceptual masking may be involved as well as the countering of innate rejection by innate acceptance (Chap. 4). Nonetheless, lemonade and tonic (quinine) water can easily be too sweet. Furthermore plain sugar solution or sugar glass or crystals are very boring without at least an acidulant and, better, a familiar fruit aroma and/or colour. (Even rats probably prefer a little bit of quinine in a saccharin solution: Booth 1972.) We learn to like a particular level of sweetness as one element only in a balanced flavour complex.

It follows from all this that the effect of sweetener on selection amongst familiar foods will be much like the effect of any other dietary or contextual stimulus: as an integral part of a *Gestalt* or holistic percept (Lockhead's 1972 "blob").

Thus the role of sweetness in food selection can be effectively investigated only by methods that use test foods and test situations that are familiar to the subjects. The psychophysics of sugar–water intake, selection or hedonic ratings is liable to mislead the study of appetite if the innate reflex is largely irrelevant.

Indeed, something wrong with the formulation of even a familiar food may cause reversion to a crude sweetening strategy. The innate reflex "breaks through" the normal preference peak ("breakpoint" or ideal) established as the acquired influence of sweetness on food selection: the assessor's preference does not peak at a particular sweetener concentration, but increases monotonically, or at least to a level considerably higher than usual. This happens when a test drink is not as concentrated as is habitual for a particular assessor (Rawle et al., unpublished observations). Flavour quality or balance may be poor (Moskowitz 1972), or the test material may be difficult to regard as a real drink or food at all (e.g. sugar water; cf. Thompson et al. 1976).

The Linear Sweetener/Food-Selection Function

Food Sweetness Ideal Points and Rejection Points

On the above theory, anyone faced with what they recognise as a food will normally have a personally familiar and balanced formulation that they like the most. Perceptible deviation of a constituent from the ideal formulation is liable to reduce acceptability. The stronger the influence of a constituent on that person's appetite for that type of food, the more likely is a barely detectable deficit or excess to have an effect on selection among variants. The mechanisms reviewed above ensure that sweeteners are like any other perceived constituent in having peak preferred levels.

The other extreme of the effects of a sensory characteristic on selection is a deficit or excess so far from the person's ideal point that that particular variant is unacceptable, even though the formulation may be perfect in every other way. Even sweetness may be excessive in a real food and thus elicit rejection.

Objective Sensory Distances from Ideal to Rejection

A well-known physical parameter of sucrose (and probably other sweeteners) in the ranges commonly used in food has a very simple relation to the discriminability of different levels of sweetness and therefore possibly to the effect of sweetener level on food selection. For plain aqueous solutions of sucrose this parameter is concentration: the minimum differences in concentration that can be distinguished (by their sweetnesses) are proportional to the lower (or higher) concentration, except for the most dilute or concentrated solutions (Schutz and Pilgrim 1957). This proportion of a just noticeable difference (JND) to the level from which it is measured (both expressed in physical units) is known as the Weber ratio. If, when the sweetener is included in a food, concentration is still the physical phenomenon that meets this perceptual condition of a constant Weber ratio, then an equal-ratio (logarithmic) scale of concentration should be proportional to the strength of the sweetener's perceptual effect, provided the tests are carried out in a way (Booth et al. 1983) that avoids various sources of bias demonstrated by experimental psychologists (Poulton 1968, 1979). This is not the statement of a psychophysical "law" (Stevens 1957), nor are we "forcing" the data to fit a semi-log function (Thurstone 1927). We are merely using a perceptual fact about sucrose concentrations to provide a model for the design and analysis of precise measurements of individual performance (Booth et al. 1983, 1986), avoiding arbitrary assumptions about psychological (Stevens 1969) or physical "scales" (Weiss 1981; Myers 1982).

Individuals' bias-minimised ratings of normal variations of sucrose (or salt, whitener, etc.) in familiar foods and drinks do indeed fall rather precisely on a straight line against logarithmic concentration of the perceived constituent (Booth et al. 1983, 1986). So far, for several perceptually fairly simple tastants, aroma substances and colourings of common foods and drinks, there appear to be sensory influences on preference that are unitary above and below the ideal point in a good formulation (Booth et al. 1986).

The Tolerance Triangle

When a sweetener's Weber ratio in concentration units is constant in the region of the sweetener level that a person most prefers in a food, then concentration ratios just discriminably higher or lower than that ideal point will be spaced equally on either side of it (Fig. 10.2).

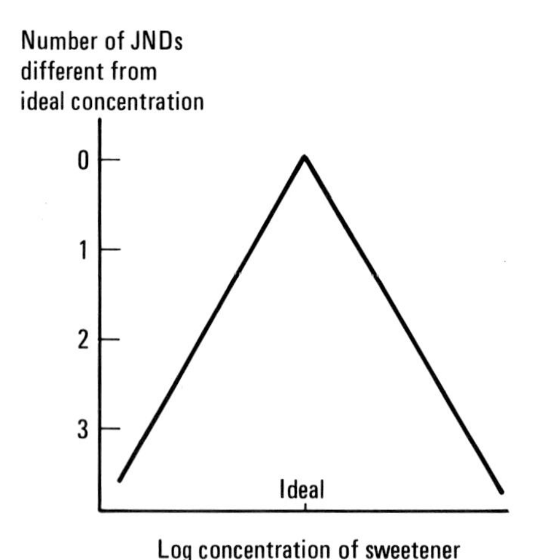

Fig. 10.2. The sweetener tolerance triangle for an individual for whom the sweetener has a constant ratio of the concentration discriminability to concentration, around the most preferred concentration in a food.

This triangle should be reproduced by an acceptance rating task, if performance has not been biassed. In fact any rating can be used by an assessor to perform at this perceptual limit, if the task demands. The perceptual nature of the descriptor "sweetness" gives it no inalienable merits as a discriminator of differences in sweetener concentration. Attribute descriptors such as "pleasantness", "liking" or even "healthiness for the teeth" can be used to make finely discriminated choices, so long as the rater's attribution is substantially enough affected by sweetener concentration.

Nevertheless, psychophysical acceptance rating, analysed on a multiple-stimulus, multiple-choice model, is liable to be less sensitive than the two-alternative forced-choice model used to calculate traditional JNDs. So the "Weber ratio" of "accept/reject ratings JND" to concentration (which we shall call the discrimination ratio: DR) is liable to be larger, i.e. the slope in Fig. 10.2 lower than a scale of traditional JNDs. Furthermore, difference tests in complex foods should not be assumed to be based on a single perceptual dimension (Frijters 1979; Ennis and Mullen 1985). When an acceptability response is taken to refer to the food overall, it is likely to be based on reaction to sensory influences on selection in addition to the sweetener concentration, although this remains to be demonstrated for test sessions in which sweetener is the only constituent varied.

Even if less sensitive, the DR is unlikely to vary around the ideal if the Weber ratio does not. In any case the data can be tested for variation in DR. In the range of constant DR, the degree of acceptability will decrease linearly on each side of the ideal point, reproducing the isosceles triangle of Fig. 10.2. That is, the indeterminate inverted U, or even asymptotic curve obtained from hedonic ratings of sweet stimuli in the traditional uncontrolled manner, should become a mathematically determinate inverted V when the ratings are obtained under bias-free conditions from a well-formulated food. So instead of guessing where the real peak is in a rounded, hand-drawn curve (and being influenced mainly by only the few data near the peak), we can calculate an exact estimate of the ideal level in that food for that person by interpolation from all the data on the tolerance triangle.

Characterised Tolerance Functions

Intensity increases monotonically with physical amount, from too weak to too strong, if there is a peak preference. Thus, if the test is free from range-frequency bias and other distortion, the ratings of a sensory characteristic which has a constant Weber ratio in the test range should "unfold" the isosceles triangle (Fig. 10.2) into a semilog function which has the same slope above and below ideal concentration (Fig. 10.3).

It follows that intensity ratings can substitute for log concentration and still give the triangular function. Frijters and Rasmussen-Conrad (1982) call this intensity-acceptability plot the "psychohedonic function". If both ratings suffer from the same biases (Poulton 1979), there is a possibility that they will cancel out. The theory can be used to interpret biassed intensity-hedonic data by regarding the usually asymmetrical and rounded hedonic peak as a triangle distorted by the conditions of the experiment (e.g. Fig. 10.4).

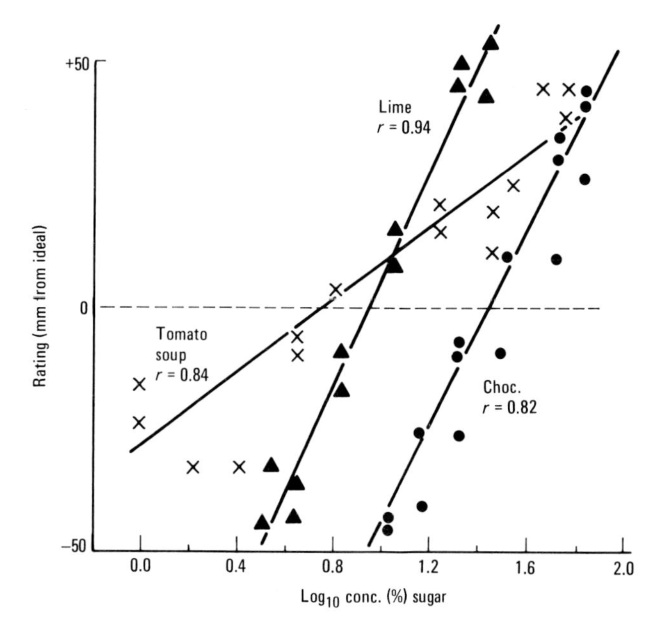

Fig. 10.3. Linear function of an individual's single session of sweetness intensity ratings on logarithm of sucrose concentration in a bias-minimised set of samples of chocolate baker's coating, lime drink or tomato soup. Note the choice of test formulations to equate response ranges on each side of the mid-category and to avoid close approach to the end-categories.

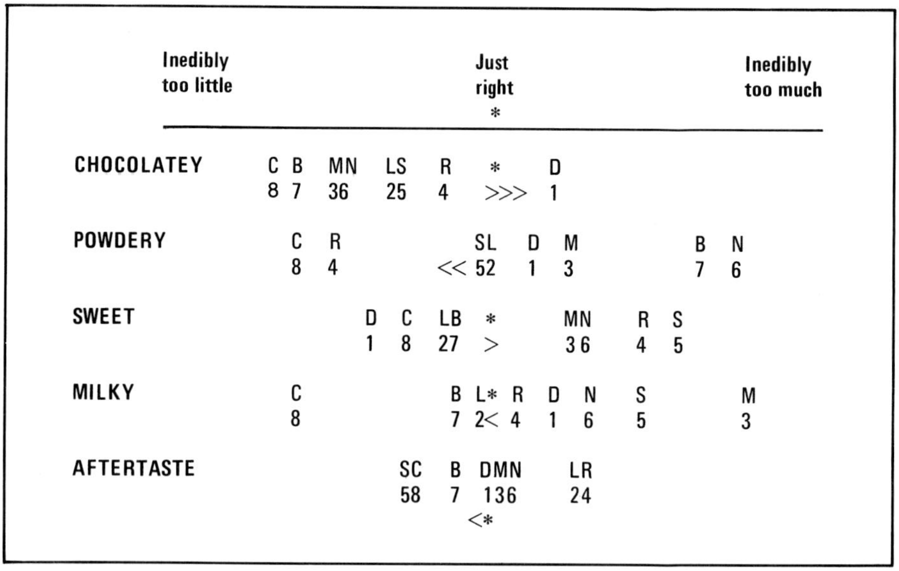

	Inedibly too little						Just right *							Inedibly too much
CHOCOLATEY	C 8	B 7	MN 36		LS 25	R 4	* >>>		D 1					
POWDERY		C 8	R 4				SL << 52	D 1		M 3			B 7	N 6
SWEET				D 1	C 8	LB 27	* >			MN 36	R 4	S 5		
MILKY		C 8				B 7	L* 2<	R 4	D 1	N 6	S 5			M 3
AFTERTASTE					SC 58	B 7	DMN 136 <*			LR 24				

Fig. 10.4. A British young woman's blind ratings of eight milk-chocolate brands (*B*, *C*, *D*, *L*, *M*, *N*, *R*, *S*) on freely described off-ideal characteristics (two-session means of positions on the response format given at the *top*). *Numbers* below the rating positions are liking ranks. The *arrow* on the ideal-response position indicates the range bias that these brands seem to be inducing in her ideal point for that characteristic. (Conner et al., to be published)

Sensory Affect Quantification with Free Descriptors

This theory of appetite psychophysics can be extended to the usual situation where several or even many sensory characteristics may be relevant to selection. Segment-representative individual consumers can give precise acceptability-composition data from test variants formulated to be close to a multifactor preference peak identified by Simplex optimisation or response surface methodology (Chap. 17). The physical basis of a descriptor does not even have to be known initially; indeed, the most accurate way of identifying the factor is the full psychophysical acceptance analysis subsequently. The individual's own freely characterised deviations from ideal in test samples whose physicochemical constitutions are not systematically controlled can provide estimates of bias: one of the errors in conventional group-analysed sensory spaces.

An example of use of the theory when there was no control over food sugar concentration or any other aspect of formulation is given in Fig. 10.4 — an untrained assessor's selection ratings for a set of milk chocolates, using free descriptions of perceived deviations from ideal. In this way, even without a bias-free experiment, a producer could vary composition at least semiquantitatively to match this individual's type of multidimensional ideal.

In a study of vending-machine coffees with controlled compositions, we avoided range bias (McBride 1982) and end-effects in three free, ideal-deviant descriptors simultaneously and identified distinct sources of the descriptors in levels of sugar, whitener and coffee solids (Table 10.1). Even though sometimes less discriminating, uncharacterised acceptance/rejection ratings gave narrower distributions of ideal points than did ideal-relative intensity ratings (Booth et al. 1986), an indication of validity in the "straight-through" model of selection without conscious sensation (Fig. 10.1). The bias-minimised ratings of sweetness and whiteness relative to ideal were also directly validated psychometrically as measures of food selection behaviour: when combined according to any of several preference models, the ratings accurately predicted the assessors' usual selections among the coffee variants from the vending machine.

Rating Anchors for Acceptance Psychophysics

Any rating requires two anchor points (or one anchor and a subjective difference or ratio "unit", which may amount to the same thing). Rating is discriminating

Table 10.1. Median (range) percentage residual variance of mean linear regressions between log concentrations of constituents and ideal-relative intensity ratings on freely described off-ideal characteristics of hot coffee drinks, in nine habitual drinkers of sweet white coffee

	Constituent		
	Sugar		Whitener
Sweetness ratings	3 (18)	*	48 (93)
	*		*
Whiteness ratings	23 (81)	*	2 (3)

*$P < 0.01$, one-tailed Wilcoxon test.

among stimuli in terms of responses referred to those anchors. Instructions, category labels (verbal, numerical or graphic) and standard stimuli (even if only the first test stimulus and a notional comparison) provided by the experimenter can influence the rater's task, but their effect cannot be assumed and can be determined objectively by analysis of relationships between stimuli and ratings.

The Ideal Point

In a familiar food, the memory of the preferred sweetness can serve as a standard stimulus that is not presented. Sweetness magnitudes can then be rated relative to this ideal point, above or below ideal, unlike acceptability ratings. Ideal-relative intensity rating is well recognised in psychology and in the food industry (Chap. 17). Moskowitz (1972) expounded some of its merits, e.g. ease of analysis and economical precision, compared with the more widely used rating of ideal intensity at the end only of a series of intensity ratings. However, its use to help minimise biases and so gain precision and validity in preference psychophysics was only recently set out (Booth et al. 1983).

There is some evidence that construction of this "internal standard" of comparison from long-term memory of preferred percepts of a familar food can be at least as precise as medium-term memory of the perception of a formulation that has been presented as an "external standard" for comparison. McBride and Booth (to be published) found that the JND for normal concentrations of a breakfast drink (containing sucrose, acidulant and fruit flavouring, with colour masked) was the same with a traditional successive two-alternative forced-choice JND determination and a JND determination by forced placing of a single presented sample on one side or other of the simultaneously remembered ideal concentration. Marie (1986; S. Marie et al., unpublished results) found that the linear regression of intensity on log aroma concentration was no less reliable when the ideal point was used as a rating mid-point than when a physically presented standard was used to anchor the mid-point.

The Rejection Points

In the Marie (1986) experiment, the necessary second category was an end-category of zero intensity. There was an unlabelled end-response on the other side of the standard-anchored mid-category, and assessors were found to use this as an anchor at twice the standard intensity. Ideal-relative intensity ratings have generally used vaguer intensity labels for the end-categories, invoking elements of selection motivation, e.g. "not nearly sweet enough" and "much too sweet". In an attempt to estimate the effects of the rated constituent on selection among food variants, we now use as end-categories the sweetness magnitudes that are regarded as sufficiently weak or strong to be rejected in such a choice situation (Fig. 10.4).

However precisely or vaguely the wording of the end-categories specify rejection behaviour (and, indeed, the context for the choice), end-effect bias must be avoided, by using the individual assessor's first two or more ratings to avoid presenting an intolerably low or high level of sweetener. Even if an unbounded response dimension is provided (e.g. numbers with instructions to respond in ratio), avoidance of intolerable sweetness keeps the experiment more relevant to everyday behaviour and avoids the bias suffered by numerical ratings that range over two or more orders of magnitude (Poulton 1968).

Perceived Sweetener Differences Between Ideal and Rejection

Rated distances of sweetness between maximally acceptable and rejected concentrations (Fig. 10.4) should then be a measure of the effect of sweetener levels on selection among variants of that food. The slope of a psychophysical function (the exponent of the power function if data are plotted on log–log coordinates) is commonly taken to be a measure of the sensitivity of the ratings to stimulus differences. However, such a parameter has no objective meaning: we cannot distinguish perceptual performance from response style, for example, and there is no fundamental justification for the common practice of combining ratings from different people, even after standardisation. If there are enough ratings of each of several stimuli close enough together to be imperfectly discriminated from one another, then the higher half of the ratings can be collapsed into one judgement of stimuli "greater than" the mid-concentration and a JND calculated (cf. Schutz and Pilgrim 1957, for example). The ratio of a JND to the stimulus level at which it is estimated (the Weber ratio) is a measure of the sensitivity of the ratings that is independent of arbitrary scoring principles. A similar objective measure of sensitivity (a discrimination ratio: DR) can be derived from the slope of the psychophysical regression line and the variance of the ratings around it, using the same JND calculation principle (Conner, Clifton, Griffiths and Booth, to be published).

Someone who is not very concerned about the particular sweetness of a food will not pit their ratings against perceptual limits. In other words motivational factors are liable to flatten the objective "slope" of the psychophysical function (Booth et al. 1983). Our data so far indicate that a majority of people are sufficiently intolerant of non-preferred magnitudes of sweetness to give a distribution of DRs across assessors which is skewed to a sharp low limit, at perhaps double the two-alternative forced-choice Weber ratio for sweetness.

Relations Among Psychophysical Acceptance Parameters

We find that individuals' ideal points for sucrose concentration in a number of foods are log-normally distributed across panels. The ratio of low to high rejected concentrations (rejections ratio: RR) appears to be normally distributed. The number of JNDs in the rejections range (the JRR) may therefore be skewed like the DRs. The correlations among assessors' DRs, RRs and JRRs depend on the wording of the end-categories, thus reflecting the actual psychological structures in the ways that assessors carry out the rating task with different category labels.

Differences in Sweetness Acceptance Between Individuals

Stability of Measurement

The sensory precision of bias-minimised sweetness ratings relative to ideal is remarkably high for unidentified repeat samples of foodstuffs: variances as low as 10% within sessions and 20% between sessions a few days apart.

Individuals vary with time relative to each other in their ideal points for food sugar (or salt), up to a total variance of 25%–50% over intervals of a few days or weeks, including the sensory error. This could represent real changes in preferences arising from ingestive experiences between test sessions.

Variation Among Individuals in Average Ideal

Analysis of variance among ideal sugar concentrations in tomato soup, dark chocolate and a lime drink shows a highly significant between-subjects effect: i.e. people differ in the level of sugar they prefer on average across these diverse foods.

The liking for high sweetness averaged over the three foods correlates with choice of carrot over celery and flavoured over unflavoured milk. This provides support for the full construct of a "sweet tooth", some people both liking food sweet and preferring sweet foods to others. Olsen and Gemmill (1981) also reported a correlation between peak-preferred sugar level and some choices of sweet foods, in children.

However, this subjects effect and a foods effect in sugar ideal points also interacted significantly in ANOVA. In other words there are idiosyncrasies in the pattern of sweet preference, even in foods other than those to which some people do and some do not add sugar, such as hot drinks. In the middling range of sweet preference there appear to be sub-groups with markedly different relative sugar concentration preferences between chocolate and lime. Variation in reactions to bitterness, sourness etc. (Chap. 4) may have more to do with observed variation in sugar preference than does variation in the sweet preference itself. Complexity is also indicated by a correlation between the average ideal point and choice of soda water over sweet and carbonated lemon and quinine drinks in response to questioning without actual samples.

Roles of Sweetness in Food Selection

Rapid Objective Measurement of Sweetness Selection

Bias-minimised scaling of an individual's ideal point for sugar in a food, and other characteristics of the acceptance/rejection function, can be achieved with only three or four tastes, taking a minute or two (Conner, Haddon and Booth, to be published). The precision and validity of the objective preference method therefore makes it practicable for mass screening for epidemiological or market data, wherever the range of samples to cover the population's ideal points can be provided.

Weight Control Problems: Sweet Preference or Abnormal Satiation?

Relationships between sweetness intensity rating and sugar concentration, and even the detectability of a minimum level of sweetener, have frequently been assessed in people who are suspected of having disorders of appetite or of the satiating effects of food, e.g. the obese, anorexic or cachectic. The logic is obscure, for what is at issue is peculiarity in the relation between their acceptance of real foods or drinks and the range of sweetener concentrations normally used. Thus acceptance, not sensation, let alone detection, is what needs testing, at least at first. Also it would be better to focus on people who have identifiable problems in restricting their intake rather than mere excess weight, which can often accompany normal appetite (Booth, 1978).

In the case of obesity a satiety deficit has been suggested as widely as an excessive liking or sweetness. However, the same point applies: satiety is a decrease in eating motivation, not a decrease in food perception. Potentially effective ratings of acceptance (namely, how pleasant consumption of the food would be) have indeed been used in tests of satiability in obese subjects, although sometimes the prima facie motivational decrease in the pleasantness of consumption has unwarrantably been assumed to be more a loss of emotionally experienced sensual pleasures from oral stimulation (Cabanac 1971). Hedonic ratings of the plain sugar water generally used in these acceptance and satiety tests are most unlikely to be expressing sensual pleasure (particularly perhaps in French subjects!). More to the point, on the theory of acquired acceptance presented here, ratings of a sweet non-food are unlikely to have anything to do with the role of sweetness in appetite/satiety, i.e. the acceptance of real (familiar) foods. Indeed, further evidence that hedonic ratings of unfamiliar mixtures may not relate to normal food selection was provided by Drewnowski's (1985) finding that pleasantness of sugar–cream mixtures was not affected by food deprivation, as appetite should in principle be affected and as the pleasantness of smelt foods (Ducleaux et al. 1973) or named staple foods (Booth et al. 1982) is in fact affected. Indeed, the pleasantnesses of eating several ordinary foods measure appetite/satiety (the level of acceptance of foods in general), and its changes from several sources, with greater sensitivity than any other measure, including amount eaten (Booth et al. 1982).

Nobody has yet used validated food acceptance ratings to test for a satiety defect in people who have difficulty in avoiding obesity. Unsuccessful dieters have been shown to be poor at learning both the dietary and the bodily cues to conditioned satiety (Booth and Toase 1983). However, there is no reason to think that this acquired insatiability is peculiar to sweet tastes more than any other sensory characteristic. Furthermore distinguishing high palatability from low satiability, physiological, sensory or cognitive, is an objective scaling problem entirely beyond the usual rating methods. We need objective measures to determine whether satiety generally involves increased selection of sweet over non-sweet foods, and whether this or any other failure to satiate or persistence of palatability varies among people in a way that might contribute to difficulty in weight control.

Food selection does relate to satiety in a manner that may be relevant to weight control, but it has nothing specific to do with sweetness. Readily assimilated energy kills hunger, as indicated by reductions of both food intake and the rated pleasantness of eating named staple foods; however, the satiating effect is only transient, for the hour or less while the energy is being rapidly absorbed (Booth 1981b). Concentrated sugars produce this transient suppression of appetite, even when sweet-specific satiety (Wooley et al. 1972) has been controlled out (Booth et al. 1970); indeed low-glucose maltodextrins have the same effect (Booth et al. 1976, 1982). Therefore refined carbohydrate (and alcohol and perhaps fats) taken between meals and at the ends of meals will suppress appetite at a time when there is no food intake to be suppressed in compensation for that snack or dessert energy. Such a mechanism appears to explain the incomplete compensation of intake-energy observed when sucrose was replaced by aspartame in people who are encouraged to take lots of soft drinks (Porikos et al. 1977, 1982). That is to say, if sugar contributes to obesity the problem is likely to be "inefficient" satiation rather than high palatability or undersatiability: low intake suppression per calorie depends on timing, not sweetness. Thus any energy-containing food or drink that is commonly selected as a snack or dessert may be a problem. A dieting strategy of cutting out sugar, regardless of when eaten, appears not to be effective (unlike cutting down on fats) in weight loss or its maintenance (Finer 1985; Lewis and Booth 1986). Exclusion of all energy selectively from drinks and foods frequently selected as a snack or desserts might, with other behavioural changes, be a more effective dietary contribution to self-maintenance (Booth 1985b) of weight loss, at the same time increasing intervals between cariogenic challenges.

Normal Role of Sweetness in Food Selection

The biological function of the human preference for sweetness is an unsolved mystery, yet to be tackled by behavioural ecologists. What is practically relevant nowadays is that sweetness can be used in a more or less sophisticated manner by rich and technologically advanced cultures to encourage self-selection of a varied, nutritious diet, as long as sugary snacking is kept infrequent. A precise and valid measure of the effect of a sweetener on an individual's selection of a food should help both education in self-management of healthy eating and engineering of food products for market success.

References

Beauchamp GK, Moran M (1982) Dietary experience and sweet taste preference in human infants. Appetite 3: 139–152

Birch LL (1980) Effects of peer model's food choices and eating behaviors on preschoolers' food preferences. Child Dev 51: 489–496

Birch LL, Marlin DW (1982) I don't like it; I never tried it: effects of exposure on two-year-old children's food preferences. Appetite 3: 353–360

Birch LL, Zimmerman SI, Hind H (1980) The influence of social-affective context on the formation of children's food preferences. Child Dev 51: 856–861

Booth DA (1972) Taste reactivity in satiated, ready to eat and starved rats. Physiol Behav 8: 901–908

Booth DA (1974) Acquired sensory preferences for protein in diabetic and normal rats. Physiol Psychol 2: 344–348

Booth DA (1977) Satiety and appetite are conditioned reactions. Psychosom Med 39: 76–81

Booth DA (1978) Acquired behavior controlling energy intake and output. Psychiatr Clin North Am 1: 545–579

Booth DA (1979) Preference as a motive. In: Kroeze JHA (ed) Preference behaviour and chemoreception. IRL Press, London Washington, pp 317–334

Booth DA (1981a) Momentary acceptance of particular foods and processes that change it. In: Solms J, Hall RL (eds) Criteria of food acceptance: how man chooses what he eats. Forster, Zurich, pp 49–68

Booth DA (1981b) The physiology of appetite. Br Med Bull 37: 135–140

Booth DA (1985a) Food-conditioned eating preferences and aversions with interoceptive elements: learned appetites and satieties. Ann NY Acad Sci 443: 22–41
Booth DA (1985b) Holding weight down: physiological and psychological considerations. Medicographia (Servier) 7(3): 22–25, 52
Booth DA (1986) Cognitive experimental psychology of appetite. In: Boakes RA, Burton MJ, Popplewell DA (eds) Eating habits. Wiley, Chichester New York
Booth DA, Toase A-M (1983) Conditioning of hunger/satiety signals as well as flavour cues in dieters. Appetite 4: 235–236
Booth DA, Campbell AT, Chase A (1970) Temporal bounds of post-ingestive glucose induced satiety in man. Nature 228: 1104–1105
Booth DA, Stoloff R, Nicholls J (1974) Dietary flavor accceptance in infant rats established by association with effects of nutrient composition. Physiol Psychol 2: 313–319
Booth DA, Lee M, McAleavey C (1976) Acquired sensory control of satiation in man. Br J Psychol 67: 137–147
Booth DA, Mather P, Fuller J (1982) Starch content of ordinary foods associatively conditions human appetite and satiation, indexed by intake and eating pleasantness of starch-paired flavours. Appetite 3: 163–184
Booth DA, Thompson AL, Shahedian B (1983) A robust, brief measure of an individual's most preferred level of salt in an ordinary foodstuff. Appetite 4: 301–312
Booth DA, Conner MT, Marie S, Griffiths RP, Haddon AV, Land DG (1986) Objective tests of preference amongst foods and drinks. In: Leitzmann C, Diehl JM (eds) Measurement and determinants of food habits and food preferences (Euro-Nut Report 7). University Department of Human Nutrition, Wageningen, pp 87–108
Cabanac M (1971) Physiological role of pleasure. Science 173: 1103–1107
Conner MT, Clifton VJ, Griffiths RP, Booth DA (to be published) Sensory optimisation of salt level in white bread.
Conner MT, Haddon AV, Booth DA (to be published). Very rapid, precise assessment of effects of constituent variation on product acceptability: consumer sweetness preferences in a lime drink. Lebensmittel-Wiss-Technol
Drewnowski A (1985) Sweet tooth reconsidered: taste responsiveness in human obesity. Physiol Behav 35: 617–622
Ducleaux R, Feisthauer J, Cabanac M (1973) Effets du repas sur l'agrément d'odeurs alimentaires et non alimentaires chez l'homme. Physiol Behav 10: 1029–1033
Ennis DM, Mullen K (1985) The effect of dimensionality on results from the triangular method. Chem Senses 10: 605–608
Farell B (1984) Attention in the processing of complex visual displays: detecting features and their combinations. J Exp Psychol [Hum Percept] 10: 40–64
Finer N (1985) The use of nutritive and non-nutritive sweeteners in the treatment of diabetes and obesity. Clin Nutr 4: 207–214
Fishbein M, Ajzen I (1976) Belief, attitude, intention, and behavior: an introduction to theory and research. Addison-Wesley, Reading MA
Frijters JER (1979) The paradox of discriminatory nondiscriminators resolved. Chem Senses Flav 4: 355–358
Frijters JER, Rasmussen-Conrad EL (1982) Sensory discrimination, intensity perception and affective judgment of sucrose-sweetness in the overweight. J Gen Psychol 107: 233–248
Harris G, Booth DA (1985) Sodium preference in food and previous dietary experience in 6-month-old infants. IRCS Medical Science 13: 1177–1178
Holman EW (1975) Immediate and delayed reinforcers for flavor preferences in rats. Learn Motiv 6: 91–100
Lewis VJ, Booth DA (1986) Causal influences within an individual's dieting thoughts, feelings and behaviour. In: Leitzmann C, Diehl JM (eds) Measurement and determinants of food habits and food preferences (Euro-Nut Report 7). University Department of Human Nutrition, Wageningen 7, pp 187–208
Lockhead GR (1972) Processing dimensional stimuli: a note. Psychol Rev 79: 321–328
Marie S (1986) Measurement of aroma perception from food in the mouth. PhD Thesis, University of Birmingham
McBride RL (1982) Range bias in sensory evaluation. Food Technol 17: 405–410
McBride RL, Booth DA (to be published) The method of constant *hedonic* differences: using classical psychophysics to measure preference. Food Technol
Moskowitz HR (1972) Subjective ideals and sensory optimization in evaluating perceptual dimensions in food. J Appl Psychol 56: 60–66

Myers AK (1982) Psychophysical scaling and scales of physical measurement. Psychol Bull 92: 203–214
Olsen CM, Gemmill KP (1981) Association of sweet preference and food selection among four and five year old children. Ecol Food Nutr 11: 145–150
Pliner P (1982) The effects of mere exposure on liking for edible substances. Appetite 3: 283–290
Porikos KP, Booth G, Van Itallie TB (1977) Effect of covert nutritive dilution on the spontaneous food intake of obese individuals. Am J Clin Nutr 30: 1638–1644
Porikos KP, Hesser MF, Van Itallie TB (1982) Caloric regulation in normal weight men maintained on a palatable diet of conventional foods. Physiol Behav 29: 293–300
Poulton EC (1968) The new psychophysics: six models for magnitude estimation. Psychol Bull 69: 1–19
Poulton EC (1979) Models for biases in judging sensory magnitude. Psychol Bull 86: 777–803
Schutz HG, Pilgrim FJ (1957) Differential sensitivity in gustation. J Exp Psychol 54: 41–48
Stevens SS (1957) On the psychophysical law. Psychol Rev 64: 153–181
Stevens SS (1969) Sensory scales of taste intensity. Percept Psychophys 6: 302–308
Thompson DA, Moskowitz HR, Campbell RG (1976) Effects of body weight and food intake on pleasantness ratings of a sweet stimulus. J Appl Psychol 41: 77–82
Thurstone LL (1927) Psychophysical analysis. Am J Psychol 38: 368–389
Weiss DJ (1981) The impossible dream of Fechner and Stevens. Perception 10: 431–434
Witherley SA, Pangborn RM, Stern JS (1980) Gustatory responses and eating duration of obese and lean adults. Appetite 1: 53–63
Wooley OW, Wooley SC, Dunham RB (1972) Calories and sweet taste: effects on sucrose preference in the obese and nonobese. Physiol Behav 9: 765–768
Zellner DA, Rozin P, Aron M, Kulish C (1983) Conditioned enhancement of human liking for flavor by pairing with sweetness. Learn Motiv 14: 338–350

Commentary

Drewnowski: Marketing research surveys and sensory panel testing, far from "simply polling opinions", often use very sophisticated methods of experimental psychology (e.g. Moskowitz 1983).

Reference

Moskowitz HR (1983) Product testing and sensory evaluation of foods. Food and Nutrition Press, Westport CT

Booth: For the reasons indicated in the paper, I cannot agree. The books on advanced methods of food testing that are now coming out illustrate the point well. They emphasise or even are exclusively concerned with statistical manipulation of aggregates of verbal responses from sets of people. This amounts to group frequency data (the poll), which can be a valuable social and market measurement, or it is frequency data corrupted by unfounded weightings (from the response grades, at best standardised). A psychological measurement, by contrast, yields a parameter describing the relationship between the individual's response and the conditions of the experiment, be it an interview item or a difference test. To be experimental psychology, the group statistics, like response surfaces, need to be done on such individualised objective data: about food acceptance, food composition, or (the present topic) the relation between them.

Drewnowski: Ratings referring to stimulus magnitude are called "objective" because they can be objectively checked against veridical stimulus concentration or against other properties of food as measured by instruments (e.g., texturometer). Measures of preference can be checked for inter- and intrasubject reliability, but proving their intrinsic validity is next to impossible. (For all we know, someone may *like* 2 M sodium chloride.) As a result, acceptability ratings are commonly treated as affective or attitudinal variables. This does not imply that acceptability data are necessarily more variable than sensory data; nor does it mean that acceptability measures are in any way less "scientific".

Booth: We agree, except that it is far from impossible to validate preference ratings, as we illustrate in our paper (cf. Booth et al. 1982; Marie 1986).

Rozin: What is the justification for using JNDs as measures of psychological distance? As I understand it, most psychophysicists have abandoned this idea, for a number of reasons.

Booth: You appreciate that we are not attempting to measure JNDs, whether by classical or new methods, and then advocating "indirect scaling" by concatenation of JNDs. Indeed, so far as I can see, the objective approach destroys the distinction between direct and indirect "scaling", by questioning that conventional use of the word "scale". We are simply using the classical calculation model of a just noticeable difference, or difference threshold, in a two-alternative forced-choice task to generate a useful measure of discriminability. Signal detection or any preferred model would do, if applicable to rating data. The point is that it is a measure of the perceptual performance achieved in a set of food sweetness selection (acceptance/rejection) ratings.

Rozin: Why is the biological function of sweetness preference an unsolved mystery?

Booth: The main reason we have for thinking that preference for sweetness might have had survival value is that it is one of the few overt behaviours that does not require postnatal experience (see Chap. 9). I know of no qualitative evidence, let alone any quantitatively established theory that the human race would have died out without our congenital responsiveness to sugars. Why should hungry primates need a taste preference to get them to sample and eat the energy in ripe fruit?

Beauchamp: Is it possible that sugars may be associated in nature with something else in addition to calories? Desor (personal communication) has suggested that since naturally sweet foods (e.g. fruits) tend to be rich in certain minerals or vitamins which themselves occur in amounts too small to be detected, sweetness might serve as a cue for such micronutrients.

Rozin: It is interesting that red, a colour associated with ripe (sweet) fruit, enhances the perception of sweetness (see Chap. 4). This might be an arbitrary cultural phenomenon, but if it had an ecological basis the effect might also be seen in fruit-eating primates. Of course, if they did show it we would still have to ask whether it was innate or they learned about the sweet–red association.

Booth: Another idea provoked by Bartoshuk's review of interactions between sweetness and bitterness is that survival value of sweet preference lies in its amplification of the suppressant effects of sweetness in sweet–bitter compounds such as some amino acids. A sweet receptor near the bitter receptor might serve to make sources of readily assimilated protein like meat, pulses and milk palatable, despite nitrogenous groups in amino acids and peptides that in other molecules signal toxicity and should be aversively bitter. Perhaps purely sweet molecules also serve to suppress the sourness that might otherwise dominate even ripe fruit. In neither case would the preference for sweetness by itself have been under selective pressure.

Chapter 11

Sweetness and Satiety

Barbara J. Rolls

Introduction

Before discussing sweetness and satiety, definitions are required. Sweetness encompasses the sweet taste associated with the various sugars, intense sweeteners, and heavy metal halides. Satiety according to the common, dictionary use refers to a glutted, unpleasant state following consumption of too much of something. However, satiety may also refer to a positive state which occurs as hunger is relieved. In scientific studies satiety generally refers to the graded termination of eating. As eating proceeds individuals start to feel full and that they have had enough, i.e., they are satiated. The preference or liking for the food being consumed declines, but interest in food can often be revived by offering another food. Thus satiety may refer to both a change in the palatability of foods, i.e. the hedonic response to foods, and to the satisfaction of the physiological need for food. In this review I will first discuss the changing hedonic response to sweetness and will then discuss the ability of different sugars and intense sweeteners to satisfy hunger and reduce food intake.

Alliesthesia: Physiological Usefulness and the Changing Hedonic Response to Sweetness

Cabanac (1971) has suggested that the pleasure derived from various sensations (alimentary, thermal, etc.) depends on their physiological usefulness. This theory was applied to eating behavior because he found that the pleasantness of food-related tastes and smells apparently depended on a person's state of repletion. Cabanac called this changing hedonic response alliesthesia, or "changed sensation."

In relation to sweetness Cabanac (1971) found that as subjects gradually consumed a sucrose solution (they drank 10 g in 50 ml of water approximately every 3 min) the subjective pleasantness of the taste of the sucrose started to decline after about 30 min, when approximately 400 kcal had been consumed. Tasting the sucrose without swallowing it had no effect on the palatability of sweet solutions. Cabanac et al. (1973) also found that the pleasantness of the taste of sucrose was decreased by 25-, 50-, or 100-g preloads of glucose delivered directly into the stomach. Because these hedonic changes occurred slowly and were seen with intragastric sugar, it was thought that sensory stimulation by the preloads had little influence on the changes, which were presumed to be due to an alteration in the physiological need for sugar.

Although the studies of alliesthesia are of historical importance, the conclusions that can be drawn from studies in which small numbers of subjects consumed or were injected with sweetened water are limited. It is likely that the postabsorptive and metabolic effects of sugars influence the hedonic response to the sweet taste, and more controlled studies addressing the role of physiological usefulness in the changing response to sweetness are required.

Sensory-Specific Satiety

Although it is not understood how the postingestive effects of sugar consumption affect palatability, a number of studies show that the sweet taste can affect the hedonic response to foods. Studies using low energy sweeteners indicate how the sweet taste affects satiety. Cyclamate was found to be as effective as glucose in reducing ratings of the palatability of sucrose solutions (Wooley et al. 1972). Since as stated above, it is not clear how relevant studies using sweet solutions are to normal eating, we conducted a study using sweet foods which had similar sensory qualities but which differed markedly in energy density because they were made with either sucrose or aspartame.

The standard test procedure was to have moderately hungry normal weight, non-dieting individuals rate the pleasantness of the taste of small portions of nine foods. Following these initial ratings, one of the foods was offered as a test meal and subjects ate as much of it as they wanted. At intervals over the hour after this test meal, subjects again tasted and rated the pleasantness of the nine foods sampled initially. In this experiment the test meal consisted of low (0.09 kcal/g, sweetened with aspartame) or 'high' (0.54 kcal/g, sweetened with sucrose) calorie orange gelatin dessert ('jello'). Subjects, who were not informed of these caloric differences, ate the same weight of the desserts so that the mean intake of the jello sweetened with aspartame was 20 kcal while that of jello sweetened with sucrose was 140 kcal. The time course and the magnitude of the decrease in the pleasantness of the taste of the jello were not affected by the calories consumed (see Fig. 11.1; Rolls et al. 1986). In studies of 2- to 5-year-old children Birch and Deysher (1986) have also found a similar decrease in preference for pudding following consumption of puddings sweetened with either fructose or aspartame.

Thus it is not necessary for sweet foods to contain nutritive sugars to activate the subjective, hedonic component of satiety. We have called this changing response to foods "sensory-specific satiety". It differs from alliesthesia in that it is not produced primarily by the state of repletion or the physiological need for particular nutrients. The palatability of sweet foods decreases following the ingestion of non-nutritive sweeteners and the changes in palatability occur within 2 min after the termination of eating. It is therefore likely that the changes depend on the sensory properties of foods, or on some cognitive process which assesses that enough of a particular type of food has been consumed (Rolls 1986).

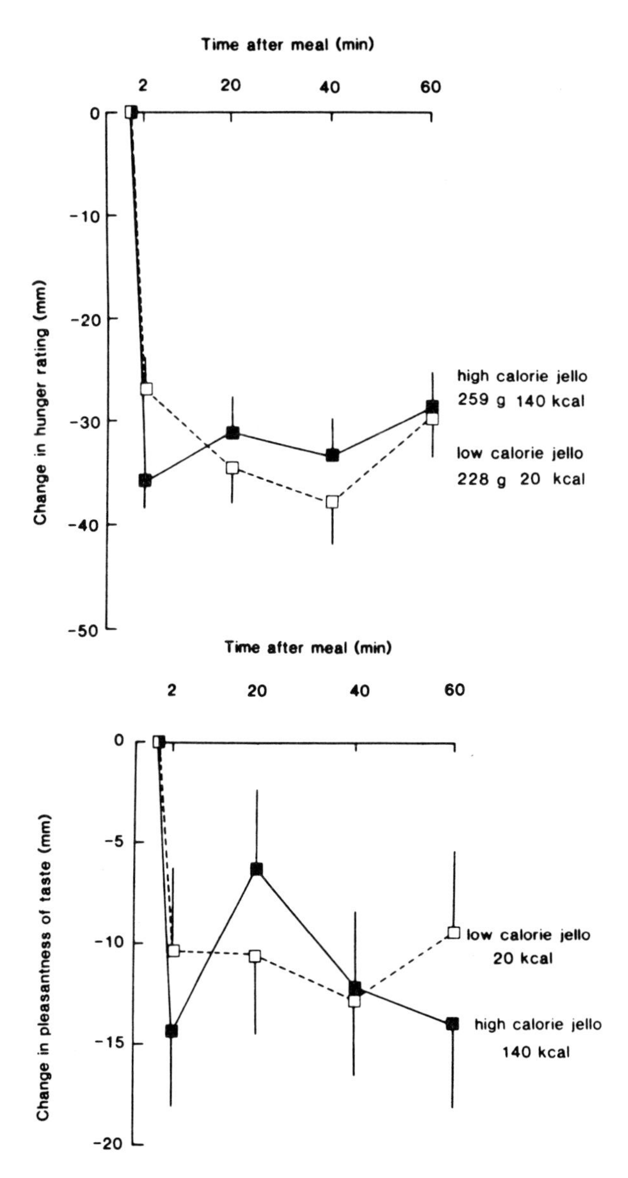

Fig. 11.1. *Above:* mean (±SEM) changes in the hunger ratings following high or low calorie orange jello test meals. No significant difference between the two conditions was observed. *Below:* mean (±SEM) changes in the pleasantness of the taste of the high and the low calorie orange jello. No significant difference between the two conditions was observed. (Rolls et al. 1986)

How Specific Is Satiety?

Cabanac and Duclaux (1970) found that the ingestion of sugar solutions decreased the pleasantness of other sugar solutions but did not affect the palatability of salty solutions and vice versa. They attributed this to the satisfaction of specific physiological needs which in turn affected hedonic responses. Since it is not clear that responses to solutions reflect normal ingestive behavior, we have looked at the effects of eating one type of food on the hedonic response to that food and other foods. Although the largest changes in palatability that occur after eating are for the food which has been consumed, there are some interactions between foods so that some other foods show a decrease in palatability (Rolls et al. 1984, 1986). Such interactions could occur because the foods have similar sensory properties, or because the foods are considered to be of the same type, or perhaps because the foods have the same macronutrient content.

If other foods are similar in taste to the eaten food, they may decrease in palatability. For example, after the consumption of a sweet food, other sweet foods declined in pleasantness, but savory (i.e. salty, meaty, piquant but not sweet) foods were unaffected, whereas the consumption of savory foods decreased the pleasantness of other savory foods, but not sweet foods (Rolls et al. 1984). Although such interactions between foods on the basis of taste can occur, of other sensory properties of foods are very different, such interactions may not be seen.

Thus sensory-specific satiety will limit the consumption of one type of food within a meal. Changing hedonic responses to foods promote the consumption of a healthy variety and make it unlikely that individuals will subsist only on highly palatable sweet foods.

Sweetness and Monotony

During a meal sweet foods decline in palatability as they are consumed. What happens to the palatability of sweet foods when they are consumed repeatedly? Is

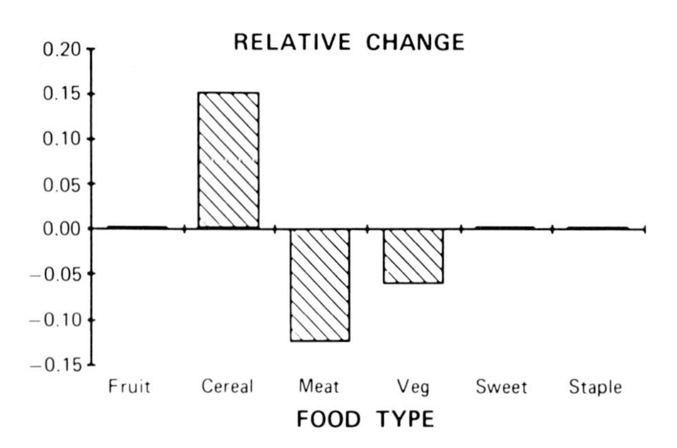

Fig. 11.2. The effect of repeatedly eating particular types of foods over 5 weeks. The relative change in palatability was calculated by dividing the difference in palatability ratings on days 9 and 37 by the number of times the foods were served. (Drawn from data in Schutz and Pilgrim 1958)

it possible that a persistent decrease in palatability will be seen? In studies of military personnel who were fed only 41 food items over 5 weeks there was no change in the palatability of the sweet desserts or canned fruits (Fig. 11.2). In contrast, the meats and vegetables did decline in palatability (Schutz and Pilgrim 1958; see also Chap. 18). We also found in a study in which confectionery was eaten every day for 3 weeks that there was no decline in the palatability of the product. Thus it appears that sweet foods are relatively resistant to long-term changes in palatability, so they can be enjoyed day after day. However, it is unlikely that the maintenance of palatability with repeat consumption is unique to the sweet taste since cereals and staple foods also remained palatable (Fig. 11.2).

Differences Among Sugars in Their Effects on Satiety

The magnitude and time course of changes in plasma insulin and glucose vary with the type of sugar ingested. Since elevated insulin and lowered glucose levels are associated with increased food intake, Spitzer (1983) hypothesized that the type of sugar ingested could affect intake in a later meal. To test this she compared the effects of equicaloric preloads of glucose and fructose on hunger and food intake 2.25 h later. Fifty grams (200 kcal) of glucose or fructose were consumed in a 500-ml lemon drink. Since fructose is sweeter than glucose, an added control for differences in palatability was to test glucose made to taste as sweet as fructose by adding aspartame. The caloric intakes in the test meal are shown in Table 11.1. Two and a quarter hours after the preloads fructose was the most effective in reducing subsequent food intake.

The different effects of the sugars on subsequent food intake could be due to the different time course of the changes in plasma glucose and insulin (Rodin and Reed 1986). Ingestion of glucose causes a rapid, sharp increase in plasma glucose and insulin which is followed by a steep decline so that 2–3 h after a preload plasma glucose levels would be below fasting levels and insulin would remain elevated. On the other hand fructose leads to slower and more moderate changes so that 2–3 h after a preload plasma glucose levels do not fall below fasting levels and insulin levels have returned to baseline. Thus 2.25 h after the preloads the hypoglycemia and elevated insulin levels associated with glucose ingestion could explain the higher food intake.

Despite these results, fructose should not be thought to be a sugar with a clear use in weight control. As Spitzer (1983) and Rodin and Spitzer (1983) have suggested, the timing of the effects on subsequent intake would be critical and indeed it would be predicted from the glucose and insulin curves that had the meal been presented 30 min after the preloads, the intake would have been higher after fructose than after glucose. Moran and McHugh (1981) found this to be true in

Table 11.1 Total number of calories eaten from a buffet 2.25 h after the preloads (data from Spitzer 1983)

Type of preload			
Fructose	Water	Glucose	Glucose–aspartame
869.2	1095.1	1401.1	1294.6

studies in monkeys. They observed that fructose produced more rapid gastric emptying than xylose or glucose. Subsequent intake was affected so that with fructose food intake was greater than with glucose or xylose preloads at first, but by 2–4 h after the preload was significantly less.

These studies suggest that sugars may differ in satiating efficiency, but more work is necessary before this conclusion can be regarded as definitive. Kissileff (1984) has stressed that to assess the satiety value of particular nutrients a dose-response paradigm is required. Also since the effects of preloads are dependent on when the next meal is offered, a systematic variation in the timing of the next meal is required. If the effects of nutrients are to be attributed to physiological changes, these changes must be measured in the behavioral studies. Also the studies should examine the effects of sucrose and non-nutritive sweeteners.

Sugars should be studied in foods as well as in drinks, since the form of the load could influence satiety. Differences between sugars on subsequent food intake of a mixed meal were found when the sugars were in solution and offered as drinks (Spitzer 1983). Recently, Rodin and Reed (1986) have reported that differences between fructose and glucose did not hold up as well when these sugars were presented in solid form and mixed with other substances. It remains to be determined the extent to which this reduced effect was due to the fact that the sugars were diluted in food as opposed to drink, or to the specific constituents of the foodstuffs with which the sugars were mixed.

Sweetness in Drinks

Most studies of the effects of sugars or the sweet taste on appetite had subjects drink sweet preloads. This complicates the interpretation of the findings since it is not clear whether preloads consumed as drinks are processed by hunger or thirst mechanisms. Since these mechanisms are different, this issue is critical.

Are sweet solutions foods or drinks? Experiments on rats indicate that they can be treated as both. Rats made thirsty with injections of hypertonic saline drink a saccharine or glucose solution avidly. However, when rats are hungry, but not thirsty, and refuse to consume water, they will consume sweet solutions (saccharin or sugar) in amounts which reflect the degree of food deprivation (Mook and Cseh 1981; Teitelbaum and Epstein 1962). Thus we do not know in the studies in which drinks were used as preloads whether they were treated as food or fluid. If they were treated as fluids, it is not surprising that sensory-specific satiety was not seen since changes in palatability appear to be food or fluid specific in that drinks decrease the palatability of other drinks but not foods, and foods decrease the palatability of other foods but not drinks (Rolls et al. 1983). Thus in preloading studies in which the goal is to study the controls of food intake, the preloads and the test meals should be foods not drinks.

The Sweet Taste Can Override Physiological Satiety Signals

Before considering further the issue of sweetness and caloric regulation, the effects of low energy sweet drinks on satiety should be discussed. When rats are given palatable, sweet solutions they drink in the absence of need. For example, non-deprived rats drink large quantities of sweet solutions such as saccharin, sucrose, and glucose. As long as the kidneys function normally this polydipsia does not matter since excess fluid will be excreted. We were interested to know what happens to consumption of sweet drinks when the formation of urine is reduced by injections of antidiuretic hormone (Rolls et al. 1978). The interesting question was whether in the face of the ensuing dilution of the body fluids, the compelling sweet taste would still stimulate fluid intake? It was found that the rats injected with antidiuretic hormone continued to drink a saccharine solution even when they were severely overhydrated. These results indicate that satiety signals from plasma dilution must be relatively ineffective against the consumption of palatable sweet solutions. Taste is overriding the homeostatic control of fluid balance.

Clinical reports indicate that humans may have only weak inhibitory thirst signals and rely on the kidneys to excrete excess fluid. Some psychiatric disorders are associated with excessive fluid intake. If this polydipsia is accompanied by high levels of antidiuretic hormone, water intoxication and hyponatremic convulsions may occur (see Rolls et al. 1978). It should be emphasized that normally people do not have problems with fluid balance when they consume large volumes of low calorie sweet drinks; however, if they have difficulty with fluid excretion, palatable drinks could maintain intake despite ensuing hyponatremia.

There is also an indication that the sweet taste can override food-related satiety signals. Two peptides, CCK and bombesin, which reduce food intake, lost potency when rats sham drank sucrose solutions (Gibbs et al. 1983). The higher the concentration of the sucrose, the less effective were the peptides in reducing intake.

Low Energy Sweet Drinks and Body Weight

Drinks in which the sugar content has been replaced with low calorie sweeteners frequently form part of a weight control program. The efficacy of these drinks in weight control has not been established. There are several studies in rats which if applicable to man could have disturbing implications for the inclusion of diet drinks in weight control programs. When rats were given saccharine solution to drink, they ate more laboratory chow than when given only water (Tordoff 1985). The sweet taste activates the cephalic insulin response and therefore increases plasma insulin, which could increase appetite. However, this did not appear to be the explanation of the rats' hyperphagia since celiac vagotomy which attenuated the cephalic insulin release did not reduce food intake. Other possible explanations for the hyperphagia are that the plasma dilution resulting from the increased fluid intake (Rolls et al. 1978) could stimulate food intake since in rats there is usually a constant ratio between food and fluid intake.

We also found that consumption of a saccharine solution which was freely available in addition to water over a 12-week period led to a significantly increased rate of body weight gain compared with rats consuming just water along with the chow (Fig. 11.3).

The effects on body weight of drinks made with intense sweeteners remain controversial in that Porikos and Koopmans (1985) have recently reported that

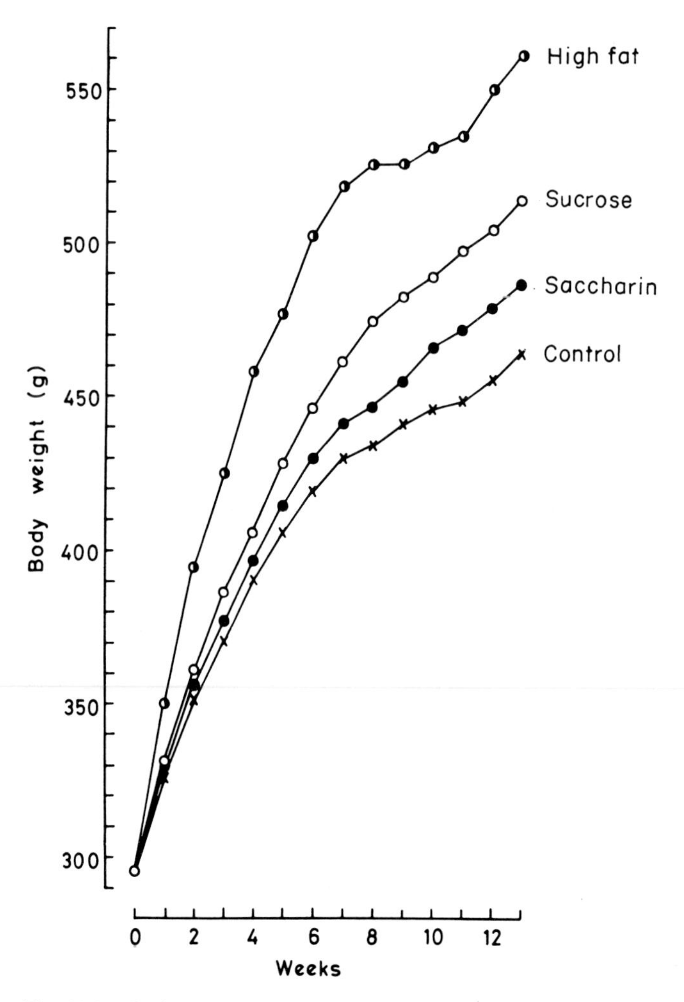

Fig. 11.3. Body weight changes in male rats over 12 weeks. All groups had continual access to laboratory rat chow. The saccharin rats ($n = 9$) also had free access to drinking water plus 0.01 M saccharine solution. The sucrose rats ($n = 10$) had water plus 0.5 M sucrose solution. The high fat group was offered a variety of palatable supermarket foods. The saccharin group gained weight at a significantly faster rate than the controls ($P < 0.01$).

rats drinking a solution sweetened with aspartame and saccharin weighed the same as controls drinking water at the end of 16 weeks, whereas those consuming 10% sucrose weighed 30% more. Rats switched from sucrose to the low energy drink lost weight and those switched from the low energy drink to sucrose showed a weight gain. They concluded that consumption of low energy sweet drinks instead of calorically sweetened beverages prevents weight gain and promotes weight loss. However, the interpretation of this experiment is complicated by the possibility that aspartame (and also cyclamate) apparently does not taste sweet to rats (see Sclafani and Nissenbaum 1985).

It is not clear why the results of the studies differ nor is it clear that the rat studies have any relevance to human behavior. They do, however, indicate that the effects of caloric dilution and the use of low calorie sweeteners may not always be straightforward. There is a clear need for long-term controlled studies of the effects of low energy drinks on food intake and body weight in humans.

Sweet Foods and Caloric Regulation

Although there are few studies of the influence of non-caloric drinks on food intake and weight control, there have been a number of studies of liquid diets where the sugar was replaced with low energy sweetener. The results of these studies are contradictory, with some showing compensation for the caloric dilution and some showing no compensation. Porikos et al. (1982) suggested that the variable results were due to the cognitive cues available to the subjects during the tests. They suggest that when subjects could not see the amounts of diet being consumed they would rely on bodily signals and compensate for missing calories. However, when subjects could judge the amounts consumed they relied on preconceived notions about appropriate amounts to eat and ate fixed volumes regardless of calories. Birch and Deysher (1986), in studies comparing caloric compensation in children and adults, also concluded that preconceived notions about the calorie content of food determine intake in adults. They found that 2- to 5-year-old children were better at compensating for caloric differences in pudding made with either sugar or aspartame than were adults. This could be because adults eat portions which they have learned are appropriate while children, who are relatively inexperienced with foods, rely on bodily signals.

We found in a study of normal weight adults who were offered jello sweetened with either sucrose or aspartame that they ate a constant weight of food and despite the difference in calories consumed, hunger was suppressed equally by the low and high calorie foods over the next hour (see Fig. 11.1), and there was no compensation for the caloric difference when cheese and crackers were offered an hour later (Rolls et al. 1986). The subjects were not informed of the caloric manipulation. Large quantities of the foods were presented in bowls so the subjects could assess the amounts they were eating. In these individuals who had never been on diets and who had no problems with body weight, calories had little impact on intake during a meal. Whether calories will be more important in dieters who are more familiar with appropriate caloric intake rather than appropriate volumes of food is being tested. We are also determining whether compensation occurs when the difference in caloric intake exceeds a critical level. Such short-term studies of the effects of caloric dilution during a meal must be regarded as a prelude to long-term studies of energy regulation.

Porikos et al. (1977, 1982) have examined the effects of replacing the sugar in a number of foods and drinks with aspartame. Subjects were both normal weight and obese individuals who were confined to a hospital room for the duration of the study. They were not told food intake was being measured and that the calorie content of the foods and drinks was being manipulated. Foods were served from platters with large portions available and subjects were encouraged to consume sweet drinks. Over the first 6 days of the test, the foods and drinks were sweetened with sucrose. With the free availability of a variety of foods, both obese and

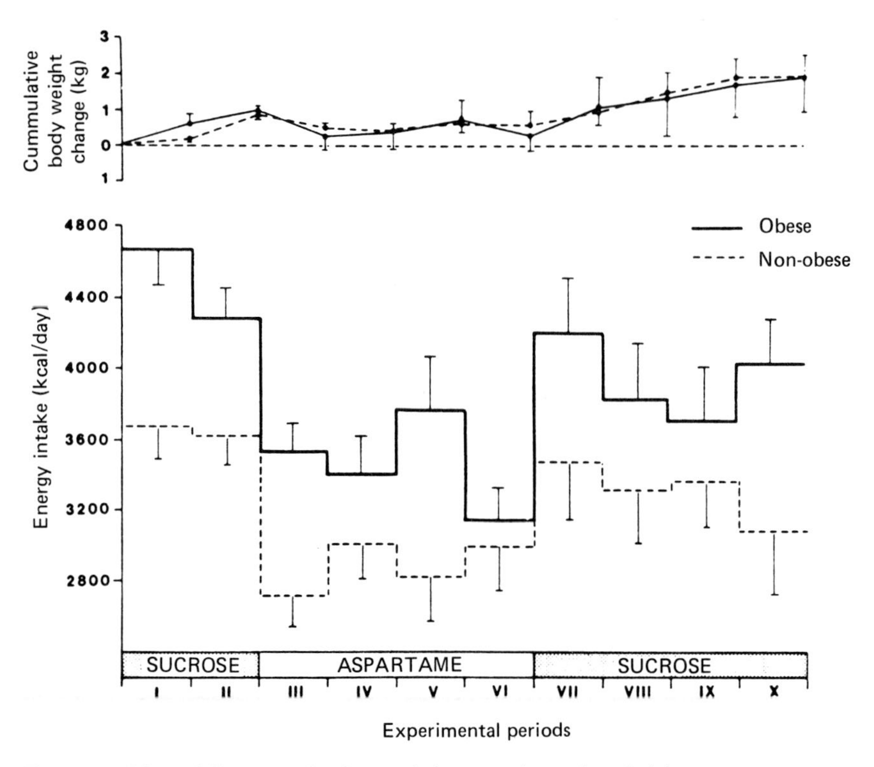

Fig. 11.4. Mean daily energy intake per 3-day experimental period (sucrose or aspartame) and cumulative body weight change (±SEM). The *solid lines* represent the data for five obese subjects, and the *dashed lines* the data for eight non-obese subjects. (Porikos and Pi-Sunyer 1984)

normal weight subjects gained weight (Fig. 11.4). However, when the aspartame replaced the sugar, this weight gain stopped. The subjects continued to eat the same volume of food that they had been consuming during the sucrose baseline period and therefore showed poor caloric compensation. On the aspartame diet the obese subjects ate 21% fewer calories and the normal weight subjects ate 26% fewer than at baseline. When at the end of the study sucrose was again consumed, energy intake rose back to the level eaten in the initial baseline period (Fig. 11.4). Although the authors suggest that low energy sweeteners may have some efficacy in weight control, more work is needed before this can be said definitively. Studies must be carried out for a longer period of time. Also, since in most situations dieters are aware of the calorie content of the foods they are consuming, tests of reduced calorie foods must be conducted in individuals who know the calories are being manipulated. It is possible that greater compensation will occur when individuals know how many calories they are consuming.

Conclusions

We still have only a rudimentary knowledge of the many factors involved in satiety, or the termination of eating. In some ways it could be considered quite contrived to consider just sweetness and satiety, since it is not clear that sweetness has special satiating properties. On the other hand, sweetness and satiety can be considered as unique because it is possible to consider sweeteners with a wide variety of postingestive and metabolic consequences.

The sweet taste is very compelling, but consumption of sweet foods may be limited during a meal by their decreasing palatability. The taste of sweet foods appears to be involved in this changing hedonic response since such "sensory-specific satiety" is seen with foods sweetened with low energy sugar substitutes. However, more studies are required before the mechanisms of such sweetness-induced satiety can be understood. Although low energy sweeteners obviously do not affect satiety by providing energy, they can induce some of the same physiological responses as sugars, i.e., they cause stomach fullness and they can stimulate the release of insulin. Also, the interpretation of the majority of the studies on sweetness and satiety is complicated by the use of drinks rather than foods so that body fluid changes accompany ingestion.

In the few studies in which foods have been used to study sweetness and satiety, it is clear that low energy sweeteners can be as satisfying as sugars during a meal, particularly when the subjects are unaware of the caloric manipulation. In the longer term there is some compensation for the missing carbohydrate, but this compensation is incomplete. More long-term studies are required before firm conclusions about the efficacy of low energy sweeteners in weight control can be made.

Because of the wide variety of sugars and the availability of palatable low energy sugar substitutes, the study of sweetness and satiety promises to be among the most informative of future areas of research in shedding light on appetite, hunger, and satiety.

References

Birch LL, Deysher M (1986) Caloric compensation and sensory specific satiety: evidence for self regulation of food intake by young children. Appetite (in press)

Cabanac M (1971) Physiological role of pleasure. Science 173: 1103–1107

Cabanac M, Duclaux R (1970) Specificity of internal signals in producing satiety for taste stimuli. Nature 227: 966–967

Cabanac M, Pruvost M, Fantino M (1973) Negative alliesthesia for sweet stimuli after varying ingestions of glucose. Physiol Behav 11: 345–348

Gibbs J, Bernz JA, Smith GP (1983) Sweet taste inhibits peptide-induced satiety. Proceedings of the American Psychiatric Association

Kissileff HR (1984) Satiating efficiency and a strategy for conducting food loading experiments. Neurosci Biobehav Rev 8: 129–135

Mook DG, Cseh L (1981) Release of feeding by the sweet taste in rats: the influence of body weight. Appetite 2: 15–34

Moran TH, McHugh PR (1981) Distinctions among three sugars in their effects on gastric emptying and satiety. Am J Physiol 241: R25–R30

Porikos KP, Koopmans H (1985) Effects of long-term use of aspartame on body weight in rats. Int J Obes 9: A85

Porikos KP, Pi-Sunyer FX (1984) Regulation of food intake in human obesity: studies with caloric dilution and exercise. Clin Endocrinol Metab 13: 547–561

Porikos KP, Booth G, Van Itallie TB (1977) Effect of covert nutritive dilution on the spontaneous food intake of obese individuals: a pilot study. Am J Clin Nutr 30: 1638–1644

Porikos KP, Hesser MF, Van Itallie TB (1982) Caloric regulation in normal-weight men maintained on a palatable diet of conventional foods. Physiol Behav 29: 293–300

Rodin J, Reed D (1986) The relationship between plasma glucose and insulin changes and food intake after preloads of fructose or glucose. Proc Eastern Psychol Assoc 57: 41
Rodin J, Spitzer L (1983) Differential effects of fructose and glucose on food intake. Abstracts of Fourth International Congress on Obesity. New York
Rolls BJ (1986) Sensory-specific satiety. Nutr Rev 44: 93–101
Rolls BJ, Wood RJ, Stevens RM (1978) Palatability and body fluid homeostasis. Physiol Behav 20: 15–19
Rolls ET, Rolls BJ, Rowe EA (1983) Sensory-specific and motivation-specific satiety for the sight and taste of food and water in man. Physiol Behav 30: 185–192
Rolls BJ, van Duijvenvoorde PM, Rolls ET (1984) Pleasantness changes and food intake in a varied four-course meal. Appetite 5: 337–348
Rolls BJ, Hetherington M, Burley VJ, van Duijvenvoorde PM (1986) Changing hedonic responses to foods during and after a meal. In: Kare MR, Brand JG (eds) Interaction of the chemical senses with nutrition. Academic Press, New York, pp 247–268
Schutz HG, Pilgrim FJ (1958) A field study of food monotony. Psychol Rep 4: 559–565
Sclafani A, Nissenbaum JW (1985) On the role of the mouth and gut in the control of saccharin and sugar intake: a reexamination of the sham-feeding preparation. Brain Res Bull 14: 569–576
Spitzer L (1983) The effects of type of sugar ingested on subsequent eating behavior. Unpublished Doctoral Dissertation, Yale University
Teitelbaum P, Epstein AN (1962) The lateral hypothalamic syndrome: recovery of feeding and drinking after lateral hypothalamic lesions. Psychol Rev 69: 74–90
Tordoff MG (1985) Sweet drinks increase food intake and preference in rats. Abstracts of a meeting on Mechanisms of Appetite and Obesity, San Antonio, Texas
Wooley OW, Wooley SC, Dunham RB (1972) Calories and sweet taste: effects on sucrose preference in the obese and nonobese. Physiol Behav 9: 765–768

Commentary

Blass: Rolls' examples of sensory-specific satiety and its constraints fit into the broader context of control exerted by physiological changes that occur within a particular behavioral context. Control is specific to the class of substances that cause the physiological change, as Rolls demonstrates. Sensory-specific satiety is also due to the context in which the change occurs. Physiological alteration outside of its normal context is a much less potent control over ingestive behavior (Blass and Hall 1974; Nachman and Valentino 1966; Baile et al. 1971).

References

Baile CA, Zinn W, Mayer J (1971) Feeding behavior of monkeys: glucose utilization rate and site of glucose entry. Physiol Behav 6: 537
Blass EM, Hall WG (1974) Behavioral and physiological bases of drinking inhibition in water deprived rats. Nature 249: 485–486
Nachman M, Valentino DA (1966) Roles of taste and postingestional factors in the satiation of sodium appetite in rats. J Comp Physiol Psychol 62: 280

Scott: The replacement of sugar by aspartame in the diets of hospitalized subjects, and the failure of these subjects to increase intake to compensate for the caloric loss, is mentioned as evidence that aspartame may be effective in a weight control program. I am concerned about the subjects having spent 6 days in the hospital, presumably rather inactive, during sucrose consumption and the effect of this protocol on caloric need. Appetite could decrease because of the unappealing setting, lack of normal social influences on intake, lack of physical activity, or the increase in glucose or lipid levels associated with the weight gain which occurred in the first 6 days. At the least, the order of presentation should have been alternated (first aspartame, then sucrose) to see whether subjects would then make a compensatory *reduction* in intake.

Booth: The experiments on the satiating effects of sweeteners in foods are very important. The sucrose may be too dilute to be measurably satiating, especially by measures as insensitive as hunger ratings or even perhaps intake. Nevertheless, the patterns of means are exactly what I would expect: "cognitive satiety" effects occur at 2 min which have disappeared from 20 min onwards. There may even be an interesting "disappointment" effect: the subjects recover some appetite because they expected more "physiological satiety" than materialized.

This "cognitive satiety" is of course part of the appetite reaction to the Gestalt that foods evoke. It may have been so thoroughly learned that a new reaction may not be adequately relearned in weeks. Nevertheless we have demonstrated learning and relearning of adult cognitive satiety in one or two "satiety conditioning" snacks or meals. Toase and Booth (in preparation) have further found that dieters who succeed in losing weight are much better than unsuccessful dieters at learning both the dietary and the bodily cues to satiety. I suggest therefore that cognitive satiety might vary between successful and unsuccessful dieters, as well as being liable to reversal.

Section V

Sweetness and Obesity

Chapter 12

Sweetness and Obesity

Adam Drewnowski

Introduction

Increased prevalence of obesity in the United States has been paralleled by increased consumption of refined sugars and fat (Drewnowski et al. 1982b; Page and Friend 1978). According to USDA estimates for 1985, the average American consumed a total of 127.4 lb. (57.8 kg) of nutritive sweeteners, including 67.5 lb. (30.6 kg) of sucrose, consumed chiefly in processed foods and beverages. There is a popular belief that sugar is uniquely fattening. Overindulgence in soft drinks, sweets, and desserts has long been thought to be a causal factor in the development of obesity, adult-onset diabetes, and coronary heart disease (Yudkin 1966). To the obese, it is claimed, many of the most attractive, almost irresistible foods are those that are rich in carbohydrates, especially sugar (Yudkin 1983).

Does sugar play a role in the development and maintenance of obesity? Popular beliefs aside, there is little direct evidence that the diet of obese individuals is rich in sweetened foods. On the contrary, epidemiological surveys of different populations have repeatedly observed an inverse relationship between reported sugar consumption and the degree of overweight. One large-scale British study (Keen et al. 1979) reported a negative correlation between body weight and total energy, total carbohydrate, and sucrose intake. A nationwide survey of almost 1000 U.S. children and adolescents (5–18 years) found no significant relationship between body fatness and self-reported intakes of sugar and other snack-type foods (Morgan et al. 1983). Data from the Ten-State Nutrition Survey (1972) for 4907 adolescents found no relationship between measures of triceps skinfold and reported intakes of sugar-containing foods, including jams, honey, candies, and soft drinks (Garn et al. 1980). Survey studies on the link between sugar consumption and the incidence of adult-onset, non-insulin-dependent diabetes (Walker 1977; West 1978) show that the risk of diabetes is inversely related to carbohydrate and sugar consumption, and positively related to fat consumption and to total calories (Bierman 1979). Since obesity is the chief risk factor for adult-onset diabetes, it may be that obese individuals overeat not sugar but fat (Gonzales 1983).

Accurate assessment of the role of sugar in obesity and diabetes depends on sophisticated dietary intake methodology. It should be noted that most survey studies of dietary intake have used a single measure of self-report — the 24-h food recall. The validity of this measure as applied to the obese population is questionable: the obese may underestimate daily calories by as much as a third (Lansky and Brownell 1982), selectively underreporting the size of the evening meal and the consumption of sugar-containing snacks, candies, and soft drinks (Beaudoin and Mayer 1953). The obese may also engage in bouts of dieting: a single day's food intake, even if correctly reported, may not reflect the habitual food intake of the individual. Overweight subjects examined as part of the National Health and Nutrition Examination Surveys (NHANES I and II) reported daily intakes that were several hundred calories less than those of lean individuals (Carroll et al. 1983), thus showing an unlikely inverse correlation between overeating and overweight.

Clinical studies on dietary intakes of ambulatory obese outpatients, using food frequency questionnaires, 3- and 7-day food records, and dietary interviews have produced inconsistent results, depending on the nature of obesity studied, procedures used, or the skill of the interviewer (Beaudoin and Mayer 1953; Krantzler et al. 1982). Some studies reported increased consumption of sugar, sweets, and high calorie snacks among obese teenagers and adults (Grey and Kipnis 1971), while other studies reported opposite results (Gates et al. 1975; Lansky and Brownell 1982). Experimental studies of food intake under controlled clinical or laboratory conditions have generally failed to document excessive intake of sweet foods among obese individuals. Because of various social biases, denial of hunger, and a documented reluctance to overeat in public, obese subjects seldom eat more than normal-weight controls, even in studies utilizing deception, where care is taken to disguise the true nature of the testing procedure (Spitzer and Rodin 1981). Laboratory studies on other aspects of food consumption, including eating styles, rates of consumption, or response to caloric dilution with non-nutritive sweeteners have not observed major obese/normal differences (Drewnowski 1983).

Existing data from survey, clinical, and laboratory studies generally do not support the postulated link between obesity and excessive sugar consumption. Yet there are persistent reports of a carbohydrate-craving among obese women that may be brought about by a "sweet tooth" or elevated pleasure response to sweet taste. Given the difficulty of accurately assessing sugar intakes, several investigators have focused on taste responsiveness in obesity as a potential index of food consumption. It is generally assumed that elevated pleasure response to sweetness is reflected in greater acceptability and increased consumption of sweet foods (Birch 1979; Shrager et al. 1984). There is evidence to suggest that the obese are over-responsive to the pleasurable or hedonic aspects of food, preferring foods that are spicy or intensely flavored (Schiffman et al. 1979). Obese subjects have been observed to eat more of palatable as compared with unpalatable foods relative to normal-weight individuals (Spitzer and Rodin 1981), are reported to eat mostly for pleasure in the absence of "real" hunger (Yudkin 1983), and overeat a large variety of foods (Vlitos 1983). However, it is unclear whether this over-responsiveness to sensory aspects of food is a consequence of obesity or whether it can serve as a potential psychobiological marker for the pre-obese state. In other words, does sweetness response contribute to the development of obesity in persons not yet obese? No information is as yet available on this point.

Laboratory studies on the relationship between obesity and sweet taste have addressed three aspects of taste responsiveness: sensitivity, suprathreshold sweetness perception, and hedonic response to sweetness. Any one of those might discriminate between obese and normal-weight individuals. Early laboratory studies (Grinker 1978; Grinker and Hirsch 1972) focused on sensory and affective aspects of sweet taste response, and examined the perception of sweetness and the judged pleasantness of sweet solutions in massively obese patients and lean controls. Hedonic response to sweetness was further examined as a function of weight loss, caloric restriction, or caloric loading (Cabanac 1971; Cabanac and Duclaux 1970, 1973; Wooley et al. 1972). More recent studies examined sensory responsiveness to sweetness and fat as a function of caloric deprivation and body weight status in obese and formerly obese women (Drewnowski et al. 1985). To complete the picture, developmental studies have examined responsiveness to sweetness at birth or in infancy as a function of birth weight or risk for developing childhood obesity (Grinker et al. 1980; Milstein 1980).

Perception of Sweetness

Studies on sweet taste psychophysics using threshold detection, stimulus recognition, and magnitude estimation procedures have for the most part failed to show abnormalities in sensory functioning among overweight subjects. In early studies (Grinker et al. 1972), massively obese subjects at 100% or more above ideal body weight (IBW) and lean controls received small quantities (3 ml) of sucrose solutions (0.057–0.57 M sucrose or 1.95–19.5% wt/vol) at room temperature. The subjects were required to sip each solution without swallowing, rinse, and then identify or rate the intensity of the stimulus. The individual ability to detect low concentrations of sucrose (0.175% wt/vol) was measured using a signal detection procedure. Magnitude estimation techniques were used to monitor the perception of sweet solutions at concentrations above threshold. There were no differences in the detection or perception of sweet taste between obese and normal-weight subjects. Both groups exhibited the same degree of cognitive bias in rating solutions containing red food coloring as sweeter than colorless ones (Grinker and Hirsch 1972; Grinker et al. 1972).

Further studies on detection and recognition thresholds for sweet, sour, salty, and bitter failed to find differences between obese subjects and lean controls (Malcolm et al. 1980). Comparable sucrose thresholds (range: 0.0037–40 mol/liter) were observed for obese and normal-weight women (Frijters and Rasmussen-Conrad 1982). Magnitude estimation or category scaling of sweetness intensity for sweet Koolaid (Grinker et al. 1976; Witherly et al. 1980), chocolate milkshakes (Rodin 1975), or different mixtures of milk, cream, and sugar (Drewnowski et al. 1985) revealed no significant differences in sweetness perception between different groups of obese men and women and normal-weight controls. The relationship between sucrose concentration and perceived sweetness intensity followed the logarithmic function $I = \log S + k$ (Drewnowski et al. 1985) or, with magnitude estimation techniques, the log–log function (Frijters 1983). Comparable exponent values were found for obese and for lean subjects (Frijters 1983).

Additional clinical and laboratory data suggest that the perception of sweetness intensity is not affected by short-term caloric restriction, such as overnight fasting (Drewnowski and Greenwood 1983), and does not change following weight loss (Grinker and Hirsch 1972; Grinker et al. 1976). Sensory perception of sweet stimuli is unrelated to body fatness and is not influenced by changes in body weight or metabolic status.

Hedonic Response

The affective response to sweet taste typically follows a different psychophysical function than the perception of sweetness intensity (Moskowitz et al. 1974). Hedonic functions first increase, with sugar concentration reaching a maximum at ideal point or breakpoint, and then decline as increasingly sweet stimuli are judged as less pleasant. The typical hedonic function follows an inverted-U shape, with the breakpoint located in the range of 7%–10% sucrose. Respondents characterized by a "sweet tooth" might show an elevated response to sweetness in general, or might show a breakpoint at a different location, preferring stronger to weaker sucrose concentrations.

Cabanac and Duclaux (1970, 1973) reported that moderately obese subjects liked intensely sweet sucrose solutions more than did normal-weight controls. However, most subsequent studies reported either no link between sweetness preferences and body weight, or a significant relationship in the *opposite* direction. Grinker et al. (1972) reported that normal-weight subjects rated most sucrose solutions as neutral, showing optimum liking for solutions in the 9% sucrose range. In contrast, obese patients disliked increasingly sweet solutions, with massively obese patients (> 200% IBW) showing the greatest decline. Overweight teenagers rating the pleasantness of cherry Koolaid of varying degrees of sweetness liked solutions at breakpoint less than did normal-weight subjects (Grinker et al. 1976). Comparable findings of reduced preferences for sweet taste in obesity were reported in other studies (Underwood et al. 1972; Johnson et al. 1979). Enns et al. (1979) observed an inverse relationship between sucrose preferences and a measure of body fatness: overweight subjects preferred weaker to stronger sucrose concentrations.

Other studies using sucrose solutions (Wooley et al. 1972), sweetened Koolaid (Witherly et al. 1980), and milkshakes (Rodin 1975) as stimuli showed no consistent relationship between body weight and sweet taste preferences. Only one study (Rodin et al. 1976) reported a positive correlation between sweetness preferences and body weight, with obese women showing greatest liking for 3.0 M sucrose. Large-scale consumer surveys typically found no relationship between body weight and hedonic preferences for varying concentrations of sugar in actual foods such as apricot nectar, canned peaches, lemonade, and vanilla ice cream (Pangborn and Leonard 1958; Pangborn and Simone 1958; Pangborn et al. 1957; Valdes and Roessler 1956; Witherly et al. 1980). The diversity of the obese subject populations and the variability of individual response may account in part for some of the seemingly discrepant results that have been reported in the literature.

Individual Differences

Individual differences in the pattern of taste responsiveness are a well-established phenomenon (Pfaffmann 1961; Pangborn 1970, 1981). Pfaffmann (1961) has reported a modal response of increasing liking for sweet solutions among normal-weight subjects, with a minority of subjects reporting a dislike. In contrast a majority of subjects disliked solutions of increasing saltiness with a minority showing increased liking. The inverted-U shape of the hedonic function for sweetness may thus be a composite of widely divergent individual responses.

A variety of response types have been observed among obese and non-obese subjects. Thompson et al. (1976) classified individuals on the basis of their hedonic response profile to increasing sucrose levels. Type I response was characterized by rise and decline (inverted-U), while type II was defined by rise followed by a plateau. Seemingly, most obese subjects were classified as type I respondents. Witherly et al. (1980) distinguished four separate hedonic response patterns. In addition to type I and II responses, type III response represented a monotonic decline in hedonic preference, while type IV represented no change in response with increasing concentration of sucrose.

The observed variability of response is not altogether surprising. All too often the obese have been assumed to represent a homogeneous group. The assumption that a single "obese" taste response profile can be identified reveals a fundamental misconception about the nature of human obesity. There is mounting evidence that human obesities are of multifactorial origin, and range from those that are genetically determined to those that may be diet-induced (Sclafani 1982). Excess body weight is merely the common end point for many potentially diverse individuals, while the etiology of obesity may differ from one patient to another. The contribution of sensory factors to food acceptance and diet selection may well vary with age, sex, age of onset of obesity, stable or dynamic weight status, as well as the degree of overweight. As a result, different patterns of response to sweetness can be expected for different obese individuals. Similarly, normal-weight controls should be carefully screened for dieting behaviors, potential avoidance of sweet foods, and previous problems with body weight. The common practice of classifying obese respondents solely on the basis of Metropolitan Life Insurance criteria ($> 120\%$ IBW) can no longer be justified. Consequently past results obtained using averaged group data should be interpreted with caution.

There is a need for new techniques to assess sensory response to sweetness on an individual basis and to correlate individual preference profiles with body weight or other indices of body fatness. The group hedonic profile has no psychological meaning, especially for so diverse a group as the obese, and alternative psychometric approaches are discussed elsewhere in this volume (Chaps. 5, 10). Recent studies show that the peak of hedonic preference to sugar solutions in water may be a continuous variable log — normally distributed over individuals (Frijters 1983). Past studies have found no consistent relationship between peak hedonic response to sugar solutions and body weight (Frijters 1983; Thompson et al. 1976). Recent studies (Drewnowski and Greenwood 1983; Drewnowski et al. 1985) do in fact show a modest relationship between hedonic responsiveness and body fatness as measured by the body mass index (weight/height^2). However, the stimuli were not sucrose solutions in water, but a more complex system containing both sugar and fat.

Complex Systems

Sensory preferences for sweetened stimuli may vary depending on the composition of stimulus mixtures. Laboratory studies on obesity and sweet taste have been largely restricted to sucrose solutions in water, which may limit their interpretation (see Weiffenbach 1977 for representative studies). Laboratory reports that obese subjects dislike the taste of intensely sweet solutions (Grinker 1978) contrast with anecdotal responses of increased liking for sweet desserts (Yudkin 1983). However, it should be noted that sugar in dessert-type foods is often consumed in combination with dietary fats. Calorie for calorie, many of the palatable foods that are supposedly irresistible to the obese — ice cream, chocolate, and other desserts — often contain more fat than sugar. The so-called sweet tooth, far from implying a craving for carbohydrates, may in fact reflect enhanced liking for foods containing both sugar and fat (Drewnowski et al. 1985; Gonzales 1983).

In recent studies (Drewnowski and Greenwood 1983), sensory preferences for sweetness versus fat were examined in a complex model system containing different proportions of milk, cream, and sugar. Twenty different stimuli, incorporating five levels of fat (0.1%–52.7% fat w/w) and four levels of sugar (0%–20% w/w) were employed. Sixteen normal weight subjects rated the perceived intensity of the stimuli and assigned a hedonic rating to each sample, using separate nine-point category scales (Peryam and Pilgrim 1957). The subjects were tested twice: once following lunch (Fed) and following an overnight fast (Fasted). As shown in Fig. 12.1 (top panels), intensity estimates of sweetness rose as monotonic functions of sucrose concentration at all fat levels, and no mixture phenomena were observed. In contrast, hedonic responses to the same stimuli (bottom panels) were characterized by a region of optimum preference or breakpoint, the location of which depended on the proportions of sugar and fat. The composition of the best-tasting mixture, as determined with the use of a mathematical modelling technique known as the Response Surface Method (Drewnowski and Greenwood 1983), was 20% fat and 9% sucrose. Neither intensity estimates nor hedonic ratings were affected by overnight fasting.

Further studies (Drewnowski et al. 1985) examined sensory response to the same sucrose- and fat-containing stimuli as a function of stable body weight status in obese, formerly obese, and normal-weight women. As shown in Fig. 12.2, the three groups of subjects did not differ in their sensory perception of stimulus sweetness, fatness, or creaminess. Intensity estimates rose as logarithmic functions of stimulus concentration.

Hedonic response profiles expressed in terms of isohedonic contours and the corresponding three-dimensional projections of the hedonic response surface are shown in Fig. 12.3. For normal-weight subjects [mean weight 129.5 lb. (58.8 kg)], hedonic ratings first rose and then declined with increasing sucrose concentration (x-axis) for all levels of stimulus fat. Increasing fat concentration (y-axis) also led to higher preference ratings. Stimuli containing both sucrose and fat were liked the best; ingredient composition of optimally liked stimuli was 20% fat and 8% sugar.

Obese patients [mean wt. 211 lb. (95.8 kg)] showed a differential response to sweetness depending on the fat content of the stimuli. They disliked increasing concentrations of sucrose in skim milk but liked the same concentration of sucrose

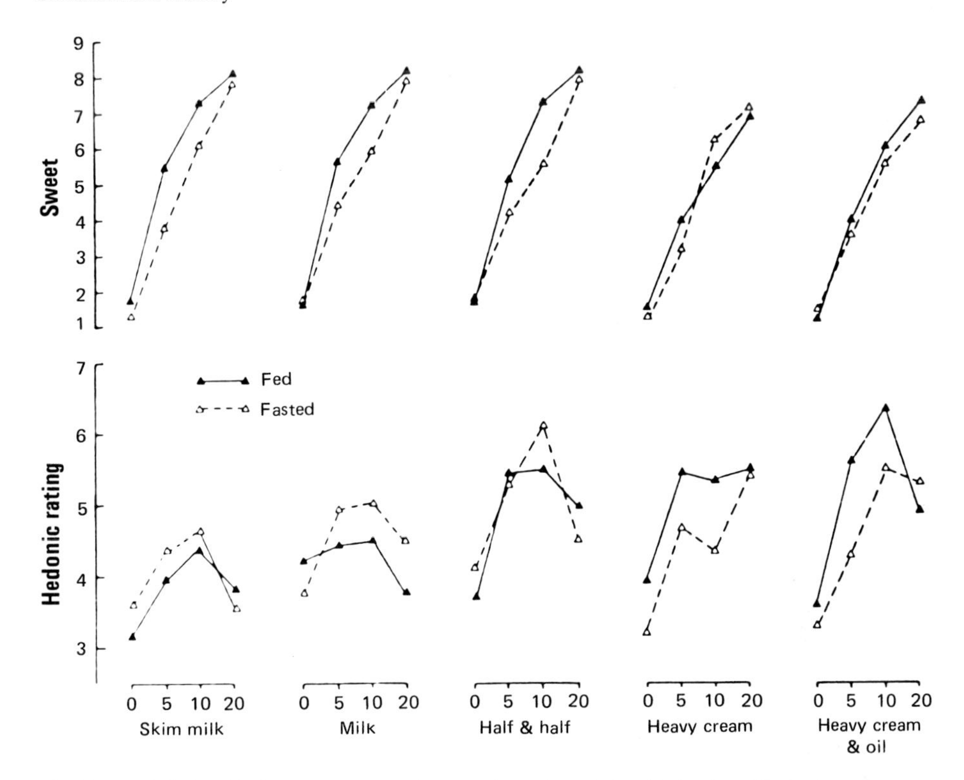

Fig. 12.1. Relationship between mean estimates of sweetness intensity (*top*) and hedonic preference ratings (*bottom*) as a function of sucrose content shown separately for different levels of fat. The data are for normal-weight subjects under Fed and Fasted conditions. (Drewnowski and Greenwood 1983)

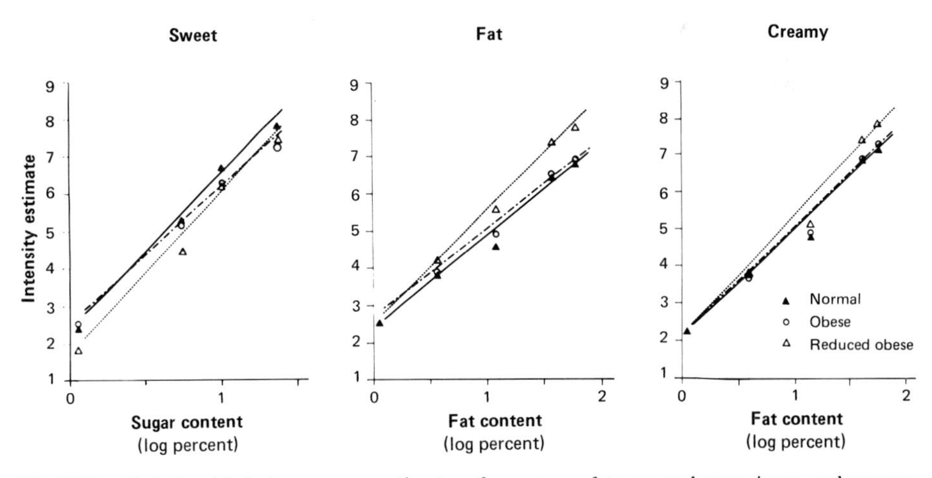

Fig. 12.2. Relationship between mean estimates of sweetness, fatness, and creaminess, and sucrose and fat content of the stimuli. Sweetness intensity scores have been averaged over the five levels of fat; fatness and creaminess scores have been averaged over the four levels of sucrose. All concentrations are expressed as log percent w/w. *Straight lines* represent least-squares fits of logarithmic functions. (Drewnowski et al. 1985)

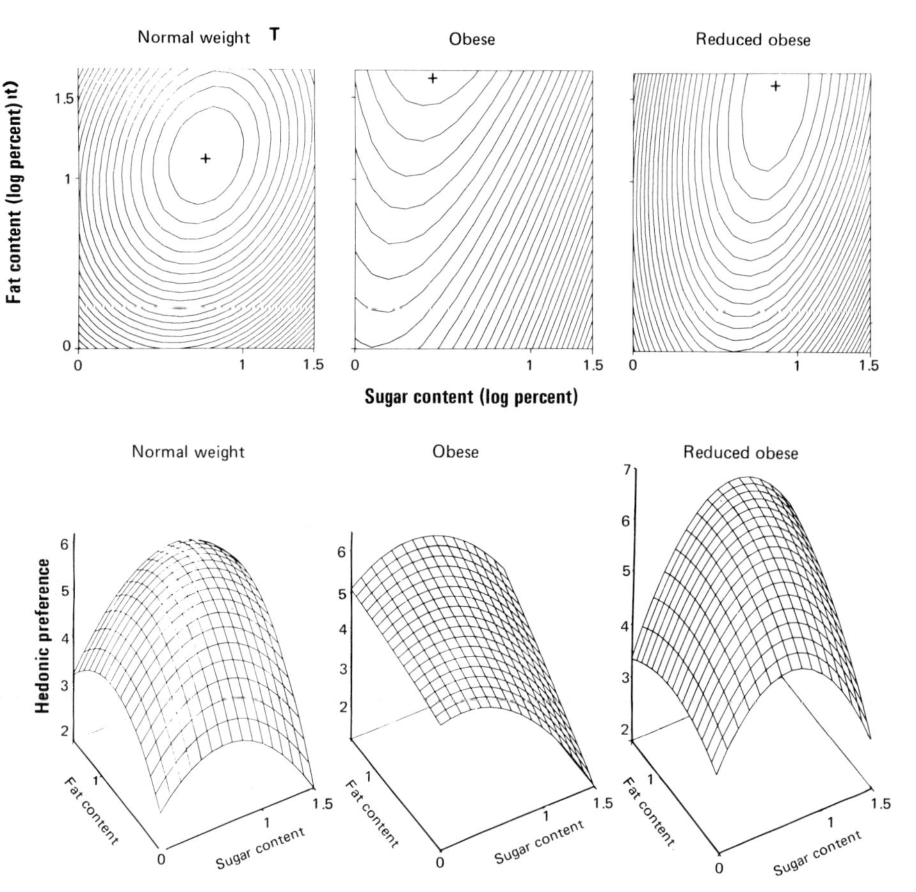

Fig. 12.3 Hedonic response profiles expressed in terms of isohedonic contours (*top panel*) or three-dimensional projections (*bottom panel*). The axes represent sucrose (*x-axis*) and fat (*y-axis*) expressed in log percentages w/w. Regions of optimal preference as derived by the Response Surface Method are denoted by + signs. (Drewnowski et al. 1985)

dissolved in heavy cream. These data appear to reconcile previous reports that obese patients disliked sucrose solutions in water (Grinker 1978) but liked sweetened fat-containing milkshakes (Rodin et al. 1976). Ingredient composition of stimuli best liked by obese subjects was 34% fat but only 4% sugar. Formerly obese subjects [mean wt. 149.5 lb. (67.9 kg)] showed elevated hedonic responses, and best liked mixtures containing 35% fat and 10% sugar.

As shown in Fig. 12.4, individual indices of hedonic responsiveness, the optimal sugar/fat (S/F) ratio, were negatively correlated with the body mass index (weight/height^2), used here as a simple measure of body fatness. However, individual responses were highly variable, and the observed relationship accounted for only 13% of the variance ($r = -0.36$; $P < 0.05$).

The issue of whether hedonic responsiveness is directly tied to some index of glucose or lipid metabolism remains unresolved. The optimal S/F ratio was not correlated with individual lipoprotein lipase levels or with the adipose cell size

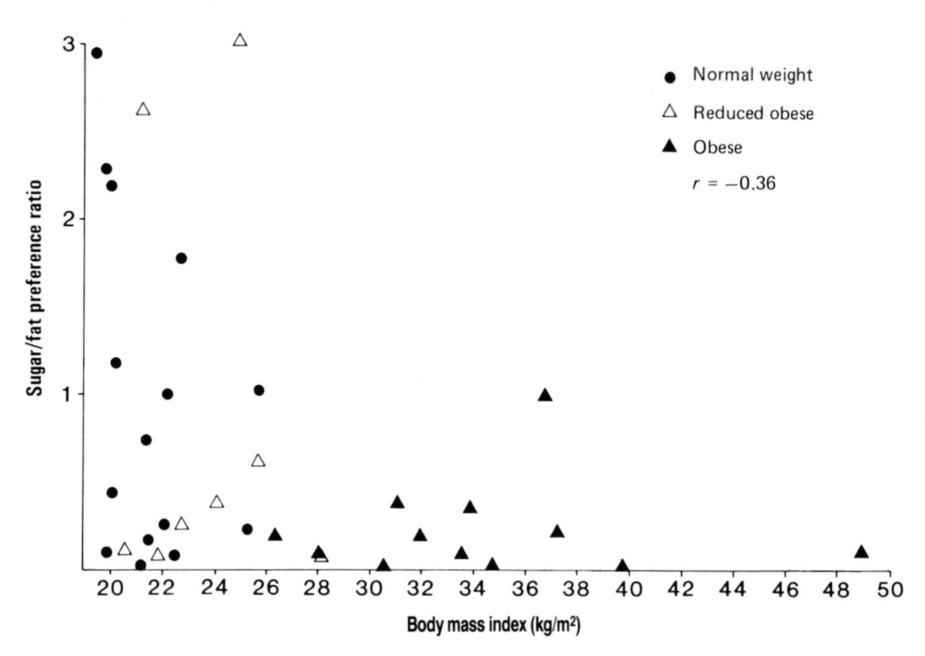

Fig. 12.4. Relationship between the optimally preferred sugar/fat ratios and body mass indices (kg/m^2) of normal weight, obese, and reduced-obese subjects. (Drewnowski et al. 1985)

(Drewnowski et al. 1985). Attempts by other investigators to link sweet taste responsiveness with hormonal or metabolic variables have also been inconclusive. One plausible hypothesis cites pre- and postabsorptive insulin release following a sugar-rich meal (Geiselman and Novin 1982). In this view, ingestion of starches leads to insulin-induced hypoglycemia, hunger, and specific taste-mediated craving for sweet foods that leads ultimately to overeating and overweight. However, direct evidence for this postulate is lacking. Further examination of possible links between metabolic status, taste responsiveness, and food selection in obesity is of great clinical and theoretical importance, and more complete indices of metabolic status, including other plasma and tissue measures, may need to be developed.

Taste and Energy Status

In sensory evaluation studies, preferences for sweet taste have been generally regarded as a stable phenomenon or individual trait. However, some psychologists believe that taste preferences for sweet stimuli are linked to body weight status, and vary directly with the energy level of the organism. In this view the pleasure response to sweetness has a role in intake regulation, since the same stimulus can be perceived as pleasant or unpleasant depending on physiological need (Cabanac 1971).

In short-term laboratory studies, sweet stimuli perceived as pleasant during caloric deprivation were found to be aversive during satiety, a phenomenon described as negative alliesthesia (Cabanac 1971). Thus normal-weight subjects showed reduced preferences for sweet solutions after drinking a caloric preload of 25 g glucose in 200 ml water (Cabanac and Duclaux 1970, 1973). Emaciated anorectic women (Garfinkel et al. 1979), dieting lean subjects below normal body weight ($n = 3$), and obese patients (Cabanac and Duclaux 1970) reportedly failed to show this satiety aversion to sucrose. These data were interpreted as showing that all three groups of subjects were in a state of metabolic need.

These data and their interpretation have long been a subject of controversy. First, the report that obese patients fail to show satiety aversion to sucrose (Cabanac and Duclaux 1970) was flatly contradicted in other studies (Grinker 1978; Grinker and Hirsch 1972). Again, there was variability in individual response: some subjects showed reduced preferences following a preload while others did not (Grinker and Hirsch 1972). Second, a shift in sweet taste preferences, even when obtained, did not appear to be directly tied to energy or metabolic status of the organism. Comparable results have been produced using glucose, sweet but non-caloric cyclamate solutions (Wooley et al. 1972), or a sip-and-spit procedure without actual ingestion (Drewnowski et al. 1982a). On the other hand, caloric manipulations, including lunch, dinner, dieting, or overnight fasting, failed to influence hedonic preferences for sweet taste (Drewnowski and Greenwood 1983; Frinker et al. 1976). It appears that preferences for sweet taste are selectively reduced by prior exposure to sweetness: a phenomenon more in line with the sensory specific satiety explanation of Rolls et al. (1981) than with the original notion of negative alliesthesia.

The demise of alliesthesia as a useful psychological concept can be linked to the decline of the once influential setpoint hypothesis. The response to sweet taste was taken as an index of physiological need, which was in turn linked to the difference between the organism's body weight and its postulated physiological setpoint. Increased liking for sweet taste among obese subjects observed in one study (Cabanac and Duclaux 1970, 1973) was thus interpreted to mean that the obese were below setpoint and in a state of energy deprivation. Increased liking for milkshakes (Rodin 1975) or Koolaid following weight reduction (Grinker et al. 1976) was also taken to be consistent with the setpoint hypothesis. However, other studies have since shown that many obese subjects actually dislike sweet stimuli (Grinker 1978; Underwood et al. 1972), while dieting and mild weight reduction have no discernible effects on taste preference.

Sweet taste preferences appear to be more resistant to metabolic manipulations than previously supposed, and their usefulness as a potential index of setpoint is severely limited. Recent reports indicate that taste response profiles of anorectic subjects do not change following dietary intervention and weight regain to target levels (Drewnowski et al., to be published). In another study former anorectic patients continued to show absence of satiety aversion to sucrose fully 1 year following return to normal body weight (Garfunkel et al. 1979).

Infancy and Childhood

Does enhanced response to sweetness promote obesity in infants of obese mothers? Such infants are often at risk for developing obesity in later life. Newborn infants respond positively to sweet solutions as noted by the facial expressions, rate and amplitude of the sucking response, and the amount of formula consumed (Engen et al. 1978). Infants can discriminate among different sugars, and are responsive to differences in sweetness concentration (Desor et al. 1973). However, there is no evidence that fatter infants or infants at risk for developing obesity show an enhanced response to sweet taste. Rather, any correlation between infant size and the amount of sweet solution consumed was due to smaller infants drinking less than both medium and larger infants (Desor et al. 1973; Milstein 1980). In another study, normal-term infants weighing less than 3 kg at birth consumed less 0.125 M sucrose than did heavier infants in 3-min brief ingestion tests held midway between formula feedings (Grinker et al. 1980). Sweetness of the feeding formula is directly related to the amount consumed. Substituting sucrose for a bland-tasting cornstarch hydrolysate (e.g., Polycose) in a formula of equivalent caloric density and macronutrient composition led to an increase in intake over a period of 4 weeks (Fomon et al. 1983) and may lead to increase in body weight in the long term.

Sweetness is an important stimulus attribute in childhood. Studies of pre-school children revealed that the principal dimensions of food preferences are sweetness and familiarity: children tend to prefer foods that are sweet and are familiar to them (Beauchamp and Moran 1982; Birch 1979). These laboratory measures of preference have been tied to intake under laboratory conditions both in children (Birch 1979) and in adults (Shrager et al. 1984). It is a reasonable assumption that preferences for sweet taste are related to the consumption of sweetened foods both in the laboratory and in real-life situations.

However, best evidence linking taste responsiveness with the development of dietary obesity is still provided by studies with animals. Laboratory rats' preferences for sucrose solutions over plain water and for greasy, fat-containing foods have been documented in a number of studies (Kanarek and Hirsch 1977; Kramer and Gold 1980). Obesity-promoting diets include a solution of 32% sucrose in water provided in addition to laboratory chow (Kanarek and Hirsch 1977) or a high-fat diet containing up to 55% vegetable oil (Schemmel et al. 1970). Perhaps the most effective obesity-promoting combination includes both sugar and fat, provided in the form of a "supermarket" diet composed of chocolate chip cookies, candy, or sweetened condensed milk (Sclafani and Springer 1976).

Conclusions

Human obesity is a disease of multiple origin, involving numerous antecedents and predisposing factors. The principal criteria for the classification of human obesities have included the degree of overweight, age of onset, adipose tissue morphology, and more recently the distribution of fat depots. Although at least some human obesities appear to be diet induced, behavioral classification criteria have been striking by their absence. No behavioral criteria are at present available that might distinguish between different classes of obesity on the basis of taste responsiveness profiles or the patterns of food selection (Drewnowski 1985).

The present data show that the perception of sweetness is not related to indices of body fatness and is not affected by shifts in energy and metabolic status.

Hedonic preferences for sweet taste show considerable individual variation and there seems to be no simple relationship between obesity and preferences for sucrose solutions in water. However, recent studies of taste responsiveness of obese women to complex stimuli containing both sucrose and fat revealed an inverse relationship between obesity and relative preference for sweet taste, as determined by the optimally preferred sugar/fat (S/F) ratio. Obese women generally preferred stimuli that were high in fat and relatively low in sucrose content. The conventional concept of obese "sweet tooth" is not supported by the present results. Hedonic response to sweetness is clearly influenced by other sensory attributes of food, including mouthfeel, viscosity, or the perceived richness or caloric density of the stimuli (Moskowitz 1978; Witherly et al. 1980).

The studies provide an encouraging sign that taste-related behaviors may be linked to body weight status. Although individual response profiles to sweet taste are highly variable, and there is no single "obese" pattern of response, the present procedures (Drewnowski et al. 1985) provide new indices of individual taste responsiveness that can be related to body weight, body fatness, or other physiological or metabolic variables. It remains to be determined whether taste factors and sugar consumption play a major role in the development of obesity in children or adults.

References

Beauchamp GN, Moran M (1982) Dietary experience and sweet taste preference in human infants. Appetite 3: 134–159

Beaudoin R, Mayer J (1953) Food intakes of obese and non-obese women. J Am Diet Assoc 29: 29–33

Bierman EL (1979) Carbohydrates, sucrose, and human disease. Am J Clin Nutr 32: 2712–2722

Birch LL (1979) Preschool children's food preferences and consumption patterns. J Nutr Educ 11: 189–192

Cabanac M (1971) Physiological role of pleasure. Science 173: 1103–1107

Cabanac M, Duclaux R (1970) Obesity: absence of satiety aversion to sucrose. Science 168: 496–497

Cabanac M, Duclaux R (1973) Alliesthesie olfacto-gustative et prise alimentaire chez l'homme. J Physiol (Paris) 66: 113–135

Carroll MD, Abraham S, Dresser CM (1983) Dietary intake source data: United States 1976–80 Vital and Health Statistics II, no. 231. Publ PHS 83-1681, National Center for Health Statistics, US Govt Printing Office, Washington DC

Desor JA, Maller O, Turner RE (1973) Taste in acceptance of sugars by human infants J. Comp Physiol Psychol 84: 496–500

Drewnowski A (1983) Cognitive structure in obesity and dieting. In: Greenwood MRC (ed) Obesity: contemporary issues in clinical nutrition. Churchill Livingstone, New York, pp 87–101

Drewnowski A (1985) Food perceptions and preferences of obese adults: a multidimensional approach. Int J Obes 9: 201–212

Drewnowski A, Greenwood MRC (1983) Cream and sugar: human preferences for high-fat foods. Physiol Behav 30: 629–633

Drewnowski A, Grinker JA, Hirsch J (1982a) Obesity and flavor perception: multidimensional scaling of soft drinks. Appetite 3: 361–368

Drewnowski A, Gruen R, Grinker JA (1982b) Carbohydrates, sweet taste and obesity: changing consumption patterns and health implications. In: Lineback DR, Inglett GE (eds) Basic symposium on food carbohydrates. AVI, New York, pp 153–169

Drewnowski A, Brunzell, JD, Sande K, Iverius PH, Greenwood MRC (1985) Sweet tooth reconsidered: taste responsiveness in human obesity. Physiol Behav 35: 617–622

Drewnowski A, Halmi KA, Pierce B, Gibbs J, Smith GP (to be published) Taste and eating disorders.

Engen T, Lipsitt LP, Robinson DO (1978) The human newborn's sucking behavior for sweet fluids as a function of birthweight and maternal weight. Infant Behav Dev 1: 118–121

Enns MP, Van Itallie TB, Grinker JA (1979) Contributions of age, sex, and degree of fatness on preferences and magnitude estimations for sucrose in humans. Physiol Behav 22: 999–1003
Fomon SJ, Ziegler EE, Nelson SE, Edwards BB (1983) Sweetness of diet and food consumption by infants. Proc Soc Exp Biol Med 173: 190–193
Frijters JER (1983) Sensory qualities, palatability of food and overweight. In: Williams AA, Atkin RK (eds) Sensory quality in foods and beverages: definition, measurements and control. Ellis Horwood, Chichester, pp 431–447
Frijters JER, Rasmussen-Conrad EL (1982) Sensory discrimination, intensity perception, and affective judgment of sucrose-sweetness in the overweight. J Gen Psychol 107: 233–247
Garfinkel PE, Moldofsky H, Garner DM (1979) The stability of perceptual disturbances in anorexia nervosa. Psychol Med 9: 703–708
Garn SM, Solomon MA, Cole PE (1980) Sugar-food intake of obese and lean adolescents. Ecol Food Nutr 9: 219–222
Gates JC, Huenemann RL, Brand RJ (1975) Food choices of obese and non-obese persons. Am J Diet Assoc 67: 339–343
Geiselman PJ, Novin D (1982) The role of carbohydrates in appetite, hunger and obesity. Appetite 3: 203–223
Gonzales ER (1983) Studies show the obese may prefer fat to sweets. JAMA 250: 579–580
Grey N, Kipnis DM (1971) Effect of diet composition on the hyperinsulinemia of obesity. N Engl J Med 285: 827–831
Grinker J (1978) Obesity and sweet taste. Am J Clin Nutr 31: 1078–1087
Grinker JA, Hirsch J (1972) Metabolic and behavioral correlates of obesity. In: Physiology, emotion and psychosomatic illness. Ciba Foundation Symposium 8, ASP, Amsterdam, pp 349–374
Grinker J, Hirsch J, Smith DV (1972) Taste sensitivity and susceptibility to external influence in obese and normal weight subjects. J Pers Soc Psychol 22: 320–325
Grinker JA, Price J, Greenwood MRC (1976) Studies of taste in childhood obesity. In: Novin D, Wyrwicka W, Bray GA (eds) Hunger: basic mechanisms and clinical implications. Raven Press, New York
Grinker JA, Marisak K, Fisher R (1980) Sweet intake as a function of infant and maternal size. In: van der Starre H (ed) Olfaction and taste, VII. IRL Press, London, pp 331–334
Johnson WG, Keane TM, Bonar JR, Downey C (1979) Hedonic ratings of sucrose solutions: effects of body weight, weight loss and dietary restriction. Addict Behav 4: 231–236
Kanarek RB, Hirsch E (1977) Dietary induced overeating in experimental animals. Fed Proc 36: 154–158
Keen, H, Thomas BJ, Jarrett RJ, Fuller JH (1979) Nutrient intake, adiposity and diabetes in man. Br Med J I: 655–658
Kramer TH, Gold RM (1980) Facilitation of hypothalamic obesity by greasy diets: palatability vs. lipid content. Physiol Behav 24: 151–156
Krantzler NJ, Mullen BJ, Comstock EM et al. (1982) Methods of food intake assessment: an annotated bibliography. J Nutr Educ 14: 108–119
Lansky D, Brownell KD (1982) Estimates of food quantity and calories: errors in self-report among obese patients. Am J Clin Nutr 35: 727–732
Malcolm R, O'Neil PM, Hirsch AA, Currey HS, Moskowitz G (1980) Taste hedonics and thresholds in obesity. Int J Obes 4: 203–212
Milstein RM (1980) Responsiveness in newborn infants of overweight and normal-weight parents. Appetite 1: 51–69
Morgan KJ, Johnson SR, Stampley GL (1983) Children's frequency of eating, total sugar intake and weight/height stature. Nutr Res 3: 635–652
Moskowitz HR (1978) Taste and food technology: acceptability, aesthetics and preference. In: Carterette EC, Friedman MP (eds) Handbook of perception, vol VIA. Academic Press, New York, pp 157–194
Moskowitz, HR, Kluter RA, Westerling J, Jacobs HL (1974) Sugar sweetness and pleasantness: evidence for different psychophysical laws. Science 184: 583–585
Page L, Friend B (1978) The changing United States diet. Bioscience 28: 192–197
Pangborn RM (1970) Individual variation in affective responses to taste stimuli. Psychon Sci 21: 125–126
Pangborn RM (1981) Individuality in responses to sensory stimuli. In: Solms J, Hall RL (eds) Criteria of food acceptance: how a man chooses what he eats. Foster, Zurich, pp 177–219
Pangborn, RM, Leonard S (1958) Factors influencing consumer opinion of canned Bartlett pears. Food Technol 12: 284–290
Pangborn RM, Simone M (1958) Body size and sweetness preference. J Am Diet Assoc 34: 924–928

Pangborn RM, Simone, M Nickerson TA (1957) The influence of sugar in ice cream. I. Consumer preference for vanilla ice cream. Food Technol 11: 679–682
Peryam DR, Pilgrim PJ (1957) Hedonic scale method for measuring food preference. Food Technol 11: 9–14
Pfaffmann C (1961) The sensory and motivating properties of the sense of taste. In: Jone MR (ed) Nebraska symposium on motivation. University of Nebraska Press, Nebraska, pp 71–110
Rodin J (1975) Effects of obesity and set point on taste responsiveness and ingestion in humans. J Comp Physiol Psychol 89: 1003–1009
Rodin J, Moskowitz HR, Bray GA (1976) Relationship between obesity, weight loss and taste responsiveness. Physiol Behav 17: 391–397
Rolls BJ, Rowe EA, Sweeney K (1981) Sensory specific satiety in man. Physiol Behav 27: 137–142
Schemmel R, Mickelsen O, Gill JL (1970) Dietary obesity in rats: body weight and fat accretion in seven strains of rats. J Nutr 100: 1041–1048
Schiffman SS, Musante G, Conger J (1979) Application of multidimensional scaling to ratings of foods for obese and normal weight individuals. Physiol Behav 23: 1–9
Sclafani A (1982) Animal models of obesity. Paper presented at the NIH Workshop on the classification of obesities. Vassar College, Poughkeepsie, New York
Sclafani A, Springer D (1976) Dietary obesity in normal adult rats: similarities to hypothalamic and human obesity syndromes. Physiol Behav 17: 461–471
Shrager EE, Drewnowski A, Greenwood MRC, Lipsky C, Starer Y (1984) Hedonic responses to mixtures of sucrose and fat in a solid food unit (SFU). Paper presented at Eastern Psychological Association Meeting, Baltimore MD
Spitzer L, Rodin J (1981) Human eating behavior: a critical review of studies in normal weight and overweight individuals. Appetite 2: 293–329
Ten-State Nutrition Survey 1968–70 (1972) DHEW Publ No HSM 72-8131, US Government Printing Office, Washington DC
Thompson DA, Moskowitz HR, Campbell R (1976) Effects of body weight and food intake on pleasantness ratings for a sweet stimulus. J Appl Psychol 41: 77–83
Underwood PJ, Belton E, Hume P (1972) Aversion to sucrose in obesity. Proc Nutr Soc 32: 93a–94a
Valdes RM, Roessler EB (1956) Consumer survey on the dessert quality of canned apricots. Food Technol 10: 481–486
Vlitos AJ (1983) Sugar and obesity. Lancet, 17 September 1983
Walker ARP (1977) Sugar intake and diabetes mellitus. S Afr Med J 51: 842–851
Weiffenbach JM (ed) (1977) Taste and development: the genesis of sweet preference. US Government Printing Office, Bethesda MD
West KM (1978) Factors associated with occurrence of diabetes. In: Epidemiology of diabetes and its vascular lesions. Elsevier, New York, pp 191–283
Witherly SA, Pangborn RM, Stern J (1980) Gustatory responses and eating duration of obese and lean adults. Appetite 1: 53–63
Wooley OW, Wooley SC, Dunham RB (1972) Calories and sweet taste: effects on sucrose preference in the obese and non-obese. Physiol Behav 79: 765–768
Yudkin J (1966) Dietetic aspects of atherosclerosis. Angiology 17: 127–132
Yudkin J (1983) Energy requirements and obesity. Lancet, 27 August 1983

Commentary

Blass: Drewnowski's point concerning the obese population is much the same as that made by Rodin concerning anorectic and bulimic populations: there are remarkable interindividual diversities among members of these seemingly homogeneous populations. I would urge the same research approach for the obese that Rodin calls in for other clinical-feeding disorders: namely to establish subtypes and (ideally prospectively) evaluate the obesity characteristics of each subtype, especially regarding etiology, rate of onset, feeding habits, etc.

This is in keeping with the approach of DSM III (see Chap. 13) and with that of experienced clinicians treating even the more homogeneous psychoses, depression, for example, where the characteristics of the individual patient can determine the nature of therapy and its direction.

Bartoshuk: The insight that the "sweet tooth" is really a "sweet–fat tooth" is fascinating. Has anyone considered testing newborn infants to see if the sweet–fat interaction is present at birth? I have a feeling that sweet preference may be hard-wired at birth but that the fat preference contribution may require experience. The reason why I suspect this is that the sensory cue for "fatness" is very vague. Sweet–fat is probably a cognitive fusion. We know of no neurons that respond to specific combinations of texture (fattiness) and taste. Thus experience would be required to make the connection between the two sensory inputs.

Blass: A possible source of the fusion may be the milk derived through suckling, especially at the breast.

Booth: There is even reason to think that preference for certain sugar–fat mixtures, like any other preferred sensory combination, is acquired from their presence in familiar foods (see Chap. 10). However, whatever difficulty there may be in analyzing the fattiness cue, it is clear that fat preference itself is not cognitive in a sense that it is peculiarly human (Hamilton 1964).

Reference

Hamilton CF (1964) Rat's preference for high fat diets. J Comp Physiol Psychol 58: 459–460

Booth: What is the hypothesis about the obese reaction to sweetness and are these experiments adequate to test it?

Drewnowski: Historically, the concern has been that any differences in sweet taste preferences among the obese might be due to the misperception of sweetness. The classic methods of taste psychophysics involved threshold detection, stimulus recognition, and magnitude estimation of suprathreshold stimuli, and these were the ones that were used (Grinker 1978). As it happened, most studies showed no deficits in sensory functioning among the obese and attention has shifted to the study of taste hedonics. However, we still include sensory evaluation of sweetness intensity in the experimental design as an appropriate control.

Reference

Grinker J (1978) Obesity and sweet taste. Am J Clin Nutr 31: 1078–1087

Booth: Self-deprived people must want nice food more, and so measures that show no such effect must be insensitive or even invalid.

Drewnowski: The present procedures are standard sip-and-spit techniques of taste psychophysics and the stimuli are evaluated without any opportunity for ingestion. Evaluating the taste of food and wanting to eat it in any amount are two different issues, and it is very likely that measures of food acceptability or food consumption would be more sensitive to deprivation than the present measure of hedonic taste preference.

Chapter 13

Sweetness and Eating Disorders

Judith Rodin and Danielle Reed

Introduction

Considerable clinical attention has been paid to the eating disorder of anorexia nervosa and more recently to bulimia nervosa, yet our knowledge about the etiology of these disorders remains remarkably limited. Biological, psychological, and family/social environment theories have all been advanced (Garner and Garfinkel 1985). A critical challenge to understanding these disorders is the heterogeneity of subtypes and perhaps of etiologies. This heterogeneity argues against undimensional models of etiology (Striegel-Moore et al. 1986).

Anorexia nervosa, which occurs most frequently in adolescent girls, has three essential diagnostic features. First is extreme weight loss, accomplished primarily by an avoidance of most foods, especially those considered fattening (Russell 1967; Garfinkel and Garner 1982). Recently two patterns of food avoidance have been described. The first is the restricting anorexic, in which the patient virtually stops eating. The second is the bulimic anorexic, in which the patient alternates between periods of restriction and periods of binging and purging. A second diagnostic feature is cessation of menstruation and the appearance of other neuroendocrine and metabolic dysfunctions (Russell 1967; Mawson 1974). A third is distinct psychopathology of which the central theme is a morbid fear of becoming fat (Theander 1970; Russell 1967, 1979; Blitzer et al. 1961). The Diagnostic and Statistical Manual of Mental Disorders of the American Psychiatric Association (DSM III, 1980) identifies five diagnostic criteria for anorexia, which are shown in Table 13.1.

Bulimia has been classified by the DSM III (1980) as a mutually exclusive diagnosis to anorexia, although there is considerable overlap in the diagnostic criteria for the two conditions. The diagnostic criteria for bulimia, suggested by the DSM III, are shown in Table 13.2. While the British diagnostic criteria for the syndrome, which is called bulimia nervosa after Russell's (1979) original description of the syndrome, differ somewhat from the DSM III, there is general agreement that the disorder is characterized by recurrent episodes of binge eating

Table 13.1. DSM III diagnostic criteria for anorexia nervosa

A.	Intense fear of becoming obese, which does not diminish as weight loss progresses
B.	Disturbance of body image, e.g., claiming to "feel fat" even when emaciated
C.	Weight loss of at least 25% of original body weight, or, if under 18 years of age, weight loss from original weight plus projected weight gain expected from growth charts may be combined to make the 25%
D.	Refusal to maintain body weight over a minimal normal weight for age and height
E.	No known physical illness that would account for the weight loss

(rapid consumption of a large amount of high calorie food in a discrete period of time), often accompanied by dysphoric affect and fear of being out of control. Many bulimics purge by vomiting, taking laxatives or diuretics, and fasting, and in this way maintain normal weight.

It is perhaps surprising that more research has not been done on the eating behavior of anorexics and bulimics per se since both syndromes represent profound disorders of food intake. Because fear of fatness and of food figure so importantly in the disorders it is reasonable to ask, however, whether differential reactions to highly palatable foods, such as those with sweet taste, distinguish these eating-disordered women from normal controls. To address this question, we will first review those studies that have investigated eating behavior and attempt to determine whether a selective preference for sweet was reported or could be inferred from the data or discussion. However, we note at the outset that in general the data have not been collected in ways that directly measured intake of sweet foods in particular. Next, the studies that directly measured sensory or hedonic differences in response to sweet will be discussed. Since it does appear that there are some differences in hedonic ratings of sweet, evidence for their role either as causes or consequences of the disorders and associated behavioral and biological events will be discussed. Finally, types of research strategy that may help to elucidate better the relationship between response to sweetness and eating disorders will be suggested.

Table 13.2. DSM III diagnostic criteria for bulimia

A.	Recurrent episodes of binge eating (rapid consumption of a large amount of food in a discrete period of time, usually less than 2 h)
B.	At least three of the following:
	1. Consumption of high calorie, easily ingested food during a binge
	2. Inconspicuous eating during a binge
	3. Termination of such eating episodes by abdominal pain, sleep, social interruption, or self-induced vomiting
	4. Repeated attempts to lose weight by severely restricted diets, self-induced vomiting, or use of cathartics or diuretics
	5. Frequent weight fluctuations greater than 10 lb. due to alternating binges and fasts
C.	Awareness that eating pattern is abnormal and fear of not being able to stop eating voluntarily
D.	Depressed mood and self-deprecating thoughts following eating binges
E.	The bulimic episodes are not due to anorexia nervosa or any known physical disorder

Types of Food Eaten

Most patients with eating disorders who binge feel that the taste or texture of the food is important at the start of a binge, although many feel that while binging they frequently eat too quickly to taste anything at all (Abraham and Beumont 1982). The amount, type, and nutritional content of a binge varies widely both within and between individuals (Abraham and Beumont 1982) but many include in their binges food that they do not allow themselves to eat at other times, which they refer to as the "fattening foods" (Rosen et al. 1986). In their survey of binge eating in identified bulimic populations, for example, Leon et al. (1985) found that the foods eaten during a binge tended to fall into two major categories: pastries, breads, cookies (biscuits), and other "soft" carbohydrates and a category of "junk food" items such as potato chips (crisps) and pretzels. Rosen et al. (1986) found that increased consumption of these two kinds of food was the only factor related to food type that distinguished binge from non-binge eating episodes among severe bulimics. Interestingly, non-bulimic college students who reported binging on occasion also reported increased consumption of these same types of food during a binge (Leon et al. 1985).

The studies just reviewed and several others with comparable results (Mitchell et al. 1981; Pyle et al. 1981; Russell 1979) have relied upon patient self-reports. Mitchell and Laine (1985) directly observed the eating behavior of six bulimic women hospitalized on a research ward. The foods selected by these subjects for binge eating were similar to those reported previously as being typical for this group of patients (Mitchell et al. 1981). The foods eaten most frequently were doughnuts, pies, carbonated soft drinks, sandwiches, and chocolate, but many other types of food were also eaten. Some subjects even incorporated fresh vegetable salads into the binge-eating episode. In non-binge eating situations these subjects consumed a mean of 451 kcal during the 24-h period. The types of food eaten during non-binge items were usually low in calories, with salads and diet beverages the foods of choice.

Taken together, the studies show a varied pattern of foods consumed by bulimics. Certainly an invariably high sweet intake by bulimics during binges appears largely unsubstantiated by the data. Similarly, the suggestion that the specific avoidance of carbohydrates is an *invariable* characteristic of the diets chosen by anorexics also does not appear supported by newer data (e.g., Beumont et al. 1981; Huse and Lucas 1984).

Early descriptions of anorexia specifically noted carbohydrate avoidance (Crisp 1965, 1980; Russell 1967). Russell actually measured food eaten by nine patients on a metabolic ward and reported an avoidance of carbohydrate compared to the general population, with relatively adequate amounts of protein and fat. Russell's work did not distinguish sweets from other forms of carbohydrate, however. Two newer and large studies suggest that there is no typical pattern of nutritional behavior evidenced by anorexics. Indeed, they suggest that beyond the inevitable caloric restriction, dietary patterns actually appear quite variable. Beumont et al. (1981) obtained 24-h diet histories from 17 anorexic patients. They found that, compared to normals, anorexics did not appear to eat proportionately less carbohydrate, while their intake of fats was lower. Again, however, sweet carbohydrates were not considered separately. Huse and Lucas (1984) evaluated diet histories of 93 patients and found that four general diet patterns could be identified, but none was specific to avoidance of sweets. In fact, there were only a few patients in each subtype reporting avoidance of sweets.

It is useful to reiterate in summary that while many bulimics appear to include sweet foods in their binges and many anorexics appear to avoid sweet foods, these patterns are far from evident in all patients. Much more work is needed in this area before firm conclusions can be drawn about the role of sweet preference or aversion in food intake in these disorders. Perhaps as we learn more about the heterogeneity of anorexia and bulimia, sweet preference or aversion will only be characteristic of certain subtypes. Also worthy of empirical evaluation is the clinical suggestion that restriction of carbohydrates, most notably sweets, may provoke binge episodes (Garfinkel et al. 1980).

Sweet Sensitivity and Hedonic Valuation

Some theories about the etiology of anorexia and bulimia suggest that sweetness takes on a forbidden quality because of its apparent significance for weight gain and the experiential meaning of this prospect. It is also clearly possible that there is some disturbed feature of sweet taste perception that may contribute to an eating disorder. This hypothesis assumes, of course, that there is selective consumption or avoidance of sweet among people with anorexia nervosa or bulimia. Before suggesting a possible causal direction, however, let us consider the strength of the evidence for an association between differential responsiveness to sweet taste and disordered eating.

Lacey et al. (1977) studied the sucrose taste sensitivity of hospitalized anorexics and controls using a two-alternative forced choice method for solutions determined to be in the "region of uncertainty" (i.e., at threshold) for most subjects. Taste sensitivity for sucrose did not differ significantly between controls and anorexics, suggesting that carbohydrate restriction, if it occurs in patients with anorexia nervosa, is not caused by any primary abnormality of sucrose taste sensitivity. These findings were replicated by Casper et al. (1980), who found no difference in sucrose sensitivity between anorexics and controls. Interestingly, the majority of anorexics did have subnormal acuity for bitter (urea) and sour (HCl) substances, while taste for salt was less affected. Lacey et al. (1977) hypothesized that sucrose avoidance may produce heightened sucrose taste sensitivity since it was evident in both those anorexics and normal controls who were currently restricting their carbohydrate intake. Indeed, refeeding anorexics with a normal diet including normal amounts of carbohydrate was associated with lessened sucrose taste sensitivity. Because of the Lacey et al. (1977) observation that sucrose sensitivity varied as a function of the amount of carbohydrate consumed in the diet, it is useful to ask whether other aspects of nutrient or metabolic status alter sucrose sensitivity in anorexics. In fact the Casper et al. (1980) study was an effort to determine whether levels of trace metals such as zinc were related to differential taste sensitivity in anorexia.

Using a direct manipulation of caloric repletion, Garfinkel et al. (1978) had anorexic and normal subjects rate the pleasantness and sweetness of a 20%

sucrose solution before and after a 400-kcal mixed lunch. Further, the lunch was designed to connote high or low calories on two different testing occasions although it was equicaloric. There was no difference between groups on sweetness intensity ratings of the sucrose solutions either over time or between the different meals. When looking at pleasantness ratings the authors found that anorexic subjects showed significantly less of a postmeal decline in their judgments of the sucrose solutions than did the non-anorexic controls. These findings were replicated in a 1-year follow up study with the same subjects (Garfinkel et al. 1979).

Cabanac (1971) has suggested that set point for body weight may be defended through acute changes in the palatability of gustative stimuli. He argued that satisfaction of short-term needs in subjects at or close to set point should render subsequent sucrose solutions less palatable, a phenomenon that has been termed "negative alliesthesia" (Cabanac and Fantino 1977). Esses and Herman (1984) examined the perceived pleasantness of sucrose solutions as rated by dieters and non-dieters, before and after the ingestion of a glucose solution. They expected their study to provide support for the Garfinkel et al. conclusion that weight restriction may reduce the occurrence of negative alliesthesia. Unfortunately only the restraint scale (Herman and Mack 1975) was used as the measure of chronic dieting. Surprisingly, subjects scoring high in restraint, whom they labelled as chronic dieters, showed negative alliesthesia after a glucose load, revealing a pattern of hedonic decline similar to the non-dieters. Moreover the dieters rated the high concentrations of sucrose solutions as less pleasant than did the unrestrained eaters both prior to and after glucose ingestion. The argument proposed by Esses and Herman (1984) is that the dieters may be denying the pleasantness of sweet taste. We find this suggestion less compelling in light of the Garfinkel et al. (1979) findings that even anorexics rated sweet taste as pleasant.

It is also possible that sweet taste preferences might vary as a function of how the sweet stimulus is presented. Traditionally, the effects of sweet taste have been measured by providing the sugar in water solutions. Recently, efforts have been made to approximate more naturalistic stimuli. For example, Drewnowski et al. (1984) evaluated hedonic ratings made by anorexic women of sweetness, fatness, and creaminess of 20 different mixtures. Hedonic ratings for sweet stimuli showed a sucrose breakpoint that was higher for anorectic patients than for normal weight controls. Hedonic ratings for stimuli of increasing fat content rose for normal weight subjects but declined for anorexics. Thus anorectic patients actually appeared to prefer intense sweetness over fat. The work of Drewnowski and his colleagues argues for assessing the sensory and hedonic properties of the complex mixtures in which sweet tastes usually appear (e.g., with fat) and for trying to use multidimensional scaling techniques to make possible assessment of individual response patterns. Given the variability in food intake reported by Huse and Lucas (1984) when studying almost 100 anorexics, efforts to identify determinants of individual variability could prove fruitful.

Because of the array of unclear and conflicting findings, and because there are almost no data from similar studies with bulimics, we cannot at present determine the extent to which the hedonics of sweet taste are altered in patients with eating disorders. It does appear, however, that there is no distinctive sensory deficit or enhancement related to the perception of sweetness that has etiologic significance in disordered eating.

Biological Mechanisms Relating Sweet Preference to Disordered Eating

While at the present time there is no clear body of literature showing a systematic relationship between response to sweetness and eating disorders, a consideration of the psychobiological variables that might mediate such a relationship could suggest future research directions. While the discussion below focuses on physiological processes, food attitudes and the cultural meaning of sweetness as something sinful and forbidden (see Chap. 7) should not be overlooked in any attempt to consider the role of sweetness in eating disorders. Indeed, cognitions about sweet, both positive and negative, may affect the food intake of humans as much as, or more than, the actual sensation of sweet. More work is needed to relate culture and cognition to the psychobiology of sweet taste, in general, and specifically in eating disorders (see also Striegel-Moore et al. 1986).

Considerable debate and disagreement has centered around the relationship beween carbohydrate consumption and hunger (e.g., Geiselman and Novin 1982; Sclafani 1982; Geiselman 1985; Vasselli 1985; Vanderweele 1985), which makes an analysis of the role of sweet preference in eating disorders more difficult. While an extensive evaluation of the metabolic effects of sugar or the metabolic effects of food deprivation on responses to sugar is specifically excluded as a charge of this paper, these processes in fact may be the primary determinants of the relationship between sweet taste responsiveness and disordered eating. Therefore let us consider briefly how other biological changes that occur in anorexia and bulimia might influence responsiveness to sweet taste.

Basic work has examined anorexics' glucose tolerance and their ability to secrete insulin in response to a glucose stimulus. Blickle et al. (1984) found that anorexics had a depressed insulin response to intravenous glucose administration; 46% of their subjects had an abnormal glucose tolerance. Similarly, Alderdice et al. (1985) found that anorexics had lower fasting plasma glucose and insulin levels than normals, and were glucose intolerant. Castillo et al. (1985) also measured glucose tolerance by the euglycemic hyperinsulinemic clamp method in anorexics and they found normal glucose disposal rates. Since insulin was secreted endogenously in the Blickle et al. (1984) and Alderdice et al. (1985) studies and provided exogenously in the Castillo et al. (1985) study, it could be concluded that glucose intolerance in anorexics stems from an inability to secrete enough insulin to sequester a given glucose load rather than from insulin insensitivity at the cellular level. This hypothesis is further supported by the work of Wachslicht-Rodbard et al. (1979), who found that anorexics show a (possibly adaptive) increase in insulin binding in erythrocytes.

The previous data suggest that anorexics are glucose intolerant but they are not insulin resistant in peripheral tissues. They may be slightly glucoprivic because they secrete inadequate amounts of insulin and, as a result, might manifest some increases in sweet preference. Insulin administration and concomitant hypoglycemia elicit an increase in carbohydrate preference (Kanarek et al. 1980). Jacobs (1958) has shown that rats learn to prefer (even after a single trial) the substrate that most effectively eliminates hypoglycemia. Cytoglucopenia was not measured directly in either of these studies, but it is generally inferred from the low levels of available glucose after exogenous insulin administration. Preliminary data in humans suggest that anorexics who restrict their food intake but do not vomit show less of a preference for sweet (as assessed in fat and sweet mixtures) than anorexics who binge and vomit (Drewnowski et al. 1985b). There are no data directly showing that anorexics who binge and purge are more glucose intolerant than restricting anorexics, although fasting has been shown to reduce glucose tolerance (Bjorkman and Eriksson 1985).

Another characteristic potentially affecting glucose availability in anorexics is their abnormally high activity level. Although no studies have specifically tested the effects of activity on sugar consumption, Collier and co-workers (Collier et al. 1969; Leshner et al. 1971), using dextrinized starch, have shown that increased energy expenditure in animals leads to increased sweet carbohydrate preference, perhaps by elevating peripheral glucose utilization. Variability in sweet preference by anorexics in the studies reviewed in this paper may be due to some studies but not others testing subjects during periods of high activity. It is also known that vigorous exercise can produce hypoglycemia (Koivisto et al. 1981), and that hypoglycema is alleviated by quickly metabolized carbohydrate such as glucose and sucrose. Thus, anorexics who exercise vigorously may crave sweet caloric foods to compensate for exercise-induced hypoglycemia.

Studying bulimics presents a different type of challenge because bulimics can be in very different physiological states depending on their body weight, whether they purge or not, and on the method they use if they do purge. A bulimic can have a substantial loss of body weight (as long as she does not meet the 25% loss which defines anorexia), she may be obese, or she may be at any weight in between. Even some normal weight bulimics appear to have adaptations that are usually found only in starvation (Pirke et al. 1984); this suggests they may not be metabolically or behaviorally comparable to normal weight individuals.

Given the extreme variation in body weight in bulimics and given that level of adiposity may be an important factor in sucrose preference (Drewnowski et al. 1985a; Enns et al. 1979; Wooley et al. 1972), it might be helpful first to limit our discussion to bulimics who maintain their normal body weights. Within this category it is also critical to distinguish between vomiting bulimics, those who use laxatives, and those who severely restrict their food intake. Vomiting, laxative use, and fasting all have potentially profound effects on sweet preference since postingestinal availability of nutrients may vary widely between these subtypes.

In contrast to anorexics, normal weight bulimics do not manifest abnormal glucose tolerance (Hohlstein et al. 1986) or abnormal fasting blood sugar (Mitchell and Bantle 1983). This may be because the abnormalities of glucose metabolism seen in some anorexics are secondary to chronic low body weight or to the acute effects of fasting. Since, based on the few available studies, bulimics do not appear to be glucose intolerant, a sweet preference based solely on chronic abnormal glucose tolerance is not a likely mechanism for differential responsiveness to sweet.

One intriguing possibility is that bulimics who vomit as a method of purging may resemble experimental animals implanted with gastric fistulas since they may be receiving insufficient postingestional cues from food. These food-deprived animals fail to become satiated when ingesting sweet substances from which they receive little caloric benefit (Mook 1963; Sclafani and Nissenbaum 1985; Kemnitz et al. 1981). On the other hand, there is some suggestion from animal work that simple sugars might be among the only nutrients that the bulimic absorbs and utilizes even if she purges. In the rat, glucose is rapidly passed through the lining of the gastrointestinal tract beginning 2 min after a carbohydrate rich meal, and this is enhanced with long periods of food deprivation (Steffens 1969). If this is true, bulimics may be experiencing all prerequisites for a phenomenon called conditioned satiety. Animals and humans that are exposed to diets that differ in flavor and amount of bland complex carbohydrate they contain come to prefer the flavor paired with the food that contains the most carbohydrate (Booth et al. 1981, unpublished work; Bolles et al. 1981). Bulimics who vomit may learn to prefer high carbohydrate meals since those are the foods from which they receive the most nutritional benefit. Drewnowski et al. (1985b) have shown that when compared to restrictor anorexics, vomiting bulimics prefer higher sucrose concentrations, which is consistent with this hypothesis.

Bulimics who purge by vomiting may also place themselves in a dysfunctional metabolic state. Cephalic release of insulin in response to the sight, smell, and taste of food is well characterized in rats and humans (Berthoud et al. 1981; Deutsch 1974; Krotkiewski et al. 1979; Louis-Sylvestre and LeMagnen 1980; Rodin 1978; Sjostrom et al. 1980). Bulimic purgers receive all of the cues (sight, smell, and taste) that prepare the body to digest larger amounts of food than it gets. If they secrete more insulin in anticipation of greater amounts of food, then they may be hypoglycemic because of surplus insulin levels after a bout of binging and vomiting. Alternatively, excess insulin may lead to an increase in cellular glucose utilization, which in turn promotes elevated sweet preference (Rodin et al. 1985). Based on these assumptions, one would expect that bulimics who vomit would have high preference for sweet tastes.

Bulimics who purge by taking large amounts of laxatives hope to avoid the caloric content of their binges by malabsorbing the ingested meals. Whether this is an effective weight loss strategy is doubtful (Lacey and Gibson 1985). In fact, one study found that over 88% of the calories eaten by laxative-abusing bulimics are absorbed (Bo-linn et al. 1983). But laxatives do have an effect that is important for our analysis of sweet preference. They speed the passage of food through the gastrointestinal tract, which may cause increased sugar preference. In an animal study manipulating the rate at which sugar was infused across the gut, it was demonstrated that the faster the sugar flowed across the intestinal wall, the more the animals ate. While the food was not sweetened, it was a high-carbohydrate food (Geiselman and Novin 1982). These findings suggest that bulimics who use laxatives for weight control may experience increased hunger, especially for high carbohydrate food, due to the increased rate at which nutrients flow across the gut (which in actuality may be secondary to rate for absorption).

The putative sweet preference of bulimics and of anorexics might also be mediated by yet another mechanism. Both groups share feelings of anxiety and stress associated with eating and fear of weight gain, stressors that could also influence sweet preference. The hormones that are secreted during stress are well known to influence glucose tolerance and glucose utilization. Adrenaline affects glucose tolerance by altering insulin secretion and increasing hepatic insulin resistance (Rizza et al. 1980; Porte 1967). Glucocorticoids also reduce insulin receptor binding and adversely affect glucose tolerance. In summary, the secretion of stress hormones acts to inhibit glucose utilization by making insulin less effective at sequestering the available glucose into cells. These stress hormones and the concomitant anxiety may increase the appetite felt by the bulimic or anorexic by decreasing the amount of available intracellular glucose (Nowland et al. 1985).

Conclusions

Descriptive research needs to be done on sweet preference in eating-disordered populations, using categories that reflect metabolic status, method of purging, and degree of depletion of adipose tissue stores. Longitudinal studies of patients at various stages of the disease and recovery process would be especially useful in clarifying how sweet preference and intensity judgments, and intake of sweet foods, change as a function of the state of the eating-disordered patient. What we know at present suggests that anorexics and bulimics do not differ from a primary deficit in sweet taste perception. Changes in the perceived palatability of sweet, when they occur, appear to be the result of changes caused by the eating disorder. We have tried to speculate, hopefully with heuristic value, on some of these changes and how they might relate to the perception of sweet. Undoubtedly we will stumble down several more blind alleys as we try to understand the nature of eating disorders and how they are influenced by the sweet tooth with which they appear, prima facie, to be associated. Conceptual and diagnostic clarity in characterizing the different etiologies and subtypes of these eating disorders will enable better research to be done, and increase our ability to identify metabolic and pathophysiological mechanisms that relate to responsiveness to sweet taste.

References

Abraham SF, Beumont PJV (1982) How patients describe bulimia or binge eating. Psychol Med 12: 625–635

Alderdice JT, Dinsmore WW, Buchanan KD, Adams C (1985) Gastrointestinal hormones in anorexia nervosa. J Psychiatr Res 19 (2–3): 207–213

American Psychiatric Association (1980) Diagnostic and statistical manual of mental disorders, 3rd edn. APA, Washington DC

Berthoud HR, Bereiter DA, Trimble ER, Siegel EG, Jeanrenaud B (1981) Cephalic phase insulin secretion. Diabetologia 20: 221–230

Beumont PJV, Chambers TL, Rouse L, Abraham SF (1981) The diet composition and nutritional knowledge of patients with anorexia nervosa. J Hum Nutr 35: 265–273

Beumont PJV, George GCW, Smart DE (1976) "Dieters" and "vomiters and purgers" in anorexia nervosa. Psychol Med 6: 617–622

Bjorkman O, Eriksson LS (1985) Presence of a 60 hour fast on insulin-mediated splanchnic peripheral glucose metabolism in humans. J Clin Invest 76: 87–92

Blickle JF, Reville F, Stephan P, Meyer C, Demangeat C, Sapin R (1984) The role of insulin, glucagon and growth hormone in the regulation of plasma glucose and free fatty acid levels in anorexia nervosa. Horm Metab Res 16: 336–340

Blitzer JR, Rollins N, Blackwell A (1961) Children who starve themselves: anorexia nervosa. Psychosom Med 23: 369–383

Bo-linn GW, Santa Ana C, Morawski G, Fordtran S (1983) Purging and caloric absorption in bulimic patients and normal women. Ann Intern Med 99: 14–17

Bolles RC, Hayward L, Crandall C (1981) Conditioned taste preferences based on caloric density. J Exp Psychol 4: 59–69

Cabanac M (1971) Physiological role of pleasure. Science 173: 1103–1107

Cabanac M, Fantino M (1977) Origin of olfacto-gustatory alliesthesia: intestinal sensitivity to carbohydrate concentration? Physiol Behav 18: 1039–1045

Casper RC, Kirschner B, Sandstead HH, Jacob RA, Davis JM (1980) An evaluation of trace metals, vitamins, and taste function in anorexia nervosa. Am J Clin Nutr 33: 1801–1808

Castillo M, Scheen A, Lefebvre J, Luyckx AS (1985) Insulin stimulated glucose disposal is not increased in anorexia nervosa. J Clin Endocrinol Metab 60: 311–314
Collier G, Leshner A, Squibb RL (1969) Dietary self-selection in active and non-active rats. Physiol Behav 4: 79–82
Crisp AH (1965) Some aspects of the evolution, presentation and follow-up of anorexia nervosa. Proceed R Soc Med 58: 814–820
Crisp AH (1980) Anorexia nervosa: let me be. Grune & Stratton, New York
Deutsch R (1974) Conditioned hypoglycemia: a mechanism for saccharin-induced sensitivity to insulin in the rat. J Comp Physiol Psychol 86: 350–358
Drewnowski A, Greenwood MRC, Halmi KA (1984) Carbohydrate or fat phobia: taste responsiveness in anorexia nervosa. Fed Proc 43: 475
Drewnowski A, Brunzell JD, Sande K, Iverius PH, Greenwood MRC (1985a) Sweet tooth reconsidered: taste responsiveness: taste preferences in human obesity. Physiol Behav 35: 4
Drewnowski A, Duberstein P, Gibbs J, Halmi KA, Pierce B, Smith GP (1985b) Taste and eating disorders: hedonic responsiveness in anorexia nervosa and bulimia. Presented at AChems Meeting, Sarasota
Enns MP, Van Itallie TB, Grinker JA (1979) Contributions of age, sex, and degree of fatness on preference and magnitude estimations for sucrose in humans. Physiol Behav 22: 999–1003
Essex VM, Herman CP (1984) Palatability of sucrose before and after glucose ingestion in dieters and nondieters. Physiol Behav 32: 711–715
Garfinkel PE, Garner DM (1982) Anorexia nervosa: a multidimensional perspective. Brunner/Mazel, New York
Garfinkel PE, Moldofsky H, Garner DM, Stancer HC, Coscina DV (1978) Body awareness in anorexia nervosa disturbance in "body image" and "satiety". Psychosom Med 40: 487–497
Garfinkel PE, Moldofsky H, Garner DM, Stancer HC, Coscina PV (1979) Stability of perceptual disturbances in anorexia nervosa. Psychol Med 9: 703–708
Garfinkel PE, Moldofsky H, Garner DM (1980) The heterogeneity of anorexia nervosa: bulimia as a distinct subgroup. Arch Gen Psychiatry 37: 1036–1040
Garner DM, Garfinkel PE (1985) Handbook of psychotherapy for anorexia nervosa and bulimia. Guilford, New York
Geiselman PJ (1985) Appetite, hunger and obesity as a function of dietary sugar intake: Can these effects be mediated by insulin-induced hypoglycemia? A reply to commentaries. Appetite 6: 64–67
Geiselman PJ, Novin D (1982) The role of carbohydrate in appetite, hunger, and obesity. Appetite 3: 203–223
Herman CP, Mack D (1975) Restrained and unrestrained eating. J Pers 43: 647–660
Hohlstein LA, Gwirtsman HE, Whalen F, Enns MP (1986) Oral glucose tolerance in bulimia. Int J Eating Disorders 5: 157–160
Huse DM, Lucas AR (1984) Dietary patterns in anorexia nervosa. Am J Clin Nutr 40: 251–254
Jacobs HL (1958) Studies on sugar preference. I. The preference for glucose solutions and its modification by injections of insulin. J Comp Physiol Psychol 51: 304–309
Kanarek RB, Marks-Kaufman R, Lipeles BJ (1980) Increased carbohydrate intake as a function of insulin administration in rats. Physiol Behav 25: 779–782
Kemnitz JW, Gibber JR, Lindsay KA, Brot MD (1981) Preference for sweet and the regulation of caloric intake by *Macaca mullata.* Am Soc Primatol Abstr
Koivisto VA, Karonen S, Nikkila EA (1981) Carbohydrate ingestion before exercise: comparison of glucose, fructose and sweet placebo. J Appl Physiol 51 (4)
Krotkiewski M, Sjostrom L, Sullivan L (1979) Early biphasic peripheral insulin reponse to cephalic stimulation in obese and lean women. In: Vague J, Vague PH (eds) Diabetes and obesity. Excepta Medica, Amsterdam, pp 47–52
Lacey JH, Gibson E (1985) Controlling weight by purgation and vomiting: a comparative study of bulimics. J Psychiatr Res 19: 337–341
Lacey JH, Stanley PA, Crutchfield M, Crisp AH (1977) Sucrose sensitivity in anorexia nervosa. J Psychosom Res 21: 17–21
Leon GR, Carroll K, Chernyk B, Finn S (1985) Binge eating and associated habit patterns within college students and identified bulimic populations. Int J Eating Disorders 4: 43–57
Leshner AI, Collier GH, Squibb RL (1971) Dietary self-selection at cold temperatures. Physiol Behav 6: 1–3
Louis-Sylvestre J, LeMagnen J (1980) Palatability and preabsorbtive insulin release. Neurosci Biobehav Rev 4: 13–15
Mawson AR (1974) Anorexia nervosa and the regulation of intake: a review. Psychol Med 4: 289–308

Mitchell JE, Bantle JP (1983) Metabolic and endocrine investigations in women of normal weight with the bulimic syndrome. Biol Psychiatry 18: 355–365
Mitchell JE, Laine DC (1985) Monitored binge-eating behavior in patients with bulimia. Int J Eating Disorders 4: 177–183
Mitchell JE, Pyle RL, Eckert ED (1981) Binge eating behavior in patients with bulimia. Psychiatry 138: 835–836
Mook DG (1963) Oral and postingestional determinants of the intake of various solutions in rats with esophageal fistulas. J Comp Physiol Psychol 56: 645–659
Nowland NE, Bellush LL, Carlton J (1985) Metabolic and neurochemical correlates of glucopriovic feeding. Brain Res Bull 14: 617–624
Pirke KM, Pahl J, Schweiger U, Warnhoff M (1984) Metabolic and endocrine indices of starvation in bulimia: a comparison with anorexia nervosa. Psychiatry Res 15: 33–39
Porte D (1967) A receptor mechanism for the inhibition of insulin release by epinephrine in man. J Clin Invest 46: 86–94
Pyle RL, Mitchell JE, Eckert ED (1981) Bulimia: a report of 34 cases. J Clin Psychiatry 42: 60–64
Rizza R, Cryer PG, Haymond M, Gerich J (1980) Adrenergic mechanisms for the effects of ephinephrine and glucose production and clearance in man. J Clin Invest 65: 682–689
Rodin J (1978) Has the internal versus external distinction outlived its usefulness? In: Bray GA (ed) Recent advances in obesity research, vol II. Newman, London
Rodin J, Wack J, Ferrannini E, DeFronzo RA (1985) Effect of insulin and glucose on feeding behavior. Metabolism 34: 826–831
Rosen JC, Leitenberg H, Fisher C, Khazan C (1986) Binge-eating episodes in bulimia nervosa: the amount and type of food consumed. Int J Eating Disorders 5: 255–267
Russell GFM (1967) The nutritional disorder in anorexia nervosa. J Psychosom Res 11: 141–149
Russell G (1979) Bulimia nervosa: an ominous variant of anorexia nervosa. Psychol Med 9: 429–448
Sclafani A (1982) On the role of hypoglycemia in carbohydrate appetite. Appetite 3: 227–228
Sclafani A, Nissenbaum JW (1985) On the role of mouth and gut in the control of saccharin and sugar intake: a reexamination of the sham-feeding preparation. Brain Res Bull 14: 569–576
Sjostrom K, Garrelick G, Krotkiewski M, Garrelick G, Luyckx A (1980) Peripheral insulin in response to the sight and smell of food. Metabolism 29: 901–909
Steffens AB (1969) The influence of insulin injections and infusions on eating and blood glucose level in the rat. Physiol Behav 4: 823–828
Striegel-Moore RH, Silberstein L, Rodin J (1986) Toward an understanding of risk factors for bulimia. Am Psychol 41: 246–263
Theander S (1970) Anorexia nervosa: a psychiatric investigation of 94 female patients. Acta Psychiatr Scand [Suppl] 214
Vanderweele DA (1985) Hyperinsulinism and feeding: not all sequences lead to the same behavioral outcome or conclusion. Appetite 6: 47–52
Vasselli JR (1985) Carbohydrate ingestion, hypoglycemia, and obesity. Appetite 6: 53–59
Wachslicht-Rodbard H, Gross HA, Rodbard MD, Ebert MH, Roth J (1979) Increased insulin binding to erthrocytes in anorexia nervosa. N Engl J Med 300: 882–887
Wooley OW, Wooley SC, Dunham RB (1972) Calories and sweet tastes: effects on sucrose preference in the obese and non-obese. Physiol Behav 9: 765–768

Commentary

Blass: Rodin's review makes clear the futility of a single bulimic diagnosis based on feeding patterns, preferences, or body weight status. She rightly calls for efforts to identify subtypes, possibly based upon ingestion of sweets. One possible subtype might be patients whose binging most often occurs during periods of high stress. A number of studies in the animal literature have demonstrated enhanced feeding during and following stress, and this feeding can be blocked by administration of opioid antagonists.

Alternatively bulimia might be the most dramatic manifestation of a psychopathology that is not linked historically or exclusively to feeding conflicts and their resolutions. The multivariegated nature of the disorder and the relatively poor insight of these patients about their condition and its determinants raise the possibility that in addition to seeking subtypes, we explore alternative etiologies.

Rozin: The speculations about control of sweet preferences (including different physiological responses) in eating disorders are quite reasonable, as explanations. But the problem is there is nothing to explain. There are no consistent differences in responses to sugar in overweight, bulimic, or anorexic people. It seems to me that speculations might be put in abeyance until there is a clear phenomenon to be explained.

Bartoshuk: The taste studies done to date appear to be primarily threshold studies. Even if there were to be some really convincing differences between those with eating disorders and controls, the differences would have limited meaning for perception of real-world foods and beverages because the sweetness of these is far above threshold. Both Pangborn and I have found that thresholds are not good predictors of responses at suprathreshold concentrations in taste. This is because the way in which taste intensity grows with concentrations varies across subjects, across taste substances, etc.

Some years ago, threshold studies were believed to assess taste ability over the whole dynamic range (i.e., from threshold to very strong sensations). This was based on the implicit assumption that if a threshold was increased by x percent, then all perceived intensities of suprathreshold concentrations would be reduced by that same x percent. We now know that this assumption was wrong. Threshold elevations can occur even when perceived intensities at higher concentrations remain unchanged; thresholds can remain normal even when perceived intensities at higher concentrations are reduced. Thus any taste threshold differences found in individuals with eating disorders would have limited meaning for the perception of real-world foods and beverages.

Chapter 14

Sweetness and Performance

Edward Hirsch

Introduction

There is widespread public belief that sweet foods have adverse effects on behavior (see Chap. 6). For example, a recent *Boston Globe* (10 February 1986) interview with the Massachusetts Teacher of the Year began with the question "Do you think kids are eating too much sugar?". Her reply emphasized that she doesn't allow candy because it makes children jumpy and hyperactive. This response is typical of many parents. Many families independently undertake dietary interventions in an attempt to help their children with attention deficit disorders and the focus of many of these changes is the reduction of sugar intake (Varley 1984). In a highly publicized trial in San Francisco the lawyers for Dan White, accused of killing two public officials, maintained that his consumption of junk food made him act irrationally: the so-called twinkie defense. This negative view of the effects of sweet foods is not restricted to the lay public. In a recent survey of primary care physicians in the state of Washington 45% of the respondents suggested a low sugar diet for hyperactive children (Bennett and Sherman 1983). A large number of correctional institutions for juveniles are sufficiently committed to the view that sugar consumption promotes behavioral problems that they have revised their diet policies in an effort to reduce sugar consumption (Schoenthaler 1985).

The question of course arises as to whether there is a scientific basis for these attitudes. In this paper the studies that bear on this issue will be reviewed. Attention will be focused on three areas: the effects of sweetness on the behavior of children, on antisocial behavior in institutional settings, and on athletic performance. These rather disparate areas have attracted the most attention of researchers and provide the core of empirical data on the relationship between sweetness and behavior in humans. The fundamenetal issue under consideration is whether sweetness affects performances. Are there data demonstrating clear effects in methodologically sound studies? A secondary focus will be to ask if these studies can be dissected to establish that changes in behavior were due to sweetness rather than the consequences of consuming the sweetening agent.

Although the focus of this paper is on sweetness and behavior, studies concerned with dietary carbohydrate levels are obviously relevant to this issue, particularly in relation to athletic performance, and in some instances these types of study provide the only relevant information even if sweetness per se is not a factor. Beyond these considerations, studies of the relationship between sweetness and behavior confront several difficult methodological problems that have not been addressed adequately. These issues will be raised even if adequate solutions are not evident.

Recent scientific interest in the relationship between sweetness and performance appears to stem from at least two sources. Firstly, dietary therapies have been widely used in attempts to treat childhood psychopathology, most notably hyperactivity or the attention deficit disorder (Varley 1984). Of the various dietary regimens that have been proposed the most widely known and the most thoroughly tested is the Feingold elimination diet (Feingold 1975). This regimen eliminates both naturally occurring salicylates, found in many foods, and synthetic salicylates, found in food additives and food colorings, from the child's diet. This diet is extremely restrictive and some authors have suggested that positive outcomes on this diet are due to the dramatic reduction in sugar consumption that occurs when this diet is instituted (Arnold 1984; Prinz et al. 1980). This suggestion is supported by the results of a recent double-blind study of sensitivity to a wide variety of foodstuffs. This study found that sugar challenges produced symptoms in 16% of a sample of hyperactive children (Egger et al. 1985). A second impetus for the interest in sugar and behavior was the discovery that variations in diet could produce changes in both peripheral levels of neurotransmitter precursors and in the synthesis of these neurotransmitters within the central nervous system. Specifically, Fernstrom and Wurtman (1971) showed that blood tryptophan, brain tryptophan, and brain serotonin levels all rose after rats consumed a high carbohydrate meal. Recent interest in the effects of sugar on hyperactivity and antisocial behavior can be traced to these influences (Conners 1984; Schoenthaler 1983a).

Methodological and Conceptual Issues

There are several recent excellent reviews of general methodological issues (Conners and Blouin 1983; Rumsey and Rapoport 1983) and substantive findings (Conners 1984; Rapoport 1983) in the area of dietary effects and nutritional therapy in children. These reviews do not, however, address a number of methodological issues that may be peculiar to evaluating the effects of sweet foodstuffs on behavior.

Research Strategies

Table 14.1 outlines the four research strategies that have been employed to evaluate the behavioral effects of sweetness. This table also indicates which strategy has been applied most frequently to study the three areas under

Table 14.1. Types of experiments conducted in humans on the relationship between sweetness and behaviour

Procedure	Duration	
	Acute	Chronic
Correlational		Children's behavior Athletic performance
Experimental	Children's behavior Athletic performance	Athletic performance Antisocial behavior

discussion in this paper. This simple classification scheme raises an important conceptual issue. Chronic studies treat sugar as a long-term dietary variable whereas the acute or challenge studies force the sweetener into a testing paradigm where the independent variable is best viewed as a drug or toxin. This distinction has important theoretical and methodological implications.

From a theoretical perspective short-term challenge studies would lead the investigator to seek very different types of underlying mechanism to explain behavioral effects than would long-term dietary studies. In the acute study it is possible to link behavioral measurements closely in time to consumption of the sweet food or drink. In this manner oral stimulation can be separated from postingestive consequences in specifying the locus of a behavioral change. The rapid metabolic and hormonal cephalic responses to purely peripheral sensory contact with sweet foods (Powley and Berthoud 1985) could underlie immediate performance changes. However, if behavioral testing occurs more than an hour after consumption of the sweet food or drink, performance changes would probably be due to non-oral consequences of consumption. For example, many investigators have suggested that changes in brain serotonin may be responsible for behavioral changes following sugar consumption (Arnold 1984; Behar et al. 1984). Recent evidence shows that chronic consumption of diets that vary widely in carbohydrate content (62%–80%) do not differentially affect brain serotonin in rats (Fernstrom et al. 1985). These authors also point out acute studies of diet and brain serotonin obtain their largest effects when a protein-free, largely carbohydrate, meal is compared with one that contains some protein (e.g., Fernstrom and Wurtman 1971). However, diets of this nature cannot be used in long-term studies. Very low protein diets lead to inadequate levels of energy intake and severe weight loss (Andik et al. 1963; Harper et al. 1970). The consequences of providing diets so low in protein will severely confound attempts to test the serotonin hypothesis in long-term studies. At a bare minimum, the acute–chronic distinction should sensitize investigators to considering whether a putative mechanism could operate in both time frames and whether they can ascribe performance changes to sweetness or to non-oral consequences of consumption.

The acute–chronic distinction also has clear implications for the proper way to conduct an experimental investigation of the relationship between sweetness and performance. In the acute study it is much easier to establish whether sweetness was responsible for performance changes. As previously mentioned, in the acute study behavior can be monitored in close temporal proximity to consuming the sweet substance. It is also far simpler to administer appropriate control foods or drinks in the short-term challenge situation. For example, a study comparing the behavioral effects of a sweet sugar, a non-nutritive sweetener, and no tastant and closely monitoring performance for several hours before and after consumption would serve to separate out whether sweetness or the postingestive consequences of consumption were responsible for any performance changes. The outcome of such a study would indicate whether additional control drinks should be tested in an effort to disentangle whether one was dealing with behavioral effects that were due to sweetness, calories, or some other postingestive consequence of sugar consumption. For example, maltodextrins, which are highly soluble carbohydrates that are not sweet or hyperosmotic, have been used in animal studies for this purpose (Booth et al. 1972).

Turning to the matrix itself, it is clear that the three broad areas under discussion in this paper have tended to exploit different research strategies in their efforts to relate sweetness and behavior. Acute correlational studies, where the investigator would relate individual variation in sugar consumption in the short-term (meal) to some criterion behavior measured after the meal, have rarely been employed.The long-term correlational study has been employed in research on diet and children's behavior (Lester et al. 1982; Prinz et al. 1980). Correlational studies have not typically been used to study athletic performance or antisocial behavior.

Both types of correlational study can serve as important sources of suggestive hypotheses about the relationship between sweetness and behavior. However, correlation, no matter how robust the relationship, does not establish causality, and great care must be taken in interpreting the direction of the relationship. This caution is particular relevant to studies of sugar and behavior. Almost 30 years ago Lat (1956) showed that rats with constitutionally different levels of "un-specific excitability" construct very different types of diet in a self-selection situation. In these studies, excitability of the central nervous system is defined by the frequency of spontaneous reactions in a standard stimulus situation in a fixed period of time (Lat 1967). More excitable rats learn better in a number of testing situations and when given the opportunity chose high carbohydrate, low protein diets. Excitability can be altered by diet but the effect lasts only as long as the diet is fed. These observations provide an empirical basis for the possibility that correlations between sugar consumption and hyperactive behavior results from hyperactive children choosing more sugar rather than high sugar consumption producing hyperactivity.

The experimental studies, both chronic and acute, offer the most powerful approach to establishing a causal relationship between sweetness and behavior. Almost all the studies on diet and athletic performance fall into these cells in the matrix. It is perhaps not surprising that the literature on diet and athletic performance is able to offer clearly dietary guidelines with a supporting metabolic rationale (Buskirk 1981; Evans and Hughes 1985). Experimental investigations of diet and children's behavior uniformly fall into the acute or challenge category. It is striking that the chronic experimental category is devoid of entries with children. This type of experiment certainly appears to have the most face validity in relation to the manner in which one would expect sugar to exert an effect on behavior in the real world.

Long-term dietary studies are, of course, very difficult to conduct with an appropriate control group in a double-blind fashion. It should be noted that blinding procedures have not been routinely employed in studies of athletic performance. Until recently it seemed impossible to design a control diet that was equally palatable and could not be discriminated from the high sugar diet by the

Table 14.2. Methodological issues in studies concerned with sweetness and behavior

Acute studies	Chronic studies
1. Defining the independent variable	1. Defining the independent variable
2. Nature of the placebo	2. Nature of the experimental and control diets
3. Test population	3. Test population
4. Prevailing nutritional state	

participants. However, Porikos et al. (1982) have shown that aspartame can be successfully substituted for sugar in a long-term caloric dilution study without the subjects detecting the difference between the two diets. It thus seems possible to conduct a double-blind study of this nature. Whether one could construct a diet that was also isocaloric, isonitrogenous and would be consumed in equal amounts to an experimental high sugar diet is another issue, since these are all dimensions of the diet that one would want to be equivalent in the two groups. In addition, the question arises as to whether aspartame is an appropriate control substance in this type of study (Stegink and Filer 1984).

Studies of the relationship between sugar and antisocial behavior largely fall into the chronic experimental category. However, this work is best viewed as quasi-experimental and only approaches a proper experiment where cross-over and cross-back procedures are employed (Schoenthaler 1983a). Even when these studies employ this type of design one is reluctant to call them experiments for a number of reasons that will be developed in a later section.

Table 14.2 lists a number of methodological issues that either have not been addressed in the literature on sweetness and behavior or have served as important differences between studies with conflicting outcomes.

Defining the Independent Variable

The literature on sweetness and behavior does not offer a clearly articulated statement on what constitutes the independent variable in this research area. Various authors have been concerned with refined sugars, others have used the term sugar generically; some have focused on two of the sweet sugars, sucrose, and glucose, whereas others have been concerned with dietary carbohydrate levels and the ratio of carbohydrate to protein in the diet. The non-nutritive sweeteners like saccharin and aspartame (aspartame is typically considered non-nutritive at the levels needed to impart sweetness to a beverage) have been used in some of the challenge studies as placebo controls to mask whether subjects are receiving the challenge substance or the control drink. Logically one can view these control drinks as experimental drinks if one defines the independent variable as sweetness.

It is apparent that one's definition of the independent variable determines the type of agent that is tested and the kinds of control procedure that are required. To test for sweetness effects the design of the experiment requires a comparison of a sweet substance to a non-sweet substance. In actual practice diets of this nature are quite difficult to construct.

In chronic dietary studies one can add a non-nutritive sweetener to the carbohydrate source to create diets that differ only in sweetness. However, in behavioral studies with humans this strategy would not be successful if the investigator wanted to keep the subjects ignorant in regard to the type of diet they were consuming. To successfully mask the sweet diet a second experimental diet would be used where a sweet sugar would be substituted for other carbohydrate sources in the basal diet. The sugar–starch exchange is relatively simple to implement and diets of this nature have been widely used in animal studies to test the metabolic effects of different carbohydrates (e.g., Hallfrisch et al. 1979). It is more difficult to generate palatable isocaloric diets for humans when sweet sugars are substituted for starch in an isocaloric exchange, but studies of this nature have been successfully conducted (e.g., Reiser et al. 1978). In these studies the major difference in the sugar and the control diets is that 30% of the calories were provided as a sucrose patty or a wheat starch wafer. In sweetness studies one could add a non-nutritive sweetener to the wheat starch wafer to create three diets. In this type of experimental arrangement the three diets would be isocaloric and isonitrogenous. If the two sweet diets were indistinguishable one could detect either sweetness effects or sugar effects if the subject's expectations about sweetness did not influence their performance on the behavioral measures.

Ideally, long-term studies of sweetness and behavior should attempt to use an isocaloric exchange of sweet sugars for starch in one of the experimental diets. Actual practice deviates considerably from this ideal. What is more typically the case is that studies concerned with the effect of dietary protein in behavior (e.g., Chiel and Wurtman 1981) have found their way into the literature on sugar and behavior. These animal studies generally exchange protein for carbohydrate to keep the diets isocaloric and as a consequence low protein diets are also high carbohydrate diets. In the growing literature on sugar and behavior some authors have offered these animal studies of low protein–high carbohydrate diets as evidence that the dietary carbohydrate/protein ratio can affect behavior (e.g., Behar et al. 1984; Lester et al. 1982; Rumsey and Rapoport 1983). Long-term animal studies of the behavioral effects of low protein–high carbohydrate diets should be interpreted cautiously due to the reduction in food consumption that is observed on these diets (Collier and Squibb 1967; Beaton et al. 1964). For example, Beaton et al. (1964) showed that rats given a 20% casein diet and pair fed with a 5% casein group showed comparable levels of running wheel activity, suggesting that the behavioral changes were a consequence of the weight loss (Collier 1970). For our present purposes the important point is that extreme caution should be exercised in interpreting the behavioral effects of low protein–high carbohydrate diets or extrapolating from these studies to high sugar consumption in children.

One experimental paradigm that could be exploited to demonstrate sweetness effects on behavior involves offering rats a nutritionally complete diet and a sapid sugar solution (Kanarek and Hirsch 1977). Under these conditions rats consume about 60% of their daily calories from the sweet solution. It does not matter whether sucrose, fructose, or glucose solutions are employed (Castonguary et al. 1981; Hirsch et al. 1982a; Kanarek and Orthen-Gambill 1982). The very high levels of sugar consumption that are observed offers the possibility that dietary effects of sweetness could be demonstrated. Interestingly enough, in a study conducted for quite different reasons, Hirsch et al. (1982b) found that both male and female rats fed a nutritionally complete diet and a sucrose solution showed higher levels of running wheel activity than rats fed only the complete diet. While this paradigm appears to offer the advantages of engendering very high levels of sugar consumption in apparently healthy animals and a potential animal model for testing the effects of sweetness on behavior, it is difficult to envisage a proper control diet for studies of this nature to identify the specific factors responsible for the behavioral change.

In sugar challenge studies the issue of defining an appropriate placebo challenge is also very difficult to resolve. Most researchers have focused on disguising the placebo so that it is indistinguishable from the sugar drink. Accordingly, sugar challenge studies with children have used non-nutritive sweeteners such as saccharin (Behar et al. 1984) or aspartame at levels that were equivalent in sweetness to the sugar challenge (Wolraich et al. 1985). This procedure successfully blinds the subjects as to the nature of the drink they are receiving but does not control for the fact that the challenge drink not only contains a sugar, but at the doses typically employed provides almost 200 calories to the child. This is a substantial caloric load for a child, particularly after an overnight fast (Behar et al. 1984), and raises the possibility that any changes in behavior could be attributed to the caloric load rather than to the sugar. In order to establish that changes in behavior were due to the sugar and not to the calories consumed in the challenge drink, the placebo should also provide an equivalent caloric load. A maltodextrin might serve this function nicely. The maltodextrins are highly soluble carbohydrates that are not sweet or hyperosmotic. They are equicaloric to sucrose and glucose and are rapidly absorbed.

Another approach to resolving the problem of the appropriate control substance in acute challenge studies is to conduct a dose–response study using the challenge substance. At present the levels of sugar used in challenge tests approximate those employed in glucose tolerance tests (e.g., about 1.75 g/kg). A graded behavioral response to a sugar challenge in a dose-dependent manner would both define the level that should be employed in future studies and perhaps illuminate those aspects of the sugar that should be controlled for.

Test Population

One of the major difficulties in comparing studies of sugar and behavior is the nature of the subjects that have been tested. This problem cuts across the three behavioral areas under discussion in this paper.

The clinical origins of interest in sugar and hyperactivity have led to a situation where some studies have tested children diagnosed as hyperactive (Prinz et al. 1980; Wolraich et al. 1985), others have tested children identified by their parents as sugar responsive (Behar et al. 1984), whereas still others have tested children with mixed psychiatric diagnoses (Conners et al., unpublished work), presented at the Symposium on Diet and Behavior, sponsored by the AMA and ILSI, Arlington, Virginia, February 1985). Finally, some investigators have focused on the responses of normal children to sugar challenges (Goldman et al., unpublished work, presented at the American Psychological Association, Toronto, September 1984). Comparing results across these diverse populations may underlie the failure to produce data that are consistent across studies. To further confound this situation some studies have tested children whose psychiatric diagnoses have spanned a number of disorders and have collapsed the data across diagnostic categories (e.g., Connors et al., unpublished work, presented at Symposium on Diet and Behavior, sponsored by the AMA and ILSI, Arlington, Virginia, February 1985) or they have combined data from normal and clinical populations (Behar et al. 1984). In addition to these problems, age has varied rather widely across the sugar challenge studies, rendering comparisons problematic.

At first glance the literature on athletic performance and sweetness seems to be characterized by a similar problem. Some studies use trained athletes (e.g., Coyle et al. 1983) whereas others use healthy adults (Ahlborg and Felig 1977). Although this is potentially a problem, the consistency of the data across these two populations indicates that the dietary effects are robust and generally hold across these two groups (Buskirk 1981; Evans and Hughes 1985).

Studies of diet and antisocial behavior have focused solely on juveniles who are incarcerated (Schoenthaler 1985). Whatever conclusions these studies generate about diet and behavior should be restricted to the population that has been studied.

Prevailing Nutritional State

The effects of acute episodes of sugar consumption appear to depend critically on the nutritional status of the participants at the time of testing. For example, complex interactions have been found between the effects of a sugar challenge and the nature of the breakfast meal. (Connor et al., unpublished work, presented at Symposium on Diet and Behavior, sponsored by the AMA and ILSI, Arlington, Virginia, February 1985). High protein meals appeared to suppress any sugar effects, whereas high carbohydrate breakfast meals potentiated behavioral changes. Again, this dimension has varied considerably across the sugar challenge studies. Similarly, studies of diet and athletic performance show that the prevailing nutritional state prior to testing affects how an individual responds to sugar consumption during a bout of exercise.

Research Findings

Sweetness and Children's Behavior

Correlational Studies

Several studies have examined the relationship between habitual sugar intake and behavior in young children (Lester et al. 1982; Prinz et al. 1980; Wolraich et al., to be published). In the first major study of this type 7-day diet records were maintained by the mothers of 28 hyperactive children and 26 normal children (Prinz et al. 1980). The children were systematically observed in a standardized playroom setting. Both groups showed similar patterns of sugar consumption. In the hyperactive group there were significant correlations between total sugar intake, derived measures of the ratio of sugar products to other nutrients in the diet, and destructive–aggressive and restless behavior in the playroom setting. In the control group of normal children, sugar intake showed significant correlations with movement from quadrant to quadrant in the playroom setting. The authors interpret these carefully collected data to suggest that sugar consumption can affect the behavior of both normal and hyperactive children.

This study has been criticized on the grounds that sugar consumption was calculated on the basis of the weight of the food rather than the weight of the nutrients (Milich et al., to be published). This shortcoming could severely distort measures of intake, particularly if sweetened beverages were heavily consumed. Beyond this criticism nutrient intake should be converted into calories to reflect more accurately the composition of the diets that were consumed. In an attempt to replicate these findings, Wolraich et al. (1986) examined the correlation between sugar intake and performance on 37 behavioral/cognitive variables in hyperactive boys. Sugar consumption was not reliably related to any of the learning or laboratory tasks, examiner ratings, or observations in a "Restricted Academic" setting. Taken together these two studies suggest that chronically high sugar consumption may affect behavior in hyperactive boys.

There is another curious aspect to the Prinz et al. (1980) study. In their examination of the dietary variables only total consumption differed between the hyperactive and the normal children, with the hyperactive children consuming less food, yet this is the only dietary variable they did not correlate with the behavioral measures. One wonders what this lower intake is due to and how this dietary variable correlates with the behavioral measures across the two groups.

In another correlational study Lester et al. (1982) found that the ratio of carbohydrate to protein and the proportion of refined sugars in the diets of a group of 184 normal children aged 5–16 correlated negatively with WISC full scale IQ ($r = -0.24$). This relationship held up even when age, sex, and socioeconomic status were partialled out. This relationship is suggestive but is far from definitive. Firstly, this correlation only accounts for 5.8% of the variance in the IQ scores. Secondly, high sugar intake may be indicative of a poor dietary history and a long-term pattern of poor nutrition may underlie these correlations.

Challenge Studies

The published sugar challenge studies have all employed a rigorous cross-over procedure in which the subjects receive a challenge substance on one day and a placebo on another with the order of presentation counterbalanced across subjects. With one exception (Gross 1984) these studies have also employed a double blind procedure where neither the subjects nor the experimenters knew which substance was being tested on a particular day. As previously discussed, the similarity across studies ends here. The published studies have examined different populations, different sugars at different doses, different placebo agents (some of which may affect behavior) and testing has occurred against different nutritional backgrounds (overnight fast, after breakfast, or after lunch). In an effort to impose some order on the challenge studies the discussion will be organized around the type of children who participated.

Hyperactive Children

Gross (1984) studied a group of boys ranging in age from 5 to 17 whose mothers indicated they responded adversely to sugar. The mothers were given lemonade that was sweetened with sucrose (75 g per serving) or saccharin, and were asked to serve one-third of a jar at a time when the child could be observed for several hours. Three tests of each type of lemonade were conducted with the mother rating her child's behavior following consumption of the beverage on a scale that ranged from much worse (−5) to much better (+5). None of the children showed a consistent response to the sugar. Despite the obvious shortcomings in this study it has the virtues of being conducted in a naturalistic setting and as Milich et al. (to be published) point out, it was the mothers, who are most often the source of claims about adverse sugar effects, who did the behavioral ratings.

A much more carefully conducted naturalistic study was done with hyperactive boys in a summer treatment program (Milich and Pelham, to be published). After an overnight fast the boys received Koolaid containing sucrose (1.75 g/kg) or aspartame and their behavior was monitored during two recreational periods and an hour of classroom work for the next 3.5 h. Dependent measures included behavior in the classroom, academic productivity and accuracy, peer interactions during recreation, and measures of compliance and rule violations. An additional strength of this study was that the authors were able to document the reliability and validity of the dependent measures. The data failed to reveal differences on any of the 25 measures between sugar and aspartame days in this homogeneous sample of hyperactive boys.

There are also two well conducted laboratory studies with hyperactive children. In the first (Behar et al. 1984), an advertisement in a community newspaper sought participants who were judged to be affected adversely by sugar by their parents. Twenty-one boys, ranging in age from 6.5 to 14 years, participated in the study. Eight were classified as normal and nine were characterized as having attention-deficit disorders with hyperactivity; the remaining boys either provided insufficient information for diagnosis or had difficulties in the past that were not evident at the time this study was conducted. All participants were on a high carbohydrate diet for 3 days preceding the experiment. After an overnight fast they received glucose (1.75 g/kg), sucrose (1.75 g/kg), or a saccharin placebo. The sugar drinks also contained saccharin to disguise their natural taste. The order of the drinks was randomly assigned across subjects and each test session was separated by at least 48 h. For the group as a whole, motor activity showed a decrease at 3 h when the data for both sugars were combined. Both sugars led to a small increase in pulse rate throughout the test period. Additional measures, including observer behavioral ratings and measures of attention and learning, failed to show any effects of sugar consumption. None of the boys showed blood glucose patterns that met the criteria for diabetes or impaired glucose tolerance. With the exception of one boy who showed a prediabetic type of blood glucose curve, hormonal indices of glucose metabolism such as insulin and cortisol levels were within the normal limits. Although this study used children who were identified by their parents as sugar sensitive, the test population was quite heterogeneous, with data collapsed across both normal and hyperactive children. Until the issue of specificity or generality of sugar challenge effects is clarified, the results of this study will remain difficult to interpret.

Wolraich et al. (1985) also conducted two careful laboratory challenge studies that were restricted to hyperactive boys from 7 to 12 years of age. In both studies 16 boys were admitted to a clinical research center where they were fed a sucrose-free diet for 3 days. On day 1 baseline measures were taken. On days 2 and 3 sucrose (1.75 g/kg) or aspartame drinks (of equal sweetness) were administered. In the first study the drink was given after lunch and in the second after an overnight fast. Measures of playroom behavior, sustained attention, paired associate learning, nonsense word spelling, matching familiar figures, drawing a line slowly, and examiner ratings all failed to reveal a difference between the behavior of the boys following sugar and aspartame. In fact when the data for the two studies were combined only a measure of impulsiveness showed a significant sugar effect. The boys made fewer errors on this measure following the sugar challenge. Although a large number of measures were taken the authors carefully provided a rationale for their use and demontrated that 28 of the 37 dependent variables showed significant test–retest reliability. Beyond the problems raised earlier about what constitutes an appropriate placebo control, the only methodological limitations in this study are that behavioral measures were not systematically repeated over time and that testing ended 3 h after the challenge. In all other respects this is a well designed and carefully conducted negative study that casts serious doubt on whether there are short-term consequences of sugar consumption in hyperactive boys.

Normal Children

There are surprisingly few sugar challenge studies with normal children. Goldman and her colleagues (Goldman et al., unpublished work, presented at the American Psychological Association, Toronto, September 1984) studied eight preschool children ranging in age from 3 years 7 months to 5 years 10 months in a double-blind sugar challenge study. Following an overnight fast the children were given orange juice sweetened with sucrose (2 g/kg) or aspartame. Following sucrose the children showed a decrement in performance on a sustained attention task, with more errors occurring at 60 min, and also demonstrated more "inappropriate" behavior during free play at 45–60 min after the drink. Despite the small size of the sample and limitations imposed by the nature of the placebo, this is a provocative, well-conducted study.

Mixed Psychiatric Diagnoses

Conners and his colleagues (Conners et al., unpublished work, presented at symposium on Diet and Behavior sponsored by the AMA and ILSI, Arlington, Virginia, February 1985) have reported the effects of several sugar challenge studies in children with a broad range of psychiatric disturbances. The studies have not yet been published or presented in sufficient detail to allow critical evaluation, but they are included here to document an important methodological point.

The first study used high doses (50 g) of sucrose and fructose in severely disturbed child psychiatric inpatients. Both sugars led to an increase in activity level but teachers could not distinguish between the challenge conditions. The second study used lower doses of these sugars (1.25 g/kg), monitored food intake, and examined a much broader spectrum of behavior in children with diagnoses of anxiety disorder, conduct disorders, or attention-deficit disorder. Both sugars reduced activity level in the classroom, speeded reaction time, reduced errors on a sustained attention task, and increased evoked response amplitudes. Hierarchical regression analyses revealed that the sugar effects were quite weak relative to the composition of the breakfast meal. In the third study, breakfast composition was directly manipulated. Sugar (1.75 g/kg) or aspartame challenges were given to normal and hyperactive children after fasting or after eating a high carbohydrate or a high protein breakfast. The results indicated that the effect of the sugar challenges depended on the nature of the breakfast meal, with protein suppressing sugar effects and carbohydrate meals potentiating them.

The incomplete nature of this report makes it impossible to interpret these complex findings. However, they do illustrate the important methodological point that in acute studies the nature of the prevailing nutritive state dramatically affects the response to a sugar challenge.

Summary of Findings

These observations, with all the limitations raised earlier, suggest the following tentative conclusions:

1. Sugar challenges do not affect the behavior of children classified as hyperactive.
2. Chronic dietary studies are more likely to reveal sugar effects than acute studies.
3. Normal children are as likely to be affected by high, chronic sugar consumption as children diagnosed as hyperactive.
4. The available evidence does not indicate widespread or powerful effects of sugar on the behavior of children.

Sweetness and Athletic Performance

Studies of exercise endurance and athletic performance have focused largely on the effects of dietary carbohydrate rather than sweet sugars per se. As in the literature on sweetness and children's behavior, the studies on diet and athletic performance fall into acute and chronic categories with the acute studies further subdivided into those that study dietary influences just prior to exercise and those that are concerned with effects of foods or drinks consumed during exercise.

The well-established relationship between metabolism and exercise intensity provides the basis for the interest in the effects of dietary carbohydrate on athletic performance. The energy demands of the muscle at rest are met almost exclusively by fat (Ahlborg et al. 1967), but as exercise intensity increases an increasing amount of energy is derived from carbohydrate (Hultman 1980). Carbohydrate storage in the body is limited and found almost exclusively in muscle and liver in the form of glycogen. The amount of energy stored as glycogen is of considerable importance in endurance sports because the depletion of muscle glycogen coincides with muscular fatigue (Bergstrom et al. 1967).

Long-Term Diet Studies

A number of studies have shown that performance in prolonged demanding exercise is improved by eating a carbohydrate-rich diet in order to elevate the liver and muscle glycogen stores. For example, Bergstrom et al. (1967) showed that a high carbohydrate diet both enhanced submaximal exercise performance and led to higher levels of glycogen content than did a mixed diet or a high fat–protein diet. In the now classic one-legged ergometry study Bergstrom and Hultman (1966) showed that by depleting glycogen stores in one leg and following this exercise by a high carbohydrate diet muscle glycogen was not only rapidly replaced but the concentration rose to a level far above the normal range. This phenomenon has been labelled glycogen supercompensation and is the condition many endurance athletes strive for prior to competition. Following this type of regimen, Karlsson and Saltin (1971) showed that actual performance was improved in a 30-km race.

From the perspective of the present paper there is little systematic information on whether these effects of high carbohydrate diets on endurance activity are influenced by the nature of the carbohydrate or their sweetness. There has been some work on the effect of different carbohydrates on glycogen synthesis and resynthesis following exercise. Costill and Miller (1980) did find that starch as compared to glucose promoted somewhat greater glycogen synthesis. In a similar manner, Costill et al. (1981) compared the effects of diets containing simple sugars (glucose, sucrose, fructose) or complex carbohydrates and found that the nature of the carbohydrate did not influence glycogen synthesis at 24 h but that at 48 h the starch diet produced higher levels of muscle glycogen. Buskirk (1981), in this thorough review of nutrition and athletic performance points out that there is a need for additional information regarding the effects of different dietary carbohydrates on glycogen synthesis.

Carbohydrate Consumption Prior to Exercise

Many athletes drink sweetened beverages prior to competition in the belief that it will provide them with additional energy. This practice may reduce rather than increase their tolerance for exercise (Astrand 1967; Costill et al. 1977). Ahlborg and Felig (1977) have shown that glucose consumption prior to exercise is followed by a rise in blood glucose and a further rise in carbohydrate utilization by the exercising muscle. In addition, glucose consumption prior to exercise is associated with a decrease in hepatic gluconeogenesis (Ahlborg and Felig 1977) and a reduction in lipolysis (Costill et al. 1977). The net result of these metabolic changes is that prior glucose consumption may precipitate hypoglycemia during exercise. Ahlborg and Felig (1977) have suggested that these metabolic responses to glucose consumption are mediated by hyperinsulinemia. These observations have led some authors to emphasize that no carbohydrate should be ingested 2 h before athletic competition (Costill and Miller 1980; Foster et al. 1979).

The admonition not to consume carbohydrates prior to exercise may apply to glucose but not to other sweet carbohydrates. Several studies have shown that fructose consumption prior to exercise does not cause a significant rise in plasma insulin levels and a subsequent drop in blood glucose during exercise (Koivisto et al. 1981; Levine et al. 1983). Levine et al. (1983) also demonstrated that muscle glycogen utilization was significantly reduced after fructose administration in comparison to the glucose and no-carbohydrate conditions. Thus it appears that fructose prior to exercise may enhance endurance.

Carbohydrates During Exercise

When hard physical work extends over several hours, depletion of glycogen stores can constitute a limit to continued performance of the activity. A number of studies have examined whether providing carbohydrates during exercise serves to spare the endogenous glycogen stores and allows the duration of exercise endurance to be extended.

In one study to test this idea Brooke and Green (1974) had trained cyclists exercise until their respiratory quotients fell to 0.73 or until they could no longer cycle. At this point they were allowed to rest for 40 min and were fed either a glucose syrup drink with added salts or a tinned rice pudding with added sucrose that provided the same energy or a low energy drink with added salts that was colored and flavored to match the glucose syrup. After this rest they were exercised again. Results showed the high carbohydrate drink allowed more work to be done following recovery. Coyle et al. (1983) obtained similar results when the carbohydrate was given during exercise. Ten experienced cyclists were given a glucose polymer solution (Polycose) or a placebo 20 min after they began working on a bicycle ergometer. Fatigue was postponed in seven of the ten subjects following the carbohydrate drink. In addition, carbohydrate consumption during exercise prevented both the hyperglycemia and hyperinsulinemia that occur when glucose is consumed prior to exercise. It should also be mentioned that at lower exercise intensities glucose consumption does not improve endurance (Felig et al. 1982).

Summary of Findings

This brief overview suggests that carbohydrates promote better performance in endurance exercise when they are consumed at high levels for several days prior to the exercise, or during the exercise itself. Sweetness only becomes a factor when the carbohydrate is consumed just prior to the exercise. In this situation glucose ingestion stimulates hyperglycemia and hyperinsulinemia, hastens exercise-induced hypoglycemia, and accelerates glycogen utilization and fatigue. These effects do not occur when fructose is the carbohydrate ingested prior to the exercise.

Sweetness and Antisocial Behavior

A large number of correctional institutions have revised their diet policies in an effort to reduce sugar consumption (Schoenthaler 1985). The empirical basis for these institutional changes derives from a series of studies conducted over the past 5 years by a single author (Schoenthaler 1982, 1983a–f, 1985, to be published) using a similar experimental approach in all studies. Beyond stories in the popular press that frequently reduce to anecdotal evidence, this work provides the core of the published data on sugar and crime. Whatever the merit of these studies it is important to realize at the outset that they are restricted to a single population: juveniles between the ages of 12 and 18 who have been placed in juvenile detention facilities. In addition, the measure of antisocial behavior consists of institutional disciplinary reports filed by the staff members. This information should be kept in mind when one reads of sugar consumption causing violent crime. The American Dietetic Association was prompted to publish a position paper emphasizing that a causal relationship between diet and crime or diet and violence has not been demonstrated (1985).

Turning to Table 14.1, the institutional studies conducted by Schoenthaler

Table 14.3. Institutional dietary policy changes intended to reduce sugar consumption (adapted from Schoenthaler, to be published)

1. Rinse canned fruits packed in syrup prior to serving
2. Replace Koolaid and lemonade with iced tea
3. Serve iced tea unsweetened
4. Eliminate jelly and cinnamon sugar from diet
5. Replace soda machine with fruit juice machine
6. Replace table sugar with honey
7. Substitute molasses for white sugar in recipes
8. Replace presweetened cereals with cereals not presweetened
9. Parents requested not to send foods with high sugar content
10. Eliminate desserts high in sugar. Replace with fresh fruits, peanuts, coconut, carrots, cheese, chestnuts, sugar-free soft drinks, peanut butter on crackers, popcorn, cold cuts, cream cheese on celery

(1982, 1983a–f) can be considered long-term dietary studies and they are also experimental in the sense that an experimental manipulation is performed, but there are no control groups or control procedures. These studies are probably best described as quasi-experiments.

The general procedure followed in these studies is deceptively simple. At a specific point in time the institution modifies its food policy in an effort to reduce sugar consumption. Table 14.3 lists the changes that were implemented in the intervention studies. The dependent measure is the number of formal disciplinary actions recorded by one of the staff members on the child's official record. To correct for the fact that juveniles are in the institution for different durations, the number of incidents are divided by the length of stay to create the dependent variable, the rate of antisocial behavior per day per child. The procedure followed to create a "double-blind" in these studies is simply not informing the staff or the juveniles that the study is taking place. In addition the following individuals are omitted from the analysis to avoid bias; those incarcerated for less than 24 h, those whose rate of disciplinary reports exceeds one per day, and females when the number in the institution is small. With minor modifications and refinements this procedure was used in the entire series of studies (Schoenthaler 1982, 1983a–f).

On the basis of this series of studies, Schoenthaler (1985) claims that the 8076 juveniles confined in 12 juvenile correctional facilities showed 47% less "antisocial behavior" when the institution's diet policy was modified to reduce sugar consumption and in one study to increase orange juice consumption. This claim has important policy implications and deserves careful scrutiny.

The first question that should be addressed concerns identifying the independent variable. The titles of these papers, the rationale given for conducting them, and the statistical analyses all point to sugar consumption as the independent variable. In fact, in some of the early papers (e.g., Schoenthaler 1982) diet is identified as the independent variable. One does not have to be a nutritionist to appreciate that many of the dietary changes listed in Table 14.3 are of dubious value in limiting the availability of sugar. In addition no effort was made to measure sugar consumption. Intake data are essential to establish that these manipulations had any effect and that the independent variable was operative. The absence of any measures did not deter Schoenthaler from stating "while total sugar consumption was not measured in either group, a definite reduction was achieved by using the revised diet" (Schoenthaler 1982, p 4). Even if a reduction in sugar consumption was achieved the nature of the diets consumed during the two periods would by definition vary in composition and in the level of intake of the other nutrients, rendering it impossible to attribute behavior change to reduced sugar consumption. In his most recent paper (Schoenthaler, to be published) the independent variable is defined as diet policy. To this reader diet policy, as an independent variable, embraces much more than dietary sugar content and is probably a more accurate description of the variable that is operating in these studies.

Beyond problems in defining the independent variable, appropriate procedures were not implemented in any of these studies to ensure that change could be attributed to the dietary intervention. Although the title of the first paper in this series (Schoenthaler 1982) includes the description "double-blind," nothing in the methodology, other than the fact that neither the staff nor the juveniles were informed of the study, even suggests a blinding procedure. In his recent review (Schoenthaler 1985), the author admits that none of the institutional diet and behavior studies met the standard of double-blind research. Rather he argues that the observations of a covert participant and the fact that the effects did not diminish over time are sufficient to rule out placebo effects. Neither argument is compelling in the absence of appropriate procedures or control groups.

The meaning of this series of studies is also clouded by the failure to establish the reliability and validity of the dependent measure. At a bare minimum, a reliability study would at least establish that staff members use the serious incident report in a consistent manner and it might be possible to draw stronger conclusions about the effects of institutional policy changes on behavior within the institutions.

The third major problem with this series of studies concerns the changing nature of the population. Some of the juveniles are included in both the treatment and the control conditions, others may be in one condition and not the other, and some may be in one condition for several months and the other for several days. Schoenthaler (1983c,d) deals with this problem in two ways. Firstly, in one study he was able to follow an a-b-a format where baseline was followed by treatment and then by a return to baseline (Schoenthaler 1983c). The orderly changes in the rate of disciplinary reports clearly suggests that something is affecting the juveniles' behavior or the staff's perceptions of it. Secondly, when the population was large enough it was possible to restrict the sample to those juveniles who were in the institution during both the control and the treatment phase (Schoenthaler 1983d). Despite the limitations of specifying the independent variable and the reliability of the dependent measure, the within-subject procedure and the return to baseline study are clearly suggestive.

Finally, in reviewing these studies one is struck by the absence of methodological details. This is a serious shortcoming in trying to evaluate the data. This lack of detail would also make it impossible to independently replicate these studies.

Summary of Findings

Despite the apparent consistency of the data across studies and institutions, these studies do not provide valid evidence for the assertion that sugar consumption leads to anti-social behavior. What appears to be a more reasonable summary of this work is that institutional policies that are intended to reduce sugar consumption lead to a reduction of serious incident reports in the institution. In my view, the imposition of the sugar restriction probably engenders a whole series of changes in the behavior of the staff and the juveniles that reduces the reporting of serious incidents. Only a carefully controlled scientific study will resolve this issue.

References

Ahlborg G, Felig P (1977) Substrate utilization during prolonged exercise preceded by ingestion of glucose. Am J Physiol 233: E188–E194

Ahlborg G, Bergstrom J, Ekelund LG, Hultman E (1967) Muscle glycogen and muscle electrolytes during prolonged physical exercise. Acta Physiol Scand 70: 120–142

American Dietetic Association Position Paper on Diet and Criminal Behavior (1985) J Am Diet Assoc 85: 361–362

Andik I, Donhoffer S, Farkas M, Schmidt P (1963) Ambient temperature and survival on a protein-deficient diet. Br J Nutr 17: 257–261

Arnold LE (1984) Diet and hyperkinesis. Integrative Psychiatry 2: 188–194

Astrand PO (1967) Diet and athletic performance. Fed Proc 26: 1772–1777

Beaton JR, Felecki V, Stevenson JAF (1964) Activity and patterns of rats fed a low-protein diet and the effects of subsequent food deprivation. Can J Physiol Pharmacol 42: 705–718

Behar D, Rapoport JL, Adams AA, Berg CJ, Cornblath M (1984) Sugar challenge testing with children considered behaviorally "sugar reactive". Nutr Behav 1: 277–288

Bennett FC, Sherman R (1983) Management of childhod "hyperactivity" by primary care physicians. Dev Behav Pediatr 4: 88–93

Bergstrom J, Hultman E (1966) Muscle glycogen synthesis after exercise. An enhancing factor localized to the muscle cells in man. Nature (London) 210: 309–310

Bergstrom J, Hultman E (1967) A study of the glycogen metabolism during exercise in man. Scand J Clin Lab Invest 19: 218–228

Bergstrom J, Hermansen L, Hultman E, Saltin B (1967) Diet-muscle glycogen and physical performance. Acta Physiol Scand 71: 140–150

Booth DA, Lovett D, McSherry GM (1972) Postingestive modulation of the sweetness preference gradient in the rat. J Comp Physiol Psychol 78: 485–512

Brooke JD, Green LE (1974) The effect of a high carbohydrate diet on human recovery following prolonged work to exhaustion. Ergonomics 17: 489–497

Buskirk ER (1981) Some nutritional considerations in the conditioning of athletes. Ann Rev Nutr 1: 319–350

Castonguay TW, Hirsch E, Collier G (1981) Palatability of sugar solutions and dietary selection. Physiol Behav 17: 7–12

Chiel H, Wurtman RJ (1981) Short-term variations in diet composition change the pattern of spontaneous motor activity in rats. Science 213: 676–678

Collier GH (1970) Work: a weak reinforcer. Trans NY Acad Sci 32: 557–576

Collier G, Squibb RL (1967) Diet and activity. J Comp Physiol Psychol 64: 409–413

Conners CK (1984) Nutritional therapy in children. In: Galler JR (ed) Nutrition and behavior. Plenum, New York, pp 159–192

Conners CK, Blouin AG (1983) Nutritional effects on behavior of children. J Psychiatr Res 17: 193–201

Costill DL (1984) Energy supply in endurance activities. Int J Sports Med 5: 19–21 [Suppl]

Costill D, Miller JM (1980) Nutrition for endurance sport: carbohydrate and fluid balance. Int J Sports Med 1: 2–14

Costill DL, Coyle E, Dalsky G, Evans W, Fink W, Hoopes D (1977) Effects of elevated plasma FFA and insulin on muscle glycogen usage during exercise. J Appl Physiol 43: 695–699

Costill DL, Sherman WM, Fink WJ, Maresh CW, Hen M, Miller JM (1981) The role of dietary carbohydrates in muscle glycogen resynthesis after strenuous running. Am J Clin Nutr 34: 1831–1836

Coyle EF, Hagberg JM, Hurley BF, Martin WH, Ehsani AA, Holloszy JO (1983) Carbohydrate feeding during prolonged strenuous exercise can delay fatigue. J Appl Physiol 55: 230–235
Egger J, Graham PJ, Carter CM, Gumley D, Soothill JF (1985) Controlled trial of oligoantigenic treatment in the hyperkinetic syndrome. Lancet I: 540–545
Evans WJ, Hughes VA (1985) Dietary carbohydrates and endurance exercise. Am J Clin Nutr 41: 1146–1154
Feingold BF (1975) Why your child is hyperactive. Random House, New York
Felig PA, Cherif A, Minagawa A and Wahren J (1982) Hypoglycemia during prolonged exercise in normal men. N Engl J Med 306: 895–900
Fernstrom, JD, Wurtman RJ (1971) Brian serotonin content: increase following ingestion of carbohydrate diet. Science 174: 1023–1025
Fernstrom JD, Fernstrom MH, Grubb PE, Volk EA (1985) Absence of chronic effects of dietary protein content on brain tryptophan concentrations in rats. J Nutr 115: 1337–1344
Foster C, Costill DL, Fink WJ (1979) Effects of preexercise feedings on endurance performance. Med Sci Sports 11: 1–5
Gross M (1984) Effect of sucrose on hyperkinetic children. Pediatrics 74: 876–878
Hallfrisch J, Lazar FL, Reiser S (1979) Effects of feeding sucrose or starch to rats made diabetic with streptozotocin. J Nutr 109: 1909–1915
Harper AE, Benevenga NJ, Wohlhueter RM (1970) Effects of ingestion of disproportionate amounts of amino acids. Physiol Rev 50: 428–558
Hirsch E, DuBose C, Jacobs HL (1982a) Overeating, dietary selection patterns and sucrose intake in growing rats. Physiol Behav 28: 819–828
Hirsch E, Godkin L, Ball E (1982b) Sex differences in the effects of voluntary activity on sucrose-induced obesity. Physiol Behav 29: 253–262
Hultman E (1980) Glycogen loading and endurance capacity. In: Stull GA, Cureton TK Jr (eds) Encyclopedia of physical education, fitness and sports: training environment, nutrition and fitness. Brighton Publ. Salt Lake City, pp 274–291
Kanarek R, Hirsch E (1977) Dietary-induced overeating in experimental animals. Fed Proc 36: 154–158
Kanarek R, Orthen-Gambill N (1982) Differential effects of sucrose, fructose and glucose on carbohydrate-induced obesity in rats. J Nutr 112: 1546–1554
Karlsson J, Saltin B (1971) Diet, muscle glycogen and endurance performance. J Appl Physiol 31: 203–211
Koivisto VA, Karonen SL, Nikkila EA (1981) Carbohydrate ingestion before exercise: comparison of glucose, fructose and sweet placebo. J Appl Physiol 51: 783–787
Lat J (1956) The relationship of the individual differences in the regulation of food intake, growth and excitability of the central nervous system. Physiol Bohemoslov 5: 38–42
Lat J (1967) Self-selection of dietary components. In: Code CF (ed) Alimentary canal. American Physiological Society, Washington DC, pp 367–386 (Handbook of physiology, vol I, Sect 6)
Lester ML, Thatcher RW, Monroe-Lord L (1982) Refined carbohydrate intake, hair cadmium levels, and cognitive functioning in children. Nutr Behav 1: 3–13
Levine L, Evans WJ, Caderette BS, Fischer EC, Bullen BA (1983) Fructose and glucose ingestion and muscle glycogen use during submaximal exercise. J Appl Physiol 55: 1767–1773
Milich R, Pelham WE (to be published) A naturalistic investigation of the effects of sugar ingestion on the behavior of attention deficit disorder boys. J Consult Clin Psychol
Milich R, Wolraich M, Lindgren S (to be published) Sugar and hyperactivity: a critical review of empirical finds. Clin Psychol Rev
Porikos KP, Hesser MF, VanItallie TB (1982) Caloric regulation in normal-weight men maintained on a palatable diet of conventional foods. Physiol Behav 29: 293–300
Powley TL, Berthoud HR (1985) Diet and cephalic phase insulin responses. Am J Clin Nutr 42: 991–1002
Prinz RJ, Roberts WA, Hantman E (1980) Dietary correlates of hyperactive behavior in children. J Consult Clin Psychol 48: 760–769
Rapoport JL (1983) Effects of dietary substances in children. J Psychiatr Res 17: 187–191
Reiser S, Hallfrisch J, Michaelis OE IV, Lazar FL, Martin RE, Prather ES (1978) Isocaloric exchange of dietary starch and sucrose in humans. 1. Effects on levels of fasting blood lipids. Am J Clin Nutr 32: 1659–1668
Rumsey JM, Rapoport JL (1983) Assessing behavioral and cognitive effects of diet in pediatric populations. In: Wurtman RJ, Wurtman JJ (eds) Nutrition and the brain. Raven Press, New York, pp 101–161
Schoenthaler S (1982) The effect of sugar on the treatment and control of anti-social behavior: a

double-blind study of an incarcerated juvenile population. Int J Biosoc Res 3: 1–9
Schoenthaler S (1983a) Diet and crime: an empirical examination of the value of nutrition in the control and treatment of incarcerated juvenile offenders. Int J Biosoc Res 4: 25–39
Schoenthaler S (1983b) Diet and deliquency: a multi-state replication. Int J Biosoc Res 5: 70–78
Schoenthaler S (1983c) The Alabama diet-behavior program: an empirical evaluation at the Coosa Valley Regional Detention Center. Int J Biosoc Res 5: 79–87
Schoenthaler S (1983d) The Los Angeles probation department diet-behavior program: an empirical analysis of six institutional settings. Int J Biosoc Res 5: 88–98
Schoenthaler S (1983e) The northern California diet-behavior program: an empirical examination of 3,000 incarcerated juveniles in Stanislaus county juvenile hall. Int J Biosoc Res 5: 99–106
Schoenthaler S (1983f) The effects of citrus on the treatment and control of antisocial behavior: a double-blind study of an incarcerated juvenile population. Int J Biosoc Res 5: 107–117
Schoenthaler S (1985) Nutritional policies and institutional antisocial behavior. Nutr Today 20: 16–25
Schoenthaler S (to be published) Diet and delinquency: empirical testing of eight theories. J Quant Criminol
Stegink LD, Filer LJ Jr (eds) (1984) Aspartame physiology and biochemistry. Marcel Dekker, New York Basel
Varley CK (1984) Diet and the behavior of children with attention deficit disorder. J Am Acad Child Psychiatry 23: 182–185
Wolraich M, Milich R, Stumbo P, Schultz F (1985) Effects of sucrose ingestion on the behavior of hyperactive boys. J Pediatr 106: 675–682
Wolraich ML, Stumbo P, Milich R, Chenard C, Schultz F (1986) Dietary characteristics of hyperactive and control boys and their behavioral correlates. J Am Diet Assoc 86: 500–504

Commentary

Fischler: A future topic for social science research could center on how and why in the first place did researchers begin to address the issue of sugar and hyperactivity. A good deal of the recent work, the author of this paper implies, essentially arose from attempts to check the accuracy of beliefs common among parents, or reports from them, regarding their children's behavioral responses to sugar. Such attitudes, in turn, had apparently been influenced by previous medical research (see Chap. 6). Thus two issues are involved here. One is epistemological and relevant to the sociology of science. Were there theoretical concerns or hypotheses on which this trend of research was based? If not, what determined such enterprises? The other approach could deal with the relationship between public attitudes and science. What we have here is a chicken and egg type of problem, involving the scientific community as a profession (a social group), the media, and "the public." How exactly they interact is not clearly understood.

Booth: Porikos et al. (1982) did not measure the effectiveness of their blinding, nor the subjects' attributions to the selected diet. They substituted table sweeteners in an unspecified form. They refer to no bulking agents in the aspartame formulations.

The so-called balanced placebo design cannot work in experienced subjects. Expectations and physical effects do not add, as the design assumes. Either disparity between expectation and physiology is liable to produce its own cognitive and emotional reaction.

Reference

Porikos KP, Hesser MF, VanItallie TB (1982) Caloric regulation in normal-weight men maintained on a palatable diet of conventional foods. Physiol Behav 29: 293–300

Rolls: In the studies where rats consume carbohydrate solutions, although the diet may be nutritionally balanced, these studies surely suffer from some of the same lack of control as other studies. Not only does the carbohydrate intake go up, but intake of other nutrients goes down.

Rodin: The analysis of exercise might make a fuller distinction between aerobic and anaerobic exercise since different energy stores are depleted, and therefore the type of exercise, not only the duration of exercise, ought to interact with what is consumed.

Würsch: The results you report to have been found by Conners et al. are disturbing. Why do proteins suppress sugar effects? Can we relate the children's behavior to the blood glucose level? It is known that a high protein load can affect blood glucose response [Thornton and Horvath (1985) J Am Diet Assoc 47: 474].

Rolls: A major problem in any chronic study is that of compliance. No adequate way of ensuring compliance apart from constant monitoring on metabolic wards, etc. has been found. Diet records are not accurate and such record keeping alters dietary intake.

Section VI

Implications of Sweetness

Chapter 15

Implications of Sweetness in Upbringing and Education

Matty Chiva

Introduction

According to *The Shorter Oxford English Dictionary*, education means "the systematic instruction, schooling or training given to the young (and, by extension, to adults) in preparation for the work of life". From the point of view of nutritional education, Fidanza Alberti (1975) also mentions and emphasises the learning of "intellectual and moral principles (taking into account) the demands of the individual and of the society." It should be unnecessary to stress the fact that education involves, implicitly and explicitly, notions and methods which are cognitive and rational, as well as ideological, moral and even affective considerations. Many specialists in the science of Education underline the "art" aspect of the educational process (in the same sense as "medical art"), in other words those matters not reducible to a simple experimental process. Furthermore the definition of food, of what is considered edible, "good" for one and for the organism, also involves experience and beliefs. It is obvious in the present state of knowledge that food is not only a matter of its nutritional or sensory qualities, but is also concerned with explicit and implicit cultural rules (Douglas 1966; Lévi-Strauss 1966, 1968). Let us, however, emphasise that this has only relatively recently become obvious because the Humanities have neglected this area for a long time.

Nutrition is at the centre of the processes which allow a human group to define itself, to mark its identity and, in this way, to separate itself from others. At another level, and in the same fashion, nutrition also has an important role for the individual, through the process of identification, integration or marginalisation (Fischler 1985; Rozin 1982). The mechanisms and processes at work are complex and multidimensional. They bring in biological, sensory and metabolic aspects, as well as symbolic dimensions. Cognition and affectivity, perception and representation must be taken into account if we are to understand the links which connect the biological to the cultural, psychological and social.

This reminder is important because one must constantly bear in mind in considering the question of sweetness that there are always two aspects: that of objective facts and that originating from ideologies. Very often the former are perceived, interpreted or distorted by the latter.

Sweetness does not only involve sugar, even though improper assimilation may occur. Thus, for example, in French, the term "sucré" is used to designate the taste of a food, its flavour. However, from an etymological point of view it would be more accurate to use the word "doux", which appeared earlier in the French language ("doux" dates back to the year 1080, "sucré" to 1175); nevertheless "doux" is currently used in the sense of diminished, or weak, and "sucré" now represents one of the fundamental tastes (Chiva 1985). This wrong attribution of a flavour to a food or to a category of foods (carbohydrates) can often be discerned in the minds of many people, whether they be the general public or specialists, and greatly influences attitudes and recommended behaviour.

Sweetness has a specific place on the hedonic level in relation to the other tastes. The pleasantness is now well established. It is an innate and universal quality which is present in our species from birth (and actually in utero) and which is not specific only to man but is common to most animal species (Mistretta and Bradley 1977; Desor et al. 1977; Maller and Desor 1974; Fomon et al. 1983; Geyer and Kare 1983). This is probably how sweet taste comes to be evaluated. Thus Chamberlain (1903) says that the Algonquins use the same word to designate sweetness and the notion of good (in the antonymy good–bad). This same phenomenon also exists in certain African cultures (Deluz 1976, personal communication). In the same way current language uses a good number of terms derived from sweetness to qualify facts, people or things which have nothing to do with sweetness: "sweetheart", "honey", "sugar" or "sweet Jesus" are a few examples; so that sweetness defines, in literature and in mythology, a true ideal of humanity. "At the beginnings of time, animals were men and ate only honey" recounts a Matako myth (Lévi-Strauss 1966). In Asian religions, Paradise is fed by an unending and freely accessible stream of sweet water, whereas in Hell one is eternally deprived of sweetness, which is replaced by bitter food (Brunet et al. 1985). Is not the Promised Land the "land of milk and honey"?

At the same time we must take into consideration the notion of guilt, whether it concerns original sin (the apple is sweet) or gluttony, one of the seven deadly sins. One cannot speak of gluttony without mentioning sweetness. At the present time, in industrialised countries, in addition to the original theological guilt there is the guilt springing from medical, social, or even aesthetic sources which affects the way we eat in general, and more particularly, the way we eat sweet food. Thus to speak of the implications of sweetness in upbringing we must remember the intricacy of different levels (Foley et al. 1979; Todhunter 1979). One must also ask oneself several questions: Whose education? Education in what (education in general or nutritional education specifically)? In what framework and with what goal (education, or reeducation)?

The Child

Sweetness has a role in children's upbringing on at least two levels: (a) as an aid to and/or influence on rearing in general, (b) in the learning of certain social aspects of non-verbal communication.

If sweetness has a part to play at these two levels, as well as in the different nutritional experiences which we will mention later, it is mostly because of the deep and innate tonality of gustative perception, existing before any kind of learned behaviour. Indeed, in the case of this perceptive modality the affective tonality is deeply linked to the sensory function. In other words any perception of a gustatory stimulus is composed of two aspects which are closely linked: the identification of the stimulus and the pleasure or displeasure it causes. These properties are intrinsic to the organism and linked to the structure and function of the CNS, which is common to man and the other animals. "The direct hedonic judgement of man may reflect the same processes that in animals undergird preference or aversion," says Pfaffmann (1978). In other words the emotional and affective quality of taste is present from the beginning. Other meanings may come later and be added.

By considering the precocious function of the gustative system, already present in utero in our species, as well as the innate, strongly positive connotation of sweetness, one can better understand the role it will play later on.

Sweetness and Education in General

Physiology, psychology and observation of social behaviour all help to place the sweet tooth in a wider educational context. Very early on, the family and institutional environments will play on the child's attraction for anything which is sweet.

On the one hand it is gratifying for the outside world to be aware of the child's specific acceptance of the foods which are specially prepared for him. At the same time this acceptance of a biological origin fits into a social context which will soon use it. Sweet foods rapidly become a means of reward, and consequently a means of control over some of the child's behaviour. Thus is created a real currency, a "sugar tender" (Fischler 1979), which is widely used in the educational system of our society.

"Sugar tender" is used at first by the parents to influence the child's behaviour, which is controlled by the giving or the taking away of sweet food. The extraordinary variety of sweets, biscuits, etc. on the market, primarily intended for children, most be considered to be proof of the extent of the phenomenon. At the same time it is not only the parents who use this attraction to sweetness but also educational institutions. Thus the giving of sweet rewards or tokens is often used in kindergartens and primary schools and even beyond, something which is in complete contradiction to another kind of education, that of good nutritional habit.

The strength of this "sugar tender" is such that the educator may eventually lose the educational control he has on the child. If the child were able to have access to it directly, he could escape the control. Even worse, he may lose the child if he accepted the "tender" from unknown people. Thus the stranger, playing on a liking for sweetness, could even ravish the child, in the double sense of the term: to kidnap and to enrapture. On the educational level the consequence of this is ambivalence towards sweetness: the attraction is good if the adult uses it with discrimination and keeps it under control; it is bad, and dangerous, when it is used by a stranger out of the sight of the person responsible for the child.

Sweetness and the Social Behaviour of Non-verbal Communication

Facial expression, and more precisely the facial expression of emotion, plays a major role in the non-verbal communication of children and adults. It is experience which is part of the process of socialisation. In this regard the behaviour provoked by the sensation of taste plays an important part. Gustatory stimulation is the basis of a specific phenomenon: the gustofacial reflex (GFR), a facial response and a behavioural pattern which is inborn, reflex and stimulus-dependent. Its control is by means of the lower centres of the CNS. This facial expression is different from one taste to another, and, for the same taste, identical from one child to another (Steiner 1973; Chiva 1979, 1985). The GFR is present from birth and, as such, is involuntary; but as soon as it appears the community interprets these expressions and responds according to a common interpretation. In these conditions and in this context the expression provoked by a sweet stimulus is always interpreted as "he likes it", "it is pleasurable".

Study of the development of the GFR has shown that quite early on, at about 16 month, the child clearly integrates these expressions in a process of communication with others (Chiva 1982, 1985). The expressions detach themselves more and more from their original, strictly sensory base and acquire other meanings which are socially accepted by the reference group. We can now ask if there is a systematic link between the interpretation by the community of the GFR, the patterns of these expressions and the pattern of the expressions which will later be useful for communication.

In order to answer this question a three-step investigation was performed (Chiva 1983). First a technique for description and measurement of facial expressions was devised. It consisted of breaking down the expressions into 52 basic components. Second, this technique was applied to both GFR and evoked facial expression in one particular population; by "evoked" we mean that the subjects were asked to display expressions for "good", "happy", "nice", "nasty", etc. Finally the data were submitted to factor analysis (correspondence analysis) (Figs. 15.1, 15.2). Results showed a strong relationship between the patterns of basic components in sweetness-provoked GRF and the patterns in those expressions supposed to mean "happy", "nice", "good". Features common to sweetness-related GFRs and to "deliberate", socially meaningful expressions included mouth components (smile), the general aspect of the face (relaxed) and a certain squinting of the eyes.

Another study was conducted on the receiver's side, at the time of perception and interpretation of the expressions (Chiva, unpublished work). Results show that the child actually goes through a process of learning how to decode the expressions. With age this decoding becomes more and more consistent with the norms of the group. In this context the GFR expressions provoked by a sweet taste, when shown to a subject unaware of the origin of the stimulus, were interpreted as "satisfaction", "happiness". The phenomenon became very significantly more constant with age.

This, along with the positive hedonic connotation of sweetness, suggests that a relationship exists between what is felt and what is expressed or what is perceived in a social context. The proper learning of facial expressions, aiding communication and therefore integration into the group, seems to be made much easier by subjective sensory experience. Consequently sweetness appears to play an important role, one which is even more interesting in that it allows one to investigate certain aspects of the universality of expressions as well as the problem of the relations of the biological to the psychological and the social.

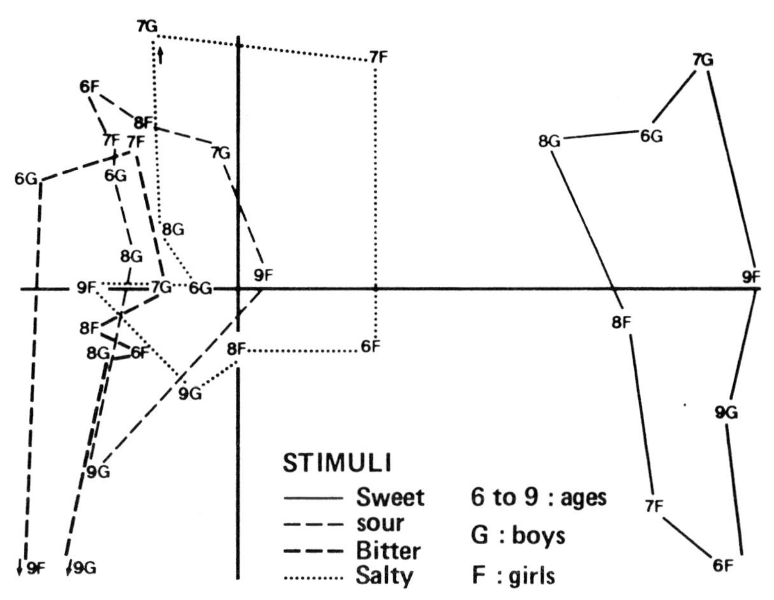

Fig. 15.1 Structure of patterns of facial expressions (GFR) in a population of children 6–9 years of age ($n = 120$).

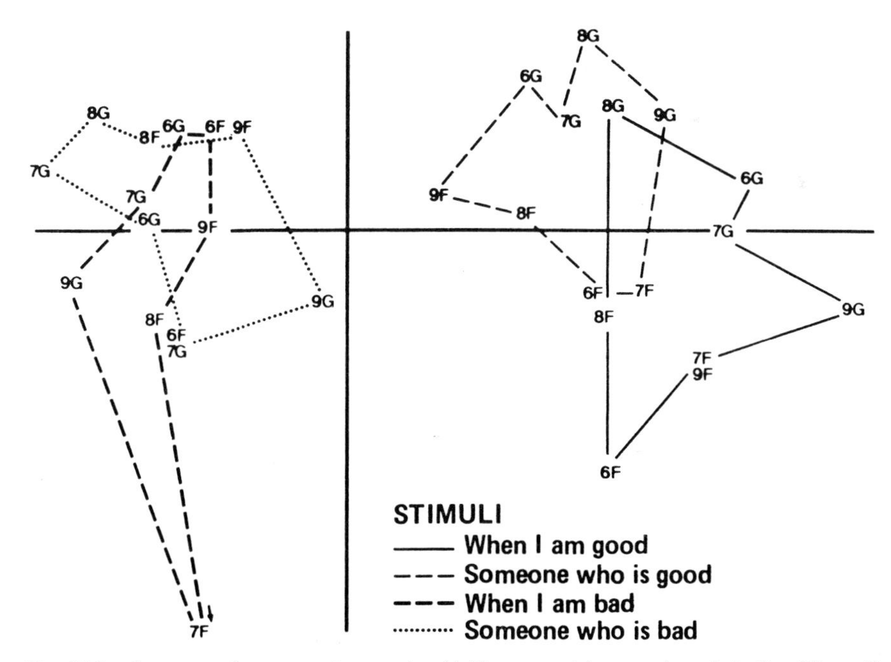

Fig. 15.2. Structure of patterns of expression (deliberate social expressions: "nice" and "nasty").

Nutritional Education

The establishment of nutrition education programmes is due to an increase in medical, biological and nutritional knowledge. It is now important in industrialised countries where, on the one hand, problems of food shortages have been overcome, and on the other hand illnesses due to overabundance have appeared. The goal of these programmes is, in general, to improve dietary habit (Caron-Lahaie 1984) or, better still, to see that good habits are ensured from birth. Two remarks are necessary here:

1. The different programmes must be developed in line with our present state of knowledge, and this is increasing quite rapidly. Nevertheless educational practice is often out of step with knowledge. This is due as much to the intrinsic difficulty of the task (the modification of behaviour which has a multifactorial aetiology) as to a resistance to change. In this way out-of-date information in this area is often to be found in our society.
2. Sweetness is often attributed to carbohydrates, and most particularly to sugar. Because of this attitudes to sweetness are influenced by those to carbohydrates. They range from avoidance to the setting up of real anti-sugar crusades.

What is most remarkable is to see the convergence of scientific attitudes with beliefs of many other origins, without any consideration for the current state of research. Thus most present-day work is in agreement that there is a lack of objective proof of the responsibility of sugars in many pathologies (Debry 1985). This does not affect the public's continuing accusation, and educational attitudes to sweetness are often influenced by these beliefs.

Sweetness and the Nutritional Education of Children

In the first months, even years of life, the child has little autonomy. As a result he becomes dependent on what is offered to him. He depends directly on parental attitudes to sweetness.

Since sugar is considered to be the source of later problems, a whole educational school of thought advises its suppression from the beginning, so that the child "will not get used to it". This is the opposite of many traditional behaviours in which reaction to sweetness is considered important in early life. In many societies, including the West, one begins by giving sweet food to the child (Jerome 1977), and in certain Asian cultures sweetness is linked to life itself: the child's appetite for sweetness, tested in the first 24 hours, measures its taste for living (Brunet et al. 1985). One might ask whether or not innate attraction to sweetness at the beginning of life is reinforced by the amount of sweet food eaten. Little research has been done in this area (see also Chap. 9). Beauchamp and Moran (1982, 1984) noticed a substantial modification in the preference and the absorption of sweet water in relation to early experience. Children who drink sweet water in early life prefer and drink more sweet water later on than those who have not had that experience. However, the authors emphasise that this early practice does not influence the preference for sweetness and its consumption in other contexts (Kare and Beauchamp 1985).

Preference for sweetness in the long run probably depends as much, if not more, on the general context and on the cultural apprenticeship as on early experience. In other words it is experience that decides not only what is a food, but also what is sweet and what is not (Young 1977) (see Chaps. 6, 7). Kaufmann et al. (1982) noted, for example, the significant differences of consumption of glucids in a population of adolescents and adults in Jerusalem, in relation to their culture: communities originally from North Africa ate more sweet foods than those who came from Europe. One might therefore deduce that early deprivation of sweet stimuli, imposed with a prophylactic and alimentary educational goal, is not very efficient and might even be useless.

We must now consider the role of taste in general, and of sweetness in particular, at another time of life: during weaning. As the child becomes more independent one might wonder whether taste may not play, in part, a controlling role when the child is faced with new foods. The available facts are quite contradictory. On the one hand Geyer and Kare (1983), considering this problem in animals, point out that even when an initial preference for sweetness exists it does not necessarily determine later food selection. Their review of the question mentions the findings of Galef (1977): adult models and the experience with foods plays a major role in young animals' learning of food preferences and rejections. Furthermore, studies of the factors which influence acceptance of a new food (in slightly older children) show that sweetness is, with familiarity, a major element in acceptability (Birch 1979a,b). However, in this case, as far as kindergarten children are concerned, one may assume an interiorisation of the model shown by adults and an integration of the educational effects of the "sugar tender". Traditional behaviour in several societies at weaning exhibits a clean break with former alimentary modes: salty or bitter flavours are ritually presented to mark the change to another status. Proverbs testify clearly to this type of custom (Loux 1978, 1979.) However, this does not mean that sweet foods are completely taken out of the child's diet.

Thoughts about certain practices whose goal is to eliminate sweet foods for prophylactic reasons converge on those concerning the beginnings of alimentary habits. The role of sensory stimuli, during weaning and afterwards, seems to have been overestimated; or, more precisely, it has been considered in too simplistic a way (Kare and Beauchamp 1985). Research facts are very rare; it would therefore be unwarranted in the state of present knowledge to take up set positions on this subject, because they would be based on convictions and not on facts.

Later, programmes of food education are directly aimed at children. Most of them first take into account the available nutritional facts. Emphasis is put on health, the nature of the foods and the balance of food intake. Taste is considered only secondarily. Sweetness is often discouraged (to diminish the intake of carbohydrates?). These programmes scarcely consider the differences in taste which come about with age. These modifications are set up as much for physiological (puberty, for example) as for psycho-social (identification models) reasons. A problem arises here: how to judge the effectiveness of these programmes? They are most often offered in a cognitive form, even if they are presented as games. Afterwards they are monitored cognitively as well, with the help of tests. However, better rational knowledge does not guarantee real behavioural modification (Caron-Lahaie 1984). The models suggested by the community, the cultural norms as well as personal motivations, play a much more important role.

Sweetness and the Nutritional Education of Adults

Most of the remarks about difficulties in children's education are also applicable to adults, with, in addition, in industrialised countries, the existence of pressure from two sources: health and aesthetic norms. Many diets and food fads combine these two preoccupations, to which are often added philosophical and religious considerations.

Without listing the details of the particular diets, it is interesting to note that mostly they can dissociate sweetness from antagonism for carbohydrates. Thus one of the most radical in this perspective, the diet of Dr. Atkins (1977), which bans carbohydrates completely, uses other sweeteners, perhaps in this way allowing the amount of pleasure necessary for the observance of the dietary advice. Let us also note that the dissociation is often justified by the fact that sweetness may have different connotations according to the food which is used: honey or fruits are "natural" and therefore good, whereas refined sugar is a source of "empty calories" and detrimental. There are good and bad sweetnesses in these alimentary guides.

Going further, it may be interesting to mention the use of sweetness during educational projects for the elderly. The problems of elderly people, especially in institutions, are well-known: disinterest, reduction of social activity, and nutritional deficiency from lack of appetite. Playing on the pleasure that sweetness continues to bring about, some people are starting to use it as an element of stimulation in the institutional environment (Baldoni 1983). Sweetness or "cake day" becomes a true periodical stimulus; the communal eating of the cake is preceded by the putting into action of the project and its preparation (means of social stimulation of this particular population).

Lastly, we must specially consider sweetness in the context of health. It is not a question of slimming diets or of nutrition education in general (see above), but of particular circumstances involving taste modifications which deserve to be communicated to practitioners. Three cases may be cited as examples:

1. Sweetness and depressive states. Modifications of food behaviour, loss of appetite and decrease of intake are well known in these states (Hopkinson 1981). Steiner and Rosenthal-Zifroni (1969) have demonstrated the changes in taste perception which occur in the depressed subject. At the beginning of hospitalisation all the thresholds are increased. Clinical improvement comes with a return to normal, the sweet threshold being the first one to regain its place. Brunet et al. (1985) mention the correlation between the good therapeutic effects of antidepressant medicine, its orexigenic characteristics and the search for sweet foods. The return of appetite for sweet foods would be a therapeutic sign of improvement here.

2. Sweetness and cancer. The perceptive and food behaviour modifications of people who suffer from cancer in its different forms are frequently noted. They are due as much to the illness as to certain consequences of the treatment. A decrease of sweet sensitivity is especially noticed (Carson and Gormican 1977; Vickers et al. 1981). This results in selective modifications in food intake, modifications which are due as much to morbid effects as to individual preferences (Trant et al. 1982). It appears, however, that sweetness is better accepted when its origin is fruit, than in products such as chocolate or sweets. The implications for the design of adequate diets is obvious, as well as during the post-cure, re-education period.

3. Several ailments and treatments can hit taste perception and, more particularly, the perception of the intensity and enjoyment of sweetness. This is frequent, for example, with zinc deficiency, and in certain diseases (such as Crohn's disease: Järnerot et al. 1983); this is also the case with many neuroleptic and psychotropic drugs. Taking this change into account produces not only greater comfort for the patient, but also, in many cases, a greater compliance with the treatment.

Conclusion

Sweetness undoubtedly has a place and plays a role in educational attitudes and behaviour. However, its real and/or supposed influence does not only come from innate appetite inscribed in the organism. We also know how great a part is played by the social and cultural context in the shaping of behaviour. At the same time perception of the role of sweetness is equally strongly attributable to the existence of principles, ideologies and guilt. Thus its use in education is much more a function of these latter aspects than of objective facts.

The wrong view that sweetness means sugar leads to attitudes which are absurd, for example the temptation to take away sweetness completely from nutrition. This is doomed to failure from the beginning, because it does not take into account the importance of the hedonic connotation, or of the symbolic aspects of the problem.

The best example to illustrate this is Doi's analysis (1982) of the formation of the personality and the social game in Japan. In the Japanese social world the structure of the person and of the interactions are defined as concentric circles. They go from the outside circle (*tanin*), which is cold and formal, to the social circle (*giri*) where everything is debt, obligation and scrupulously calculated reciprocity, and lastly, towards the interior circle (*amae*), an intimate universe of affection and indulgence. *Amae* is the gratifying world of the mother–child relationship, an interior world of permanent affective investment and of tolerance. The whole social game is to try to pass the partner from the *giri* to the *amae*, that intimate circle of well-being which is the deep aspiration and the fundamental motivation. However, it will be appreciated that *amae* is a word meaning this interior circle as well as... sweetness.

References

Atkins RC (1977) Ma cuisine diététique. Buchet/Chastel, Paris

Baldoni E (1983) Il doce nella dieta di une communità di anziani. In: I carboidrati nell'alimentazione umana: gli alimenti dolci. Perugia Edizioni Guerra, pp 223–242

Beauchamp GK, Moran M (1982) Dietary experience and sweet taste preference in human infants. Appetite 3: 139–52

Beauchamp GK, Moran M (1984) Acceptance of sweet and salty taste in 2 year old children. Appetite 5: 291–305

Birch LL (1979a) Dimensions of preschool children's food preferences. J Nutr Educ 11: 91–95

Birch LL (1979b) Preschool children's preferences and consumption patterns. J Nutr Educ 11: 189–192

Brunet M, Kamal S, Grivois H, Vitre A (1985) Le sucré et le sacré. Psychol Med 17: 2073–2078
Caron-Lahaie C (1984) Influence de l'éducation en nutrition sur le comportement alimentaire. Cah Nutr Diet XIX 229–232
Carson JA, Gormican A (1977) Taste acuity and food attitudes of selected patients with cancer. J Am Diet Assoc 70: 361–365
Chamberlain AF (1903) Primitive taste words. Am J Psychol 14: 146–153
Chiva M (1979) Comment la personne se construit en mangeant. Communications 31: 107–118
Chiva M (1982) Taste, facial expression and mother-infant interaction in early development. Baroda J Nutr 9: 99–102
Chiva M (1983) Psicologia del gusto dolce. In: I carboidatri nell'alimentazione umana: gli alimewnti dolci. Perugia Edizioni Guerra, pp 263–274
Chiva M (1985) Le doux et l'amer. P U F , Paris
Debry G (1985) Part des glucides dans l'équilibre alimentaire; évaluation de leur rôle dans certaines pathologies. Communications Economiques et Sociales, Paris
Desor JA, Maller O, Greene LS (1977) Preference for sweet in humans: infants, children and adults. In: Weiffenbach JM (ed) Taste and development. DHEW, Bethesda, 161–172
Doi T (1982) Le jeu de l'indulgence. Le Sycomore, Paris
Douglas M (1966) Purity and danger: an analysis of concepts of pollution and taboo. Routledge & Kegan Paul, London
Fidanza Alberti A (1975) Techniche per l'educazione alimentare con particolare riferimento per le scuole materne e dell'obbligo. Riv Sci Tecn Alim Nutr Um 5–6: 277–280
Fischler C (1979) Les pièges de la douceur. Le Monde de l'Education 47: 11–15
Fischler C (1985) Alimentation, cuisine et identité: l'identification des aliments et identité du mangeur. Recherches et travaux de l'Institut d'Ethnologie, Neuchâtel 6: 171–192
Fischler C, Chiva M (1985) Food likes, dislikes and their correlates in a sample of French children and youth. EURO-NUT Workshop on Measurement and Determinants of Food Habits and Food Preferences, Wageningen, pp 137–156
Foley D, Herzler AA, Anderson, HL (1979) Attitudes and food habits – a review. J Am Diet Assoc 75: 13–18
Fomon SJ, Ziegler EE, Nelson SE, Edwards BB (1983) Sweetness of diet and food consumption by infants. Proc Soc Exp Biol Med 173: 190–193
Galef BJ (1977) Mechanisms for the transmission of acquired patterns of feeding from adults to weanling rats. In: Weiffenbach JM (ed) Taste and development. DHEW, Bethesda, pp 217–231
Geyer LA, Kare MR (1983) Taste at weaning: strategies facilitating appropriate food selection by non-primate mammals. In World Rev Nutr Diet 41: 232–254
Hopkinson G (1981) A neurochemical theory of appetite and weight changes in depressive states. Acta Psychiatr Scand 64: 217–225
Järnerot G, Järnmark I, Nilson K (1983) Consumption of refined sugar by patients with Crohn's disease, ulcerative colitis or irritable bowel syndrome. Scand J Gastroenterol 18: 999–1002
Jerome NW (1977) Taste experience and the development of dietary preference for sweet in humans: ethnic and cultural variations in early taste experience. In: Weiffenbach JM (ed) Taste and development. DHEW, Bethesda, pp 235–245
Kare MR, Beauchamp GK (1985) The role of taste in the infant diet. Am J Clin Nutr 41: 418–422
Kaufmann NA, Friedlander Y, Halfon ST et al. (1982) Nutrient intake in Jerusalem: consumption in adults. Isr J Med Sci 18: 1183–1197
Krondl MM, Lau D (1978) Food habits and modification as a public health measure. Can J Public Health 69: 39–48
Lévi-Strauss C (1966) Du miel aux cendres. Plon, Paris
Lévi-Strauss C (1968) L'origine des manières de table. Plon, Paris
Loux F (1978) Le jeune enfant et son corps dans la medicine traditionnelle. Flammarion, Paris
Loux F (1979) Le corps dans la société traditionnelle. Berger Levrault, Paris
Maller O, Desor J (1974) Effects of taste on ingestion by human newborns. In: Bosma JF (ed) Fourth symposium on oral sensation and perception. US Government Print Office, Washington DC
Mavrikakis S, Lahaie L (1983) Les motifs des choix alimentaires. Communication ACFAS
Mistretta CM, Bradley RM (1977) Taste in utero: theoretical considerations. In: Weiffenbach JM (ed) Taste and development. DHEW, Bethesda, pp 51–63
Pfaffmann C (1978) The vertebrate phylogeny, neural code and integrative processes of taste. In: Carterette EC, Friedman MP (eds) Handbook of perception, vol VIA. Academic Press, New York, pp 51–124
Rozin P (1982) Human food selection: the interaction of biology, culture and individual experience. In: Barker LM (ed) The psychobiology of human food selection. AVI Publishing, Westport, Conn., pp 225–254

Steiner J (1973) The gustofacial response: observation on normal and anencephalic newborn infants. In: Bosma JF (ed) Fourth symposium on oral sensation and perception. US Government Printing Office, Washington DC, pp 254–278

Steiner J, Rosenthal-Zifroni A (1969) Taste perception in depressive illness. Isr Ann Psychiatry 7: 223–232

Todhunter EN (1979) Food habits, food faddism and nutrition. In: Nutrition and the world food problem. Karger, Basel, pp 267–294

Trant AS, Serin J, Douglass HO (1982) Is taste related to anorexia in cancer patients? Am J Clin Nutr 36: 45–48

Vickers ZM, Nielsen SS, Theologides A (1981) Food preferences of patients with cancer. J Am Diet Assoc 79: 441–445

Young PT (1977) The role of hedonic processes in the development of sweet taste preferences. In: Weiffenbach JM (ed) Taste and development. DHEW, Bethesda, pp 399–417

Commentary

Scott, Chiva: We note a clear correspondence between the dimensions that result from the application of multidimensional scaling techniques to electrophysiological data from the rat and to behavioral responses of human children. In Fig. 2.3b of their manuscript, Scott and Giza present a dimension along which a range of stimuli are ordered according to the similarity of the electrophysiological response evoked from a population of neurons in the rat's hindbrain. This is presented as a toxicity–nutrition dimension which is associated with physiological well-being, acceptability and hedonic value. In Fig. 15.1 Chiva presents a dimension along which stimuli are ordered according to the similarity of the gustofacial responses they evoke in children. A comparison between these dimensions, each transformed to the same scale, appears below. This serves as a demonstration that, while the taste systems of rats and monkeys may respond differently to physiological manipulations (as described in the text), there are, nevertheless, clear similarities between the sensory coding of taste stimuli in rats and the behavioural reactions of humans to those same tastants. It is likely that these reactions are reflexive, and mediate by neural proceses which are complete in the hindbrain.

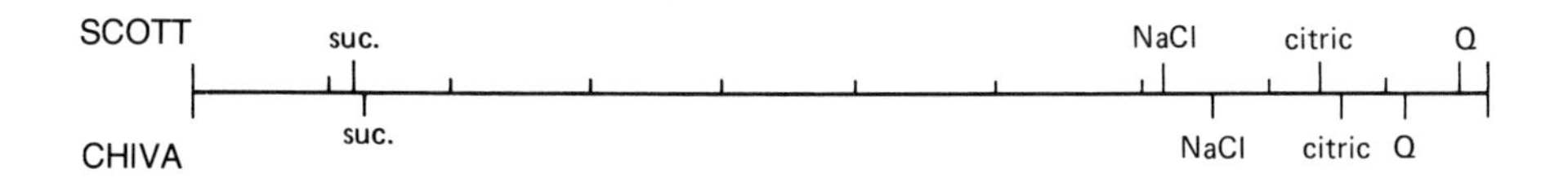

Booth: Do you see any distinction between the emotionality evoked by sweetness (pleasure) and its behaviour-directing effect (preference or pleasantness)? Could not the latter be at least partly a sensorimotor reflex, depending only on the brainstem (even in human beings), whereas the former is likely to require forebrain projections in order to be experienced subjectively, and perhaps the recruitment of the full range of diffuse autonomic and skeletal excitation?

Chiva: We have no data at present to enable us to answer this question.

Bartoshuk: I suggest that we not take the cancer data too seriously. If one scans all of the studies done there is no systematic evidence for taste loss. True, we can expect loss after radiation therapy and possibly after chemotherapy, but not from cancer per se. Similarly, although the literature abounds with anecdotal accounts of taste loss with various diseases and medications, very few of the reports are convincing to me. We must await carefully controlled studies to conclude that taste is really as vulnerable as the anecdotal literature suggests.

Rozin: A fundamental question touched on in this paper is whether early experience is of special importance in development or, more specifically, the development of preferences. There is much data in psychology that argues against special importance of early experience.

Chapter 16

Sweetness in Marketing

Howard G. Schutz and Debra S. Judge

Introduction

In general, "marketing" of any product addresses problems of product acceptance, pricing, distribution, and advertising (Nelson 1982). Sweetness may be associated with pricing to the extent that various levels and types of sweetener affect production costs and acceptability. Sweetness can have a role in product distribution in cases where potential, geographically distinct markets showed large differences in preferred levels of product sweetness. However, the most significant incorporation of product sweetness into marketing can be broken down into (a) the use of sweetness in increasing the market of a given product without specific reference to sweetness itself (i.e., sweetness as a tool to increase product acceptance), and (b) explicit marketing of the sweetness of the product (e.g., use of sweetness of the product as an advertising element). The first is more an element of product development but some methodological similarities with marketing are noteworthy. Naturally, for a specific product both of these elements could be involved to a greater or lesser degree.

Consumer Attitudes

Our topic concerns the role of sweetness in general in the marketing of products; however, for food products the identity of the specific sweetening substance is usually inextricably bound to the marketing strategy. Historically, sweetness indicated ripeness and sweet foods were relatively rare. Being scarce and pleasant, "sweet" was desirable and sweeteners were prized. Words such as "sweet," "sugar," and "honey" were assimilated into language, particularly American, as terms of endearment. In contemporary, calorie conscious, western nations, "sweet" is no longer rare and is now usually synonymous with sugar and sugar's associated negative concepts of "fattening," "high calorie," "non-nutritious," and "unhealthy." The original desirability of sweetness in food remains in our language in reference to non-food items but its desirability in food is qualified. In fact, sugar and its use appear to be, at some level, synonymous with sin and self-indulgence. At a conference on Culture and Communication (1981), Mechling and Mechling presented a discussion of sugar as a "polluter" of the body and symbolic of the pollution of the society with disorder; they noted that anti-sugar literature uses terms deriving from both the scientific world and theological ideas about "dirt." Historically, such attitudes are fairly recent. During the time of Henry VI sugar was used primarily as a medicine, and later it was sold for marzipan and sweetmeats. Throughout the sixteenth century sugar was scarce and a luxury item in Western Europe as well as in Britain (Lowenberg et al. 1979).

Because new negative attitudes of consumers regarding sugar are enmeshed in deeper belief systems about the proper ways of living in general, the marketing of sweet products must tread a fine line of attracting consumers to a product they will accept without invoking the associated cultural taboos. A 1980 survey of 2451 consumers in seven western U.S. states indicated that 87.8% of respondents favored a decrease in the amount of sugar in breakfast cereals and 70.8% agreed or strongly agreed that "sugar is responsible for many diseases in wealthy countries" (USDA Regional Research Project W-153, unpublished data). In a 1985 survey commissioned by Better Homes and Gardens, 82% of the respondents claimed to be more calorie/weight conscious than they had been 1 year previously (Anonymous 1985) and 52.2% said that the last time they read the nutritional information on a food package they were looking at sugar content (this was posed as a multiple response question and sugar was surpassed only by 72.2% looking at calories). Recent dietary guidelines advising substantial reduction in consumption of refined and processed sugars in the western industrialized nations (e.g., USDA 1980) have increased consumer awareness of sugar in products and the public has linked sugar to "sweetness."

Sweetness in Contemporary Marketing

In order to obtain contemporary information about "sweetness" in the proprietary world, we queried marketing research directors of approximately one dozen food companies (by letter) about the role of sweetness in their marketing schemes. Our selection represented companies that market a wide variety of food and beverage products. The few corporations that responded did not seem to incorporate sweetness to any significant degree; Weight Watchers Products (a subsidiary of Heinz) considers "Sweetness is one component of taste which is one of many attributes . . ." and determines "appropriate sweetness levels based on consumer preferences..." assessed "through quantitative consumer research." "Sweetness is not emphasized in advertising and we do not evaluate sweetness on marketing goals." G.D. Searle sent us a package of trade and retail publicity materials used in the promotion of Nutrasweet. A representative of Beatrice Corp. informed us that "only rarely" does Beatrice use consumer testing of products and this is generally acceptability testing. Corporate philosophy is that consumers cannot accurately assess sweetness per se; consumers can rate acceptability but their reasons for acceptability cannot be validated. Beatrice has no company philosophy regarding the marketing of sweetness in its food products. The lack of response from these companies could be interpreted as just a lack of concern, but also may be related to their sensitivity toward what is perceived as a "touchy" consumer issue.

Not surprisingly, sweetness is used in food and beverage marketing schemes in very selected ways when used at all. Food products are often advertised as containing less "sweeteners" of various forms but are never advertised as "not sweet." Specifically, "sweet" is used primarily in situations where it can be

1. Separated from the concept of high calorie
2. Distinguished as "natural" versus added sweetener, or
3. Presented as "reduced sweetening" or "lightly sweetened"

Sweetness is typically a part of the advertising mix when it is obtained through the use of high intensity, low calorie sweeteners rather than sugar. For example, Post Raisin Bran cereal boxes incorporate "Natural" into the name and carry a banner across the front of the box declaring "Unsugared Raisins." General Foods renamed their Super Sugar Crisp breakfast cereal to Super Golden Crisp without any concomitant product change because "People thought it had more sugar than it did — but it doesn't have more than the competition" (Hall 1985b). Some breakfast cereal manufacturers have attempted to retain the best of the taste and reduced sugar worlds by promoting their cereal as "lightly sweetened." Presweetened breakfast cereals have come under consumer and media criticism for the amounts of sugar incorporated into them; however, they, contribute only approximately 3% of the American dietary sugar (Franta 1985). Criticism that the amount of sugar is not easily discernible to the public and encourages the consumer to add more sugar to presweetened cereals at the table may have resulted in the addition of the fine-line printed words "Presweetened Cereal" on the front of the package of many cereal brands. An example of sweetness being used as a positive enticement (distinguishing the "natural" sweetener) is when the sweetness is inherent to a relatively unprocessed product; for example, oranges, orange juice, grapefruit, grapes, and raisins have been advertised as "sweet" or "naturally sweet." This separation consciously reverses the universal linking of sweetness to the "sins of (added) sugar." Prior to the development of aspartame, artificially sweetened products were sometimes marketed emphasizing traits other than sweetness or low calories. For example, Fresca, an artificially sweetened carbonated beverage, was marketed without emphatic reference to its low calorie nature but on the basis of its taste. Similarly, the vast majority of artificially sweetened chewing gums are marketed on the basis of flavor and non-cariogenicity.

Ingredients that connote "high calorie" to the consumer may be minimized without stating that substitute ingredients used for sweetening are also calorie dense (e.g., substitution of corn syrups for sucrose). The substantially lower cost of corn syrup products than cane and beet sugar also encourage such changes in ingredient formulations.

The discovery and use of non-nutritive sweeteners such as saccharin, cyclamates, and acesulfame K, and high intensity, nutritive sweeteners such as aspartame allowed development of food products that appealed to the "sweet tooth" without introducing large numbers of calories. Bitter aftertastes and concerns regarding safety combined to limit the use and applicability of saccharin and cyclamate. Cyclamate is no longer approved in the United States although it is used in approximately 40 other countries, saccharin has been under several consecutive moratoriums of FDA bans in the United States, and acesulfame K is approved for use in the United Kingdom, Ireland, Germany, and Belgium (Institute of Food Technology 1986). The introduction and advertising of Nutrasweet (aspartame) by G.D. Searle Co. has initiated an upswing in the use of "sweet" concepts in marketing; however, such marketing has invariably included a low calorie, non-fattening, non-sugar disclaimer. Aspartame is approximately 180–200 times as sweet as sugar, is metabolized as a regular protein, and is considered a nutritive, high intensity sweetener. Heavy public and trade marketing of Nutrasweet by G.D. Searle included development of a logo to be used on products using only Nutrasweet for sweetening and a mass mailing of millions of Nutrasweet-sweetened gumballs to consumers throughout the United States. Television advertising of Nutrasweet by G.D. Searle aims at increasing public recognition and acceptance of the sweetener and subsequently of products using Nutrasweet as the sweetening agent. Presumably, with time, use of the logo will render unnecessary the negative concept disclaimers. The result has been the inclusion of the high intensity sweetener as an important part of the marketing mix of Nutrasweet-containing products. One indication of the success of a sweetener that can be divorced from calories is the heavy campaign to discredit aspartame (including claims of health hazards and deceptive advertising) that has been mounted by The Sugar Association (Kovach 1984). Currently, aspartame is approved for use in more than 40 different countries and G.D. Searle has initiated a large-scale publicity and marketing campaign in the United Kingdom. Equal, G.D. Searle's Nutrasweet entry into the low calorie table sweetener market (Egal in France, Canderal in the United Kingdom and Europe) is sachet packaged with a carbohydrate filler and advertised as equal in sweetness to the same amount of sugar but without the calories.

A New York marketing researcher was recently quoted in the *Wall Street Journal* (3 July 1985) as saying "sweetness is seen as something that kids like. It's unsophisticated". Where sweetness is a major aspect of the product, terms such as "taste" are often substituted for "sweet." Sweetness is generally considered a characteristic appealing to children and where the term sweet is not used there is sometimes a combination of the term "taste" with use of children in advertising. For example, Kelloggs' Frosted Mini Wheats are advertised showing the uncoated side as appealing to the adult and the obviously sugar-coated (though this is not stated) side appealing to the "child inside." The same *Wall Street Journal* article notes that although Americans profess to avoid sweet drinks they are consuming increasing per capita amounts of such beverages. This dichotomy has resulted in marketing representatives who are careful *not* to promote the sweet nature of their products (e.g., Jim Beam Marketing Director, G. Lane Barnett, is quoted as saying of their new Bourbon Mixers "I wouldn't describe the taste as sweet, it's an adult taste"). Other euphemisitic terms and concepts indicate richness and self-indulgence or pampering of oneself rather than specific reference to sweetness.

Public calorie/weight consciousness has effected some changes in the food service industry as well as at the retail level. Several hotel chains have developed low calorie menus for their restaurants, using portion control, low fat entrées, items sweetened with low calorie artificial sweeteners (such as diet soft drinks), and sugar substitute table top sweeteners (Calorie Control Commentary 1986).

The discrepancy between what the public professes to want and like and actual consumption trends results in some difficulties for marketing. Marketing must determine the actual degree of sweetness required to satisfy the consumer, communicate this to the product development department of the company, and subsequently determine ways to attract the consumer to the product and retain the consumer's loyalty. To attract the consumer, the concepts of a pleasurable experience, and from consumption data and taste test panels this often means sweet, must be conveyed without using terms or concepts that go against what the consumer professes to avoid (e.g. "sweet," "high calorie," etc.). That sweetness and sweetening agents are just two of many components important to successful marketing is illustrated by the introduction of a "reduced sugar," "more fruit" jam in Norway. The jam was less sweet, faded in color more rapidly due to loss of certain properties of the sugar, and was more highly priced than its higher sugar counterpart. The introduction failed and production of the product was halted (A. M. Hjelmstad, Consumer Council of Norway, 1986, personal communication) but it is not clear which of the changes effected by the reduced sugar was most responsible for consumer rejection.

Methodologies in Marketing

Marketing managers utilize marketing research, either within the company or from outside research corporations, to determine the consumers' cognitive and affective responses to "sweetness." These techniques are also utilized at different stages in the product development process. Concept tests in which sweetness may be one component of the product concept are used in the idea exploration and screening phase. Word and/or picture descriptions rather than actual products are usually involved. The consumer may be asked to rank order, use category ratings or a variety of proprietary scales that purport to predict actual purchase potential. Occasionally, protocepts (physical products embodying basic components of a product concept) are used in consumer tests to evaluate an idea in a more concrete fashion. However, such protocepts are more typically part of focus group studies in which small groups of consumers (eight to ten individuals) are used to discuss ideas and/or products in an attempt to ascertain desirable products and their attributes. After an idea or concept is considered worthy of development, a variety of techniques can be used to determine more accurately the preferred levels and importance of product attributes. This type of research is in addition to the typical Research and Development product research and differs by providing information based on consumer responses rather than laboratory taste panels and analytical data.

There exist a multitude of relatively simple research methodologies for determining preferred levels of sweetness both with and without referencing cognitive concepts using consumers in either central location or home-use tests. In addition to hedonic judgments, simple checklist responses in taste tests of formulation variations ("not sweet enough," "too sweet," "just sweet enough") request the respondent to assess their preference for levels of sweetness. Semantic differential scales develop a more finely tuned picture of consumer response to a product by using bipolar, adjective anchored scales to assess various attributes, including sweetness.

Acceptability and preference tests ordinarily rate consumer acceptability of formula variations without specifying the attributes being tested. These two methods do not necessarily yield the same results since the former may tap into social acceptability concepts whereas the latter merely indexes preference without accessing cognitive processes. Differences in results of the two methods may recommend different marketing strategies to food product marketers. A recent and widely reported marketing effort by Coca-Cola to increase their market share against the inroads of Pepsi Cola vividly illustrates the difference between cognitive and sensory-based preference. Coca-Cola developed a modified formula which had as one characteristic a sweeter flavor than its traditional product. In thousands of blind taste tests with consumers the new formula was more popular than the old formula. However, when the new product was marketed as new Coca-Cola it resulted in a strong consumer revolt with the result that the company decided to bring back the old formula under the name "Coca-Cola Classic" and to continue the marketing of the new product. It is reported that the old product is now outselling the new one in spite of heavy advertising expenditure on the new formula. We can speculate that the cognitive characteristics of brand name associated with the changing of a deeply entrenched product image overrode any true sensory preference for the new sweeter formula. Perhaps if the new formulation had been considered a product improvement, rather than a new product, and had been marketed without reference to its sweeter taste it would have yielded the desired increase in market share.

At a more sophisticated level, use of multi-dimensional scaling techniques allows a company to determine the relative value of actual sweetness and advertising claims of sweetness in product acceptance and market success. Moskowitz (1985) has outlined the procedures for consumer testing of formulations of, and marketing concepts for, products that included rating and scaling systematic variations of product formulation, potential names, and advertising concepts. Use of multidimensional scaling techniques with consumer panels may facilitate advertising schemes that attract consumer interest and respond to consumer concerns. Product optimization techniques involving both Research and Development and Marketing departments use trained sensory evaluation panels and analytical data which can be combined with simple consumer acceptance tests at the consumer level to determine absolute and relative acceptance, the relative importance of product attributes, and the existence of market segments to which variations of formulation may appeal (Schutz 1983). This approach uses factor analysis techniques to provide for orthogonal sensory and analytical predictor variables. These factors are then used in a stepwise multiple regression to predict a measure of acceptance for the products. The resulting regression equation can then be used to calculate an optimum formula that may or may not include the characteristic of sweetness.

An example of one product optimization study referring to sweetness is a proprietary study conducted on mayonnaise. In this study the intensity of 40 sensory attributes including flavor, texture, and appearance variables in 24 mayonnaise products was evaluated by a trained Quantitative Descriptive Analysis panel (Stone et al. 1974). The 40 sensory attributes were subjected to principal component analysis from which seven factors were extracted. The seven sensory variables that weighted highest on each factor were chosen to represent the seven factors. One of these was sweet flavor.

The same 24 products were evaluated on a nine-point hedonic scale by 500 consumers representing five cities. The hedonic ratings of 500 consumers were also subjected to principal component factor analysis using their ratings for the 24 products as cases. This resulted in three major "people factors." Means for the 24 products were computed for the people heavily weighted on each of the three factors. These means made up three sets of dependent variables which were analyzed by stepwise multiple regression on the seven sensory variables derived from panel analysis.

Sweet flavor did not appear in the equation as a significant predictor for two of the "people factors"; however, sweet flavor was a predictor for the remaining cluster. In this prediction equation, sweet flavor was the second most important of four sensory attributes in the equation predicting acceptance. The four-attribute equation accounted for 81% of the variance in preference, providing a very good measure of the factors contributing to acceptance of mayonnaise for a particular segment of the population.

From this prediction equation, an optimum level of each of the four variables and the relative importance of each to acceptance could be determined. Utilizing the prediction equation and a set of optimum target values, a mayonnaise product was predicted that was superior in hedonic value to any product in the original 24-product test array.

Both of the previous methods use sensory characteristics of sweetness in a more realistic additive (and perhaps interactive or curvilinear) role with other sensory characteristics; this allows one to determine the role of sweetness in product acceptance and in marketing potential. In fact, the term sweetness as used by consumers may be the result of a combination of several sensory characteristics as measured by a trained descriptive panel. The results of marketing research described above can then be used as an integral part of the decision process in the use of sweetness as both a component of acceptance and/or as an advertised, marketing element.

References

Anonymous (1985) Healthy trends. Calorie Control Commentary 7(2): 3

Calorie Control Commentary (1986) Hotels catering to the calorie-conscious. Calorie Control Commentary 8(1): 5

Franta R (1985) Sugar and sweetener consumption trends and their significance. Presented at "Sugars and Other Sweeteners in Food Processing" Symposium, Michigan State University, October 1985 and excerpted in Food Technol 40(12): 116–128

Hall T (1985a) Consumers say they avoid sweet drinks, so why do they buy so many of them? Wall Street Journal, 3 July 1985, 1st page, Section 2

Hall T (1985b) What Americans eat hasn't changed much despite healthy image. Wall Street Journal, 12 September 1985, p 1

Institute of Food Technology (1986) Sweeteners. 3. Alternatives to cane and beet sugar. Food Technol 40(12): 116–128

Kovach J (1984) The bitter truth. Industry Week Nov. 1984, p 28

Lowenberg ME, Todhunter EN, Wilson ED et al. (1979) Food and people, 3rd edn. John Wiley, New York

Mechling EW, Mechling J (1981) Sweet talk: the rhetorical drama of sugar. Culture and Communication Conference, San Diego State University

Moskowitz H (1985) Sensory preference segmentation and implications of product development. Presented at "Sugars and Other Sweeteners in Food Processing" Symposium, Michigan State University, October 1985

Nelson JE (1982) The practice of marketing research. Kent Publishing, Boston
Schutz HG (1983) Multiple regression approach to optimization. Food Technol 37(11): 46–49
Stone H, Sidel J, Lovier S, Woolsey A, Singleton R (1974) Sensory evaluation by quantitative descriptive analysis. Food Technol 28: 24–34
USDA (1980) Nutrition and your health, dietary guidelines for Americans (Home and Garden Bull No. 232). U.S. Department of Agriculture, Washington DC

Chapter 17

Sweetness in Product Development

P. Würsch and N. Daget

Introduction

Sweetness has long been a key to the improvement of food acceptability, but until the development of the sugar industry some six centuries ago, sources were limited to fruits, honey, maple and carob syrup. Before the widespread use of sugar, food in Europe was generally highly spiced. The change in food habits, particularly towards sweet food, was intimately linked with sugar availability. Sugar from cane appeared in Europe through Venice in the early fourteenth century, but only two centuries later was the first refinery built, in the Netherlands. Beet sugar production made considerable progress in Germany, Austria and Russia in the first decade of the nineteenth century, but it was Napoleon I who assured its major advance.

World production of sucrose, which until recently has been the only cheap source of sweetness, reached 0.8 M tons in 1839, 95% being from cane. By 1900 the figure was 8 M tons, and in 1982 over 100 M tons were produced, with 63% from cane (Hugill 1979).

Starch has been converted industrially into glucose syrups, another source of sweetness, for more than a hundred years. Recently the rapid advance in science and technology in the enzyme conversion of starch, coupled with other technologies such as chromatography and hydrogenation, has resulted in a broad range of sweeteners having specific sweetness, technical properties and industrial characteristics (Table 17.1). These sweeteners are essentially used by the food industry.

The rapid increase in sweetener production has had a greater impact on the choice and availability of sweet food. Today in industrialized countries there is in general little increase in sweetener consumption, but the current major increase in consumption in the developing countries of the world will continue. A relationship between income and sugar consumption has been shown during the period 1964–1975 by compiling data from ten regions of the world (Provisional Food Balance Sheets 1972–74, FAO Rome 1977) and it has been more recently

Table 17.1. Sources of sweeteners (sweetness intensity, 10% sucrose soln. = 1.0)

Sucrose	1.0	Hydrogenated glucose syrup	0.7–0.9
Inverted sugar	1.0	Sorbitol	0.6
Glucose syrups	0.1–0.6	Mannitol	0.7
High fructose corn syrup (HFCS)	1.0–1.3	Xylitol	0.9
		Maltitol	1.0
Glucose (dextrose)	0.6	Isomaltitol	0.35
Fructose	1.2–1.8	Saccharin	250–550
Lactose	0.3	Cyclamate	30–50
		Aspartame	120–200
		Acesulfame K	100–200
		Thaumatin	1500–3000

observed in the Middle East that for every 1% rise in the standard of living, sugar consumption increased by 3%.

Many factors, including wealth and working wives, have boosted the growth of food technology in the mass processing of food, transferring food preparation from the kitchen to the factory. At the beginning of the century the use of sugar was roughly 80% domestic and 20% industrial. By 1980 the industrial use of sugar in nine European countries was between 27% and 77% (Fig. 17.1) and 64% in the United States.

These statistics do not include sweeteners derived from starch, which, especially in the United States, represented 32% of the total sweeteners consumed in 1980. As shown in Fig. 17.2, the proportion of sugar in the total sweeteners consumed has progressively decreased to a level of 50% in 1985. However, on a worldwide basis, sugar still accounts for more than 90% (Table 17.2).

Sweetness is one of the many properties of a sweetener, since it can at the same time be a bulking agent, a texture modifier, a preservative, a flavour and a fermentation substrate, depending on the physical properties of the product. In the development of a food product all of these properties have to be taken into

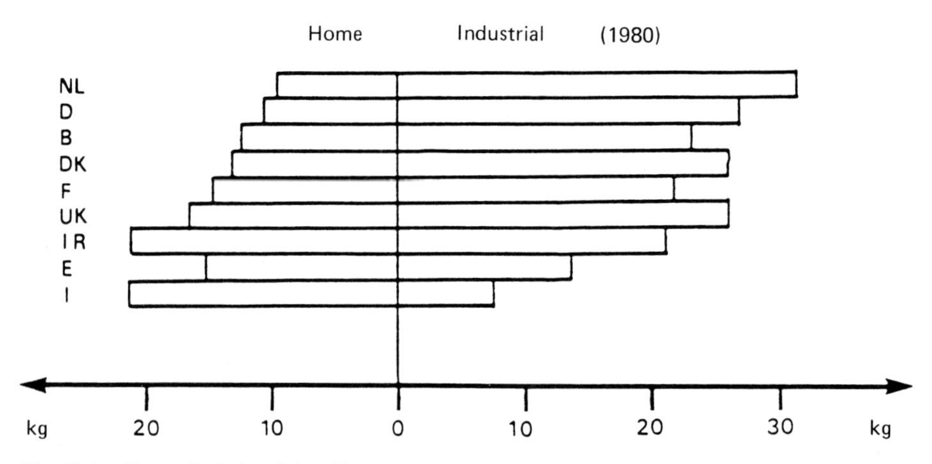

Fig. 17.1. Per capita industrial and home uses of sucrose in nine European countries (1980) (adapted from Comité Européen Fabricants Sucre, Paris, April 1985)

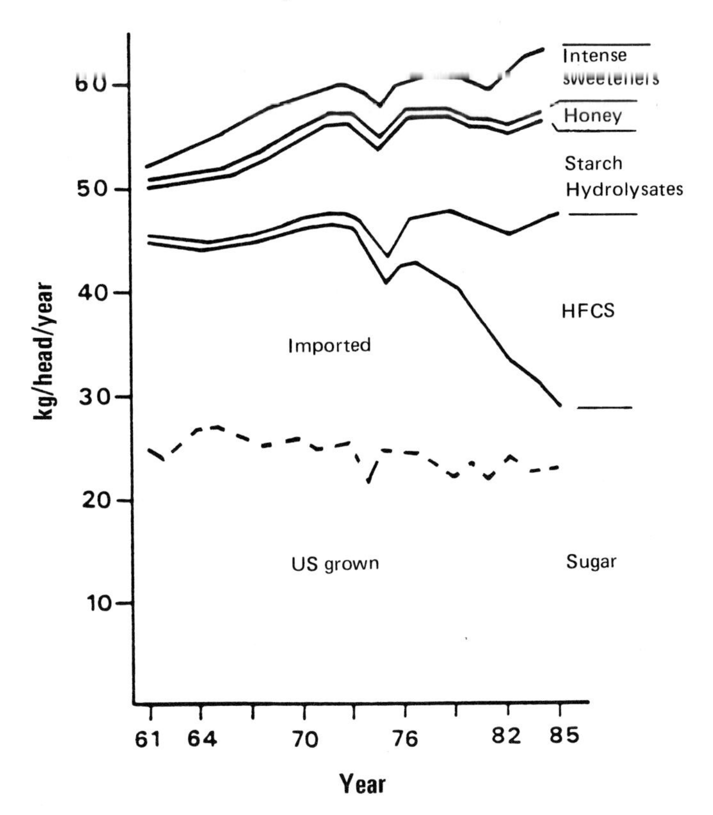

Fig. 17.2. Bulk and intense sweetener consumption in the United States. Amount of intense sweeteners is stated in terms of sucrose sweetness equivalent. *HFCS*, high fructose corn syrup. (Sugar and Sweetness Reports 1983–85, USDA, Washington DC)

account, as a change in one property will affect the others. The formulation and preparation of sweet products therefore imply an extensive knowledge of the properties and behaviour of the various ingredients used, and of the factors that can affect sweetness.

Formulation of sweet food at home mainly involves following recipes or formulae, and the housewife prepares no more food than is needed. The family plays the role of the consumer-testing laboratory: if the consumers do not like her

Table 17.2. World bulk sweetener output (Agra-Europe, issue 83.02.11, pp D2)

	1975	1982	1985
Sugar	81.6	99.0	95–105
Starch-based sweetener	4.6	9.0	12.0
High fructose syrup	0.5	4.0	5.0
Glucose syrups	3.0	4.6	5.0
Dextrose	1.1	1.2	1.2
Total	86.2	108.0	107–117

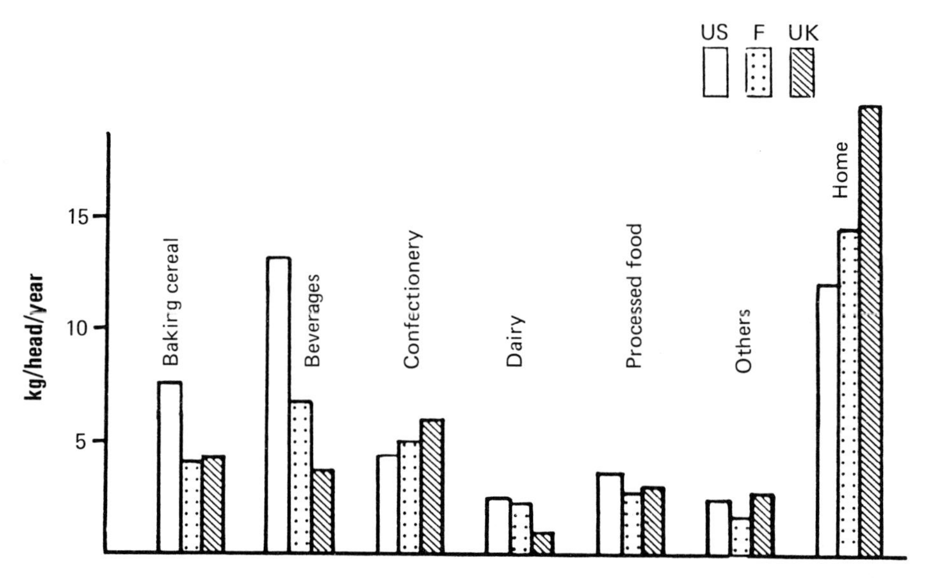

Fig. 17.3. Per capita industrial and home uses of sucrose and high fructose corn syrup (*HFCS*) in the United States (1983), France (1983) and the United Kingdom (1976) (adapted from Sugar and Sweetness Reports 1983–85, USDA, Washington DC; Centre d'Etude et de Documentation du Sucre, Paris, 1985; Nicol 1979)

creations they say so and the recipe must be corrected accordingly. The industrial food manufacturer formulates and produces a tremendous variety of products for a very diverse clientele. Further, the manufacturer must be concerned with shelf life, nutrition, wide acceptance and value for money. The researcher involved in the field of food formulation must usually work with a number of ingredients to attain these objectives. Test samples will be judged by taste evaluation panels or consumer studies, along with other more specific characteristics measured by physical and chemical tests.

The range of sweet products is generally divided into market sectors like baking, beverages, confectionery, dairy, processed food (preserved canned and frozen food) and others, which vary in magnitude from country to country (Fig. 17.3), with standard of living and with food habit.

Sweetness in the Food System

Perception of sweetness in a food system is affected by many factors which determine the amount and type of sweetener that has to be used. Table 17.3 lists some of these factors, which encompass the structural nature of the food, its technology, the interaction with other tastes, and the temperature when consumed (Moskowitz 1981). These factors may affect the intensity, quality and duration of perception of sweetness.

Table 17.3. Some factors affecting sweetness in food

Texture:	*Taste*:
Liquid	Sourness
Viscosity	Bitterness
Solid	Flavour
Aerated froth	
Homogeneity	
Consistency	
Temperature	

The most acceptable level of sweetness in liquids is attained with much less sweetener than in dry foods (Table 17.4). A concentration of 40% sugar in dry food may appear only as sweet as a 10% sugar solution. Plain chocolate contains about 50% sucrose suspended in the dry finely divided state in cocoa butter. This sweetness is needed to counteract the natural bitterness of the cocoa and to provide bulk. Fat plays an important indirect role in the taste. In melting, it produces a coating in the mouth, reducing the intensity but extending the time of sweet sensation. A decrease in sweetness would increase the bitterness of the chocolate and would require a partial replacement of the bulk provided by sucrose. Sucrose can be replaced by other sweeteners of lower sweetening power but modifies the sensory profile in chocolate, especially with respect to taste and flavour, and thus the hedonic quality of the product (Ogunmoyela and Birch 1984).

Although sweets contain more than 90% sucrose, the sweetness perceived is acceptable because of the slow dissolution rate of the sweetener and of the presence of acid in the product, which has a sweetness-depressing effect (Pangborn and Trabue 1964).

In all countries but the United States and Canada, sucrose is still the dominant sweetener in soft drinks. The attraction of sucrose lies in its excellent sweetness, mouth-feel and flavour-enhancing properties. The level of sucrose ranges generally from 9% to 14% and is governed by the type of beverage. Increasing the acid concentration in a soft drink depresses the sweetness and thus the level of sucrose has to be balanced against acidity. The level of sweetness also varies according to the flavour type and the degree of concentration of the beverage. Recent studies have shown that the most preferred level of sucrose in a lemonade at a pH of 2.75 is 10% (Pangborn 1980). There was, however, a wide variation in distribution of preference, ranging from 6% to 24% sucrose. An approximately 10.5% level of sucrose is found in carbonated soft drinks.

Table 17.4. Sweetener level expressed as % sucrose in some selected food items

Soft drinks – lemonades	9–14
Chocolate-flavoured drink	8
Yogurt with fruit	12
Ice cream	14–18
Sherbet	27
Custard	6–13
Biscuits	25
Plain cake	25–36
Chocolate	50–60

Choice of Sweetener

Food processors find it very important to look at a large choice of available sweeteners, selecting one or several in combination that offer the best functional properties, sweetness quality and level, and cost-effectiveness. Although many types of sweetener are available, each having a particular psychophysical characteristic, the two major types are still sucrose and corn-derived products (Table 17.1). There are many types and forms of sucrose and corn sweetener commercially available within these two categories (Inglett and Lineback 1982; Junk and Pancoast 1980). They are compatible in combination and achieve a wide range of properties, listed in Table 17.5. Within a given food sector, the practical choices are, however, rather limited, when considering the following general conditions to be satisfied:

1. Optimum functional and sensory properties
2. Availability/best price
3. Governmental food laws and regulations
4. Calories, nutritional value

The variety of sweet products and of sweeteners is considerable. We shall therefore limit our description of technical factors affecting sweetness to two examples.

In frozen dairy products the functionality of sugars is as important as their sensory contribution. The combination of numerous sweeteners allows one to modulate, increase or decrease various characteristics and to optimise the overall quality. For each type of frozen dairy product, however, the sweetener composition has to be treated specifically. This optimisation can be restricted in most cases to two factors: the sweetness and the decrease of the freezing point, which strongly influences the texture. The freezing point of sugar solutions being dependent upon the concentration and average molecular weight of the sugars in solution, corn syrups are used in establishing the freezing point of the mix and to prevent crystallisation of sucrose. However, one has also to take into account that sucrose will accelerate the fusion rate and glucose syrup will slow it down.

In carbonated drinks, sucrose is inverted (hydrolysed) during the routine manufacturing process and during storage. At pH = 2.5 about 50% of the sucrose is found inverted after 25 days at room temperature (Jacobs 1959). The distinct advantage of the inversion of sucrose lies in the taste of the product, which changes as a function of the degree of inversion. In particular, sweetness intensity and duration increases with increasing inversion. By inverting the sugar syrup

Table 17.5. Properties and functional uses of sweeteners (Junk and Pancoast 1980)

Sweetening agent
Bodying agent, texture, cohesiveness
Humectant
Freezing point depressor
Preservation agent
Fermentation substrate

completely during manufacturing, the final taste is produced, thus ensuring a better and consistent uniform taste in the finished product (Stone and Oliver 1969). These functional characteristics in soft drinks combined with a lower cost have favoured the progressive replacement of sucrose by high fructose corn syrups (HFCS) in the United States and Canada. In 1983, 50% of the sweetener used in carbonated beverages in the United States was HFCS. This share should increase to 67% in 1990, when only 6% will be covered by sucrose, the remainder being intense sweeteners in low calorie drinks (Beverage World Periscospe, issue 83.10.31, p 1).

The world health authorities have become increasingly concerned by excessive or unbalanced food intake, recognising its contribution to many so-called civilisation diseases. Control of nutritional intake has been focussed on fat and carbohydrates, especially sugars (Bierman 1979). Replacement of sugars has been particularly recommended with three objectives in mind: reduction in energy intake, reduction in cariogenicity of sweet food, and creation of sweet foods suitable for diabetics.

Although industry has created a variety of new bulk and intense sweeteners to replace natural sugars for medical purposes (Table 17.1), the ideal sweetener, which is identical to sucrose in taste, sweetness, functional properties and cost, and which at the same time is low in calories, non-cariogenic, and non-hyperglycemic, has still to be discovered. However, by combining different types of sweetener, along with texture modifiers and bulking agents, it is possible to create products which the consumer regards as acceptable alternatives.

Fructose is the only carbohydrate with a sweetness higher than that of sucrose on weight basis and which could give a reduction of energy content for the same sweetness in a food. In solution, however, fructose undergoes conversion to its various anomers, predominantly ß-fructofuranose, which is markedly less sweet than the pyranose form found in the crystalline state. When fructose is dissolved, equilibrium between the various anomers is achieved in 20 min at room temperature. The sweetness of a 5% fructose solution immediately after the dissolving is 1.5 (compared to sucrose = 1.0) and decreases progressively to 1.25. The sweetness level at 5°C is significantly higher, and reaches 1.8 immediately after dissolving and 1.45 at equilibrium (Hyvönen et al. 1977). Therefore the use of fructose is most advantageous in powdered products which are dissolved just before consumption, or in very cold liquids or frozen products.

One of the main advantages of certain combinations of sweeteners is that they are often synergistic, i.e., the sweetness of the combination is greater than the sum of the individual parts (Moskowitz 1974). Using the magnitude estimation method, Stone and Oliver (1969) noted synergistic effects of as much as 20%–30% when glucose at high concentration (9%–18%) was combined with fructose or sucrose. A sweetness synergy is also observed between fructose and saccharin, with the additional advantage that the carbohydrate effectively masks the unpleasant aftertaste of the artificial sweetener. The sweetness of fructose containing 0.3% saccharin is 3–4 times that of sucrose and its taste in aqueous solution cannot be distinguished from a pure sucrose solution of equal sweetness (Hyvönen et al. 1978). Further, the accentuation of fruit flavour by fructose permits reduction of the sweetener in fruit formulations with no significant change in sensory qualities.

Approval for the use of new bulk and intense sweeteners has initiated formulations of new lines of foods. However, substitution of an intense sweetener for other sweeteners implies a complete remodelling of the formulation in order to restore the original texture and taste profile. The flavour system must therefore provide an initial pleasant sensation with a quality compatible with the accepted sweetness profile.

For instance when formulating a beverage comprising artificial sweeteners, acidulants and flavouring agents, flavour–taste impression should be balanced so that one tastes a mixture of sweet flavour/acid at the same time, rather than first tasting the bitter flavour and the sour acid and only later tasting the slowly rising sweetness produced by the artificial sweetener (Larson-Powers and Pangborn 1978).

The stability of the new ingredient must also be taken into account. Aspartame-based beverages lose sweetness as a function of storage time, temperature and pH. Only about 50% of the initial aspartame remains in a cola beverage stored at 20°C for 40 weeks (Bakal 1983).

Cola beverages sweetened with aspartame and saccharin show exceptional sweetness stability with good taste acceptability. Several patents describe the synergism between aspartame, saccharin and cyclamate. Such combinations also exhibit superior taste profiles which may be associated with the flavour-enhancing properties of aspartame (Bakal 1983).

In solid foods the application of intense sweeteners is often difficult since they have no bulking properties which affect texture and mouth-feel, and thus require the use of other ingredients to provide the bulking properties of sucrose. Polyols, in combination with intense sweeteners when their own sweetness is too low, have been used with success in the formulation of desserts, baked goods and sweets of low energy content for diabetics and for the prevention of dental caries. The development and marketing of such sweet foods are, however, limited for the time being, due to legal restrictions and high cost of the ingredients.

Product Optimisation

Sidel and Stone (1983) define product optimisation as the procedure for developing the best possible product in its class. When using alternative sweeteners for nutritional reasons (diet drinks), this can be redefined as developing the product which comes closest to the standard (non-diet) commercial product.

Considering the large variety of sugars and sweeteners, their different potentialities as sweet substances, their enhancing effects in mixtures, their specific characteristics as texture modifiers, their nutritional values, and how they interact with other components, the choice of a specific sweetener or of a mixture in creating a new product has to take into account all of these aspects, as well as technological, cost and legal constraints. After having chosen a certain direction it is necessary to study the situation experimentally in order to optimise factors and to obtain the food product best accepted.

Replacing or changing the amount of sweetener in an already existing product needs the same sort of considerations, although the situation is simplified: the product has already been optimised, and many of the interactions are known, or at least the range of technically feasible samples has been defined.

Aromaticians, food technologists, marketing specialists and sensory analysts act together, adapting their approach as necessary from case to case (Moskowitz 1985; Daget 1983). The procedure includes several steps as outlined below. A preliminary selection of factors includes the ingredients, divided into those which are necessarily to be kept constant and those which could be varied in the optimisation. For instance, vitamins and minerals are fixed; sugar and acids may be changed in their relative amount. The appropriate level and the range within which concentrations may vary are established. An example is offered by the amounts of sugar and alcohol both ranging from 25% to 40% and of flavor from 0.6% to 1.02% in a liquor composition, as described in a case study presented by Schiffman et al. (1981).

Technological factors are also likely to influence the final acceptability of the product: grinding level, kneading time, cooking or sterilisation temperature and time etc. have to be taken into account. The variation of these factors results in a change of certain product features considered as dependent variables.

A selection of variables considers sensory characteristics: level of sweetness, flavor intensity or hardness, or more complex characteristics: quality or liking. It also considers physical or chemical characteristics: solubility, stability during shelf life.

The team chooses the set of variables from which they wish to obtain a response to their optimisation procedure. In the liquor case study mentioned above, two responses on "similarity between samples" and "intensity of liking" were obtained.

Establishment of Measuring Instruments

For detection and measurement of sensory characteristics the measuring instrument consists of a panel of 6 to 12 trained tasters, giving answers to questions about sweetness intensity, aroma intensity or complexity, fluidity and overall quality. Many questions can be solved using such panels, e.g. is there an enhancing effect of one sugar on another? Is there an aftertaste? Which sets of conditions maximise the sweetness perception? Which maximise the fruity aroma? Which minimise the bitterness? And so on.

The panel may also consist of a group of 30 to 300 consumers representative of the target population. Effectively, only consumers can answer questions about liking. They may score their likings or describe the negative aspects of a product using a scale from "too much" to "not enough", thus indicating which are the points to be corrected and in which direction (Chap. 16; Daget 1983).

The tasting may follow any sort of test adapted to the type of question to be asked: pair comparisons, ranking, scoring, using either category scaling or cross-modality matching between perceived intensity and length of a line traced on paper. Scoring may involve only one characteristic or several characteristics generating quantitative description analysis (for review, see Moskowitz 1983). Parallel to the panel, instrumental measurements of the products are also carried out.

Let us take the example described by Williams et al. (1983). The aim of the survey was to determine the optimum level of sweetness and acidity in apple juice. They used the ratio scaling technique (magnitude estimation scaling) with a panel of consumers representative of the target market. Seven samples of clear juices, prepared by blending juices in order to have variable sweetness and acidity, were tested against one of the samples which has been defined as standard. From the examination of the contour plot of ideal acidity and sweetness two main subpopulations emerged, one preferring sweeter and less sour and the other less sweet and sourer.

Optimisation Trials

Traditionally, technologists and product specialists establish new recipies or modify existing ones by trial and error. They often express the feeling that changing this working method would not help them to screen a greater number of factors. However, more systematic approaches can help to accelerate the optimisation of product formulation.

Such approaches use regression techniques to work out the exact amount of a factor which gives the desired intensity of response. Simple linear or non-linear regressions may be used when only one factor is tested, such as sweetness perception or liking as a function of sugar concentration in lemonades (Pangborn 1980). Multiple regression is applied when several factors are to be considered.

When simultaneous treatment of too many factors hampers the test procedure, the task is split into several subsets. Several approaches may be used, and have been extensively and nicely described by Cochran and Cox (1959). Unfortunately, very few cases are found in the literature; hence we present the cases below.

1. *The sequential approach:* In this method treatments are applied to experimental units in a definite time sequence. The process of measurement is rapid and the response to a given step is known before the following step in the time sequence is begun. The experimenter may stop at any time and examine accumulated results to decide whether to continue the trials or not. The number of cycles of the treatment sequence is not fixed in advance, as the desired results can appear rapidly. A statistical test, specially adapted for this situation, makes one of three choices: accept, continue testing the same product, or reject and therefore change formulation and then start again.

As an example, we can consider the case of replacing a sweetener in a known commercial product. A sweetener is proposed which one thinks will match the original product. The difficulty is to find out if it is more, equally or less sweet. Simple pairwise comparison does not give an assurance as to whether the response is equal or not. The trial is thus repeated, changing the level of sweetener until a match with the preference product is achieved. This method is preferred when pilot trials are very costly.

2. *The method of single factor:* The technologist first makes estimates of the optimum combination of factors and their levels. Since a series of experiments with only one factor at a time will follow, the factors must be arranged in the order of testing. It is best to start by varying the factor which is expected to give the largest response, the others being kept constant. Three to five levels may be compared to find the optimum feature. When the response is obtained, the next factor is studied (Cochran and Cox 1959). We may have, for example, a situation in which the development group wishes to make a refreshing beverage. Tasters have described such beverages as sour, slightly bitter, slightly sweet and fruity. It is assumed that bitterness is the most restrictive parameter. A series of samples

with various bitter levels and constant sweet, sour and fruity levels are prepared in order to find the optimum level of bitterness. Once found, this level is kept constant and sweetness is varied. The process is repeated for sourness and finally for fruit aroma. It is possible that at this stage the optimum level of bitterness has changed from that determined before the level of sourness was established. Consequently one has to repeat the bitterness optimisation.

This method is sometimes used in the laboratory but it is time consuming. It does not provide an estimate of the shape of the response curve and the true optimum may escape. Furthermore it does not reveal interactions.

3. *The simplex method:* The maximum is located topologically by a series of trials, each planned from the results of the preceding one. The experiment starts by fixing a minimum of three points corresponding to the factors examined. The worst point obtained is replaced by a new one obtained from reflexion, expansion and contraction procedures. The trial is repeated until the best levels are found. This operation may be represented by a SIMPLEX algorithm which has to be optimised (Debouche and Steiner 1974; Swann 1969).

Usually the space within which the results are located is limited by the technological, economic and legal constraints already mentioned. Step by step the method slowly leads to the optimal zone of desired conditions. This method can be applied in the development of an instant soluble sweet white coffee powder which should dissolve without any residue in the cup. The problem lies in the feasibility of having big enough sugar granules in a form which dissolves instantaneously. Sugar would be ground, mixed with the other ingredients and granulated with the right amount of water and at the right temperature. Since three technical factors must be controlled, four products are prepared. Solubility and taste are controlled and from the responses new conditions are set, using the results from the product just tested. The operation is continued until the best solubility is obtained within the acceptable limits of off tastes.

4. *The factorial design method:* This approach is no longer sequential and has to be planned before starting trials in order to follow a fixed design (Lowry 1979; Cochran and Cox 1959). If several factors are operating together, the effects of each of them are examined with every combination of the others. A great deal of information is obtained on factors and on their interactions. This method is applicable in an exploratory trial where one wishes to determine quickly the influence of each of a reasonable number of factors at two levels. It leads to a linear model and indicates recommendations for combinations of levels of these factors which will produce a maximum response, but the optimum may escape detection. With more than two levels, a quadratic model may be used to reveal an optimum. This directly leads to the surface response method described below.

Using a full factorial design, Moskowitz (1983, pp 372–395) presented a case study of rye bread optimisation on two factors. He examined three levels of rye and four levels of sugar, generating twelve products. A panel of consumers first evaluated the leading competitors' rye breads currently on the market in order to define their global sensory properties, and to create a base for comparing results coming from the optimisation trial. Changes in the ingredients significantly affected several of the flavour and texture characteristics. The optimum product contained no less than 6% sucrose.

5. *The surface response method:* When several factors are operating together, an optimum may be obtained for a certain set of levels of these factors, outside

which a clear drop in quality, flavour intensity or texture may result. This specific method is able to determine that optimum point (Box 1954; Cochran and Cox 1959; Vuataz 1981).

A preliminary trial is carried out to establish the range of interest for levels of each of the factors investigated, including a fringe range which may lie outside the acceptable range but can be useful in order to establish the parameters. The central value of the level is taken and other levels fixed symmetrically about it with a small number of points.

Yamaguchi and Takahashi (1984) have used this approach to optimise monosodium glutamate and sodium chloride levels in a clear soup. It could easily be transposed to the optimisation of sweetness intensity in a product containing a sweetener and an acid.

When applied to a reduced number of factors this method is more efficient than the other experimental procedures described. It cuts the time and cost required to determine the optimum product (Giovanni 1983).

Conclusions

In a food sweetness is rarely tasted alone. Rather sweetness is mixed with a range of other complex tastes along with texture and aroma. These factors can significantly modify the perception of sweetness and can ultimately determine the choice and amount of sweetener. End products of the food industry are becoming more and more complex through sophisticated processing and formulation due notably to the increased variety of available sweeteners. Therefore optimisation methods must be applied to establish the most acceptable product.

References

Bakal AI (1983) Functionality of combined sweeteners in several food applications. Chem Ind 700–708

Bierman EL (1979) Carbohydrates, sucrose and human disease. Am J Clin Nutr 32: 2712–2722

Box GEP (1954) The exploration and exploitation of response surfaces: some general considerations and examples. Biometrics 10: 16–24

Cochran WG, Cox GM (eds) (1959) Experimental designs, 2nd edn. John Wiley, New York

Daget N (1983) Understanding and interpreting consumer answers in the laboratory. In: Williams AA, Atkin RK (eds) Sensory quality in foods and beverages. Ellis Horwood, Chichester, pp 58–68

Debouche C, Steiner J (1974) A propos de deux methodes d'ajustement de modèles mathématiques non linéaires. Revue de Stat. Appliquée 22: 5–22

Giovanni M (1983) Response surface methodology and product optimisation. Food Technol 36: 96–101

Hugill JAC (1979) History of the sugar industry. In: Birch GG, Parker KJ (eds) Sugar: science and technology. Applied Science Publishers, p 26

Hyvönen L, Kurkela R, Koivistoinen P, Merimaa P (1977) Effects of temperature and concentration on the relative sweetness of fructose, glucose and xylitol. Lebensm-Wiss Technol 10: 316–320

Hyvönen L, Koivistoinen P, Ratilainen A (1978) Sweetening of soft drinks with mixtures of sugars and saccharin. J Food Sci 43: 1580–1584

Inglett EE, Lineback DR (eds) (1982) Food carbohydrates. AVI, Westport, Conn

Jacobs JB (1959) In: Manufacture and analysis of carbonated beverages. Chem. Publ., New York, p 40

Junk WR, Pancoast HM (ed) (1980) Handbook of sugars. AVI, Westport, Conn, pp 383–414
Larson-Powers N, Pangborn M (1978) Paired comparison and time intensity measurements of the sensory properties of beverages and gelatins containing sucrose or synthetic sweeteners. J Food Sci 43: 41–46
Lowry SR (1979) Statistical planning and designing of experiment to detect differences in sensory evaluation of beef loin stock. J Food Sci 44: 488–491
Moskowitz HR (1974) Models of additivity for sugar sweetness. In: Moskowitz HR, Scharf B, Stevens JC (eds) Sensation and measurements. Reidel, Dordrecht, Holland, pp 195–226
Moskowitz HR (1981) Changing the carbohydrate sweetness sensation. Lebens-Wiss Technol 14: 47–51
Moskowitz HR (1983) Product testing and sensory evaluation of foods. Marketing and R & D approaches. Food & Nutrition Press, Westport, Conn
Moskowitz HR (1985) New directions for product testing and sensory analysis of foods. Food and Nutrition Press, Westport, Conn
Nicol WM (1979) Sucrose in food systems. In: Koivistoinen P, Hyvönen L (eds) Carbohydrate sweeteners in foods and nutrition. Academic Press, London, p 152
Ogunmoyela OA, Birch GG (1984) Sensory considerations in the replacement in dark chocolate of sucrose by other carbohydrate sweeteners. J Food Sci 49: 1021–1056
Pangborn RM (1980) A critical analysis of sensory responses to sweetness. In: Koivistoinen P, Hyvönen L (eds) Carbohydrate sweeteners in foods and nutrition. Academic Press, London, pp 87–110
Pangborn RM, Trabue IM (1964) Task interrelationships. V. Sucrose, sodium chloride, and citric acid in lima bean purée. J Food Sci 29: 233–240
Schiffmann SS, Reynolds ML, Young FW (1981) Introduction to multidimensional scaling. Academic Press, London, pp 337–341
Sidel JL, Stone H (1983) An introduction to optimisation research. Food Technol 37: 36–38
Stone H, Oliver SM (1969) Measurement of the relative sweetness of selected sweeteners and sweetener mixes. J Food Sci 34: 215–222
Swann WH (1969) A survey of non-linear optimizaton techniques. FEBS Lett 2: 39–47
Vuataz L (1981) Response surface methodology. First Int. Conf. AMS: "Applied modelling and stimulation". Lyon, France, 7–11 September
Williams AA, Langron SP, Arnold GM (1983) Objective and hedonic sensory assessment of ciders and apple juices. In: Williams AA, Atkin RK (eds) Sensory quality in foods and beverages. Ellis Horwood, Chichester, pp 310–323
Yamaguchi S, Takahashi C (1984) Interactions of monosodium glutamate and sodium chloride on saltiness and palatability of clear soup. J Food Sci 49: 82–85

Commentary

Rozin: On the first page, reference is made to market segmentation in terms of sweetness. This is an issue of central concern to the topic of this paper. Could more information be presented on this subject? Some years ago, with Cines, I (1982) wrote a paper on ethnic differences in coffee and coffee product preference. We found (looking at four different American subgroups) notable differences in the way in which coffee or coffee-flavoured foods were used and prepared, with some implications for sweetness. For example, in comparison with other ethnic groups Jewish people showed a much stronger relative preference for coffee-flavoured products (soda, candy, yoghurt) than for the usually less sweet and more bitter coffee beverage.

Reference

Rozin P, Cines BM (1982) Ethnic differences in coffee use and attitudes to coffee. Ecol Food Nutr 12: 79–89

Scott: The paper provides an interesting glimpse into an area that basic researchers see rather little of. Producers and marketers are clearly caught in a quandary between the physiological benefits (and concomitant positive affect) of sweetness and the social unacceptability of it. The hint of perfidy inherent in their resolution of this quandary, while not startling, is mildly unsettling. I do not see that the producers are merely unfortunate victims of this physiological–social dichotomy, however; rather, they helped create the unacceptability of sweetness by removing sugars from their historical role of scarcity and using them in proliferation to sell products. Schutz' manuscript maintains a good balance between the views of the companies and the consumers.

Booth: Is not the criticism of presweetened breakfast cereals most ill-informed? Even with further sugar added, they are still eaten with milk and are vitamin fortified too. As well as increasing the nutrient density, the milk buffers mouth acid, and anyway that sugar is often washed off the teeth with a hot drink or later parts of the breakfast. This attack on popular convenience breakfast foods can only further discourage the eating of breakfast calories, which prevent accidents and may well improve efficiency, reduce aggression and help prevent weight gain from non-nutritious snacks later in the day. You document an impossible position for food marketers arising from muddle created by unscientific prejudice of "nutritionists".

Rozin: The recent book by Mintz (*Sweetness and Power*, see Chap. 7) is relevant to the historical materials presented at the beginning of this paper.

The lower level of sweetness in sugar in solids seems to me of fundamental importance. According to some of the data presented here, in many cultures the majority of sugar is consumed in solid form. However, in the United States, the beverage vehicle may predominate. This could have the effect of being a force to lower sugar intake.

The problem of optimal adjustment may be complicated by the issue of familiarity. As pointed out by Booth et al. and others, departure from the familiar is one basis for preference judgments. The fuss about Coca-Cola, over a very small difference in sweetness, seems to be partly in opposition to any change.

Meiselman: Dr. Würsch raises the important distinction between trained and untrained panels. This distinction is critical in the conduct of sensory experiments for product development, but little published research addresses how trained or expert panels differ from untrained or consumer panels. Although the former are supposed to perform analytic tasks and the latter evaluative tasks, the performance of each type of panel has not been thoroughly investigated.

Birch: I query the statements on the persistence of sugars. In our experience fructose persistence is greater than glucose persistence. Indeed, this difference is even more distinct than the difference in intensity between the two sugars. Typically 10% w/v fructose solution persists for about 30 s.

Chapter 18

Sweetness in Food Service Systems

Herbert L. Meiselman

Introduction

The purpose of this paper is to review the role of sweetness in food service systems. Other papers in this volume deal with sweetness as a physiochemical event, a sensory event, a physiological event, a food event within narrow definitions, and a variable affecting other aspects of behavior. This paper will attempt to extend the examination of sweetness to an examination of its impact in actual food service, situations in which foods are served, selected, and consumed. It is only in recent decades that systematic examination of food habits has advanced to the study of food habits in natural situations of food consumption.

In order to organize my material, I will use an organization of food habits which I have proposed elsewhere (Meiselman 1984) and have used in the contexts of clinical disorders (Meiselman 1986a) and flavor enhancement (Meiselman 1986b). This organization of food habits without any theoretical foundation, divides them according to the chronological sequence of eating: food attitudes, food selections, sensory evaluation/food acceptance, food intake, and food waste. I will briefly review what each food habit covers, and then will use this outline to present material on sweetness in food service.

Food attitudes are those predispositions toward foods which are more or less constant in particular eating situations. They include our preferences and aversions for foods (sweet foods in general, and specific sweet foods such as apple pie, dates, and chocolate-covered ants). Attitudes also include our perceptions about whether foods are healthful, status producing, sophisticated, etc. Their history shows many attempts to relate food attitudes to other food habits. Generally, the result has been that food behaviors are the result of more complex variables than simply attitudes. Attitudes have been shown to relate to other food habits but not in a highly predictive way.

In any actual eating environment a person is faced with food options from which he or she must make choices. These selections provide the first behavioral data in an eating environment. In commercial food service operations, food selections are often collected automatically on computerized cash registers, and provide potential data bases for the researcher. Many of the data, however, are not reported in the scientific literature. Anyone working within or with access to food service operations has potential to collect such data. At Natick we have collected selection data in schools, in universities, and in the military.

The foods selected are submitted to sensory evaluation and, based on the outcome, a food is judged to be acceptable or unacceptable. It is important to distinguish the sensory properties of foods (sweet, firm, red, cool) from the hedonic properties (good, bad). A great deal of research has gone into the perception of sweetness, and this topic is covered elsewhere in this volume. However, relatively little published research has gone into how the sensory properties of foods affect food habits in actual food service situations. As I have noted elsewhere (Meiselman 1984), most sensory research has been conducted with simple chemical solutions. While the use of more complex chemical stimuli in laboratory research is increasing (Halpern et al. 1986) and the use of foods in sensory research is also increasing (Piggott 1984), there is still very little research in eating environments relating sensory properties of foods to food habits.

The same has not been true for acceptance or hedonic research. These data have been collected during evaluation of a large number of food products (Moskowitz 1983) and food service systems (Meiselman 1984).

Based on a number of factors, including acceptability, the person consumes a quantity of the food. The method and theory of food intake has occupied a large part of the research in food habits (Krantzler et al. 1982). To date, no single method of collecting valid data on food intake has emerged. This has presented a serious problem to researchers and decision-makers who need precise data on food intake.

Whatever is selected but not consumed is food waste. This has received little attention overall as a food habit. Waste has been studied in school feeding programs because of its relationship to the economies of such systems. Little has been done relating food waste to food aversions and preferences.

Based on the above outline, the data relating food habits of sweet foods to food service can now be reviewed. I have selected three different types of food service for inclusion: feeding schoolchildren and university students, feeding the military, and feeding the elderly. These are three of the many types of institutional food service in the United States. When combined with restaurant food service, they account for the 22.5% of meals eaten away from the home and 24.9% of dollars spent each week on food. For younger people these percentages exceed 35% (Restaurants and Institutions 1985). In the discussion to follow, I have analyzed food habits for sweet foods by selecting clearly sweet food items from lists of food items. I have avoided marginally or questionably sweet items so the analyses represent a conservative criterion of sweet.

Food Attitudes

Studies of attitudes to sweet foods in food service have dealt with two issues. The first is how sweet foods compare with others in food preference. The second is general attitudes toward sweet products.

Table 18.1. Fruit and vegetable juice preferences in descending order

Military[a]	College students[b]
Orange	Orange
Grape	Grape
Apple	Apple
Tomato	Grapefruit
Pineapple	Tomato
Grapefruit	Pineapple
Vegetable	V-8 (vegetable)
Cranberry	
Prune	

[a] Meiselman and Waterman 1978
[b] Einstein and Hornstein 1970

It is the layman's view that all sweet foods are highly preferred. This is not necessarily the case. Within any food class there are foods which are highly preferred, moderately preferred, slightly preferred, and not preferred. Meiselman (1977) and Meiselman and Waterman (1978) reported on the food preferences of young enlisted personnel in the United States Armed Forces, based on a questionnaire of 378 food items. In the top ranked 50 foods, 20 were sweet, including 3 sweet beverages, 7 desserts, and 10 fresh fruits. For the bottom ranked 50 foods, there were 7 sweet foods, including fruits, beverages, and desserts. Einstein and Hornstein (1970) reported food preferences of college students, also detailing higher and lower preferred desserts. Comparison of the military and college student data yields good agreement for preferences of sweet fruit juices (Table 18.1). Lyman (1982) also noted similar ratings for desserts and other categories in two studies of college students and for military personnel.

Meiselman and Waterman (1978) also examined the distribution of ranks for each food class. Again, the layman's expectation that most desserts would be highly ranked was not confirmed. When the ranks of the 378 foods were divided into four quartiles, each quartile contained substantial numbers of desserts and fruits. There were as many low-scoring desserts as high-scoring ones. Meiselman and Waterman also examined the mean score across foods and range of scores for each food class. While fruit drinks, fresh fruit, and ice cream were among the highest scoring classes, carbonated beverages, canned fruits, and baked desserts had more moderate class averages.

Meiselman (1977) and Wyant and Meiselman (1984) have reported on differences in food preferences among different racial and gender groups within the United States military. Meiselman (1977) observed that the major food class contributing to differential preference of sweetness by blacks was beverages (Table 18.2). Data from Wyant and Meiselman (1984) confirmed this trend. Blacks had a higher preference for fruit and vegetable juices and for fruit drinks and iced tea. In the same study females showed a stronger preference for fresh fruit and for salads. Neither racial nor gender group showed a consistent preference for a wide range of desserts.

A number of studies have investigated food preferences of schoolchildren. In addition, two papers have explored what foods mean connotatively to children. Carlisle et al. (1980) studied 354 high school students, including 82 blacks (27 males and 55 females) and 272 whites (137 males and 135 females). Their

Table 18.2. Beverage preferences of blacks[a] (adapted from Meiselman 1977)

Beverages	Hedonic		Preferred monthly frequency	
	Black	White	Black	White
Cherry flavored drink	6.36	5.55	13.40	8.72
Cherry soda	6.38	5.71	12.87	9.26
Fruit punch	6.84	6.03	14.69	9.69
Gingerale	6.48	5.93	13.06	10.30
Grape flavored drink	6.84	5.91	15.41	10.08
Grape juice	6.95	6.24	16.32	12.94
Grape lemonade	6.50	5.50	13.70	8.92
Grape soda	6.92	5.77	16.04	9.75
Grapefruit juice	6.41	5.66	13.90	10.54
Grapefruit orange juice	6.81	6.12	15.98	12.97
Grapefruit pineapple juice	6.46	5.57	13.96	10.05
Lemon lime soda	6.32	5.68	13.66	10.78
Lemonade	7.21	6.73	16.17	13.00
Lime flavored drink	5.91	5.28	11.06	8.09
Orange flavored drink	6.92	6.04	16.20	10.29
Orange soda	7.14	6.04	16.97	11.34
Pineapple juice	6.37	5.69	13.32	9.23
Prune juice	4.59	3.97	7.30	4.41

[a] Based on a conservative criterion of three statistically significant comparisons among relevant population subgroups

questionnaire contained 93 food item names in random order. Among the 45 top ranked foods were four sweet beverages, nine desserts, and five fruits. Carlisle et al. compared the preferences of their racial and gender groups. Out of 93 food items there were 47 race preference differences, with blacks showing greater preference for 30. All racial preferences for six fruits showed blacks with the greater preference, and of eight racial preferences for desserts, blacks showed greater preference for six. Blacks also showed a higher preference for carbonated beverage. Meiselman (1977) had also noted that blacks had a greater number of significant racial preference differences among military personnel. The one gender preference for fruit and the four gender preferences for desserts all showed greater preference by males.

When foods were rated on connotative scales using a semantic differential technique, Carlisle et al. found that soda pop (carbonated beverage) was rated on the youthful side of a young–old scale, whereas orange juice was rated slightly on the youthful side but near neutral. Both soda pop and orange juice were rated near neutral on the male–female scale.

In an attempt to determine the dimensions of preschool childrens' food preferences, Birch (1979) had children rank order eight fruits. She did this by first asking children to point to the fruit they liked best, removing that fruit, repeating the question, and so on. Multidimensional scaling of the ranks yielded two dimensions, one labelled familiarity (29% of variance) and the other labelled sweetness (26%). In fact when adults were asked to rank order the fruits according to sweetness, the ranks correlated significantly with sweetness.

Price (1978) examined the food preference of 1009 8- to 12-year-old children, including 245 Mexican American, 263 black children, and 501 white children. Differences in preference for 58 foods using a five-point scale were negligible for the groups when sweet foods were examined. For example, sweet rolls were ranked number 9 for whites, number 8 for blacks, and number 9 for Mexican Americans. For milk products, ice cream, chocolate pudding, chocolate milk, and hot chocolate were the most popular items in a list of 17 products for all three groups. Similarly, watermelon, strawberries, bananas, and oranges were the most popular fruits for all three groups.

Price stressed the importance of individual variation in food preferences. It is probably safe to argue that individual variation will be greater than racial variation, ethnic variation, or gender variation. Pangborn (1981) has stressed the importance of examining individual variation in many food habits.

Meiselman (1977) and Kaufmann et al. (1975) have studied the food attitudes of American military personnel and Israeli high school students, respectively, of different levels of body weight. Meiselman noted relatively few differences in food preferences among sweet foods for persons of different weight as defined by life insurance tables. Fat-containing foods appeared to discriminate the weight groups better. Kaufmann et al. (1975) found that 90% of students felt that fruit should be eaten daily, but that 90% felt that "cake, sweets, and so on" were "fattening," more than any other food class, including fatty foods.

Food Acceptance of Sweet Foods

The sensory evaluation of chemical and food stimuli has been adequately covered elsewhere in this volume. I will therefore concentrate on acceptability of products within food service. By acceptance I am referring to the hedonic judgment of a product which has been sampled and evaluated. A number of studies have examined food acceptability within school food service. Head et al. (1977) compared three methods for measuring food acceptance within school food service: a five-point hedonic scale, the amount eaten, and weighed plate waste. Using the term acceptance in these ways, I believe, adds semantic confusion. In a related study, Harper et al. (1977) reported on "menu item acceptability" when actually reporting on percentage of servings consumed. While one of the goals of food habits research is to understand the interrelationships among the different levels of food habits, interchanging the terms from one level to the next is potentially confusing.

Head et al. (1977) used a two-part questionnaire. One part measured "how you liked it" and the other measured "how much you ate." The authors do not comment on the risk of yielding interdependent data sets with this combined questionnaire approach. Further, Head et al. used the scale "great, good, OK, not very good, terrible" for hedonic measurement. This scale does not meet the criteria of a good hedonic scale noted by Meiselman (1984).

Both elementary and high school students appeared to use the hedonic and amount consumed scales similarly with respect to the actual amount eaten, and this appeared to hold across different food classes. Further, Head et al. claim that over 50% of the variance in consumption data can be explained by acceptance ratings. They recommend the use of acceptance data for such predictions.

Head et al. (1982) reported on acceptability of breads and desserts in a 2-year study of elementary and high school students in North Carolina. They again used the same five-point scale as above. For overall ratings, males gave higher ratings than females. Elementary school males rated 45 out of 48 desserts higher than females, and high school males rated 33 out of 48 desserts higher than females. Black students gave more positive ratings than white students. Elementary school blacks rated 34 of 39 desserts higher; secondary school blacks rated 20 of 30 desserts higher. The acceptance data from black children are consistent with the preference data from both black children and black military. Head et al. also examined the ratings separately in those receiving free food and those who paid for their food. Those receiving free food gave higher ratings.

Ice cream desserts received the highest acceptance ratings. Cakes and biscuits followed in popularity. Acceptability of desserts varied from elementary to high school students, and from time to time. The less liked desserts appeared to be those containing raisins and prunes, those with spices, and those with a gelatin base. For both study years and for both school age groups there was a relatively narrow range of dessert scores compared with other food classes. Overall the range for dessert scores was from 3.90 to 4.49 on the five-point scale.

The United States military have studied the acceptability of their dining hall food and of their special field rations since World War II. In fact the United States Army developed the nine-point hedonic scale, which became a standard of product development technology (Peryam and Pilgrim 1957). A method has also been designed within the United States military to permit collection of valid food acceptance data in food service environments. Meiselman (1982) reported that a properly designed rating card could yield valid data. This was demonstrated by intentionally changing one dimension of a dessert item (the portion size of white cake) and asking raters to rate several dimensions of the product. Other rating cards yielded highly intercorrelated data on all dimensions. The card recommended for further use (Fig. 18.1) showed appropriate rating on the size dimension and no other correlated ratings for other dimensions.

Recent studies on food acceptance within the United States military have sought to further the earlier studies of product acceptance which utilized single

DID YOU SELECT WHITE CAKE ? ☐ YES (continue answering) ☐ NO (STOP & return card) **C**

After you have tasted the WHITE CAKE, please rate it for each of the following characteristics by checking one box in each category.

Temperature	Flavor	Portion Size	Texture
Much Too Hot ☐	Very Good Flavor ☐	Much Too Big ☐	Very Bad Texture ☐
Too Hot ☐	Good Flavor ☐	Too Big ☐	Bad Texture ☐
Slightly Too Hot ☐	Slightly Good Flavor ☐	Slightly Too Big ☐	Slightly Bad Texture ☐
Just Right ☐	Neutral Flavor ☐	Just Right ☐	Neutral Texture ☐
Slightly Too Cold ☐	Slightly Bad Flavor ☐	Slightly Too Small ☐	Slightly Good Texture ☐
Too Cold ☐	Bad Flavor ☐	Too Small ☐	Good Texture ☐
Much Too Cold ☐	Very Bad Flavor ☐	Much Too Small ☐	Very Good Texture ☐

TSA FORM 938² 30 OCT 75

Considering everything, how was the WHITE CAKE ?

Good ☐ Slightly Good ☐ Neutral ☐ Slightly Bad ☐ Bad ☐

COMMENTS:

Fig. 18.1. The feedback card developed for measurement of sensory evaluation and acceptance of products in food service (Meiselman 1982)

time testing or short-term testing. In two recent studies (Hirsch et al. 1984; Schnakenberg 1986, personal communication) large numbers of Army personnel have been fed a variety of foods for periods of up to 7 weeks. These studies have permitted detailed measurement of food acceptability and food intake over time.

Hirsch et al. (1984) did not observe any systematic changes in acceptability ratings over time for main dishes, desserts, or dehydrated fruits. There was a slight tendency for ratings of all products to increase in the first weeks of the test, probably as people became accustomed to the new rations. Schnakenberg's data also fail to show any systematic trends over time for a variety of sweet products.

Hirsch's data do not show a clearly higher acceptance of dessert items for either the traditional meals of freshly prepared foods or for the special military operational rations. For the latter, the dessert items were rated below either main dishes or fruits. For fresh food, desserts were rated the same as main dishes and also below fruits. This was consistent with opinions expressed by troops when asked about the portion sizes of their rations. For both fresh food and military rations troops said they wanted a greater portion of fruit, but that they were closer to being satisfied with their portions of main dish or dessert. Hirsch et al. were studying food intake and weight gain or loss in their troops. They analyzed whether troops who showed high weight loss gave different acceptance ratings to foods than troops who showed low weight loss. The comparisons did not uncover any pattern for foods in general, or sweet foods in particular.

Food Selection and Intake of Sweet Foods

Irrgang (1984) has collected data on dessert selections in food services outlets in Germany. These included cafeterias in hospitals and homes for the elderly, canteens at work, and restaurants. The restaurants account for 45% of all hot meals served outside the home. Utilization of some desserts was at the same level across the different types of food service, for example 7% were baked goods and cakes at all three. There were more pudding desserts in hospitals/homes for the elderly and there was more ice cream in restaurants. There is an increase in the use of fruit salad and fresh fruit for dessert.

A number of studies have examined food intakes in schools. Jansen et al. (1975) compared the nutrient content and acceptability of different lunch programs. In so doing, they determined the percent consumption of these different programs by food class. Milk beverages were the highest, with juices and entrées scoring high. Desserts scored lower than these but above salads and vegetables. These data are in agreement with the acceptance data presented above, namely that desserts are neither uniformly popular nor uniformly heavily consumed.

Jansen and Harper (1978) studied consumption of foods for fifth and tenth grade students at 23000 lunches containing 130000 food servings. For younger fifth graders, ice cream was the most heavily consumed item, followed by milk beverages and then cookies and candies. For tenth graders, milk beverages were most consumed, followed by ice cream and fish.

Tseng et al. (1983) studied the eating patterns of 10866 4th, 8th, and 11th grade students (ages 9, 13, and 16 years respectively) in California by questionnaire. Their calculation of average intake of sucrose from various sources during school

Table 18.3. Average intakes of sucrose by 8th and 11th grade students from various sources during school hours (Tseng et al. 1983)

Food source	Sucrose intake			
	Eighth grade			
	Males		Females	
	(g)	(%)	(g)	(%)
Cafeteria	3.64	25.9	3.52	23.9
Snack bar	4.15	29.5	4.68	31.8
Vending machine	0.28	2.0	0.49	3.3
Home	3.60	25.6	4.01	27.2
Other	2.38	17.0	2.04	13.8
	14.05	100.0	14.74	100.0
	Eleventh grade			
	Males		Females	
	(g)	(%)	(g)	(%)
Cafeteria	4.63	25.0	2.73	19.5
Snack bar	3.91	21.1	2.76	19.7
Vending machine	0.86	4.6	0.87	6.2
Home	5.70	30.7	3.98	28.4
Other	3.45	18.6	3.69	26.2
	18.55	100.0	14.03	100.0

hours for the older students is shown in Table 18.3. Consumption of sucrose was about 14 g for 4th and 8th grade students of both genders, and for female 11th graders. It was over 18 g for male 11th graders. For the oldest group, the source of most of the sugar intake was the home, but for 8th graders it was the snack bar. This may have been related to different types of food service for the different ages of students. Due to the relatively low sugar consumption reported in this study, caution is advised in interpreting these data.

As an interesting comparison with the studies of school food service, Camacho et al. (1984) observed the contents of bag lunches brought to school by 56 North Carolina elementary school children. The most frequently observed item was white bread, probably for sandwiches. Jelly/honey was one of the most popular sandwich fillings. Most beverages were sweet, with Koolaid and chocolate milk being most frequent. Of desserts, cookies and ice cream were most frequently observed. Fruits, especially apples, also appeared among sweet foods observed.

McLaughlin and Wickham (1982) report a comprehensive study of lunches for schoolchildren in Ireland. The study included schools which supplied a sandwich lunch and schools in which children brought a bag lunch. In addition to presenting detailed nutritional analyses of the lunches, they also provide a listing of the foods consumed. Sweet foods consumed by more than 10% of schoolchildren included apples, currant buns, chocolates, orange drink, and sweets. Interestingly, those children who went home for lunch ate more sweets and more currant buns at school than those children eating lunch at school. Soft drinks were not heavily consumed by any group at school.

Several reports are based on a study of 1135 United States children aged 5–18 using 7-day food diaries. Stults et al. (1982) reported on children's beverage consumption patterns (Table 18.4). Beverages appear in analyses of several levels

Table 18.4. Percentage contribution of various beverage groups to total beverage consumption at meals and snacks eaten by children and teenagers (Stults et al. 1982)

Beverage group	Meal/snack			
	Breakfast	Lunch	Dinner	Snacks
Coffee	4.7	0.3	0.4	1.0
Carbonated sweetened drinks	0.4	13.7	16.0	30.4
Fruit and vegetable juices	23.7	4.1	2.4	6.4
Milk and milk drinks	64.1	65.6	50.4	42.4
Non-carbonated sweetened drinks	4.7	9.5	11.1	15.7
Tea	2.4	6.8	19.7	4.1

of food habits to be a major contributor of sugar and a major differentiator between population subgroups. They found that milk and milk drinks were very frequently consumed and that carbonated and non-carbonated sweetened drinks were consumed by large numbers but at low frequencies. The same pattern carried over to the average amount consumed in 1 week, with milk being by far the highest, followed by tea, and then by both carbonated and non-carbonated sweet drinks. Finally the authors present the percentage contribution of the various beverages to total beverage consumption at both meals and snacks. Milk is the largest percentage contributor at every meal and snack, with over 50% of mealtime beverages being milk. Juices contribute heavily only to breakfast beverages, and are replaced by carbonated and non-carbonated sweetened drinks for lunch, dinner and snacks. For snacks, both sweetened drinks total more than the percentage of milk. Fromantin (1985) has reported on beverage consumption at lunch by French military personnel. He found that 87% of the beverages were sweetened drinks, including 50% Coca-Cola.

Utilizing the same data base as Stults et al., Morgan et al. (1983) computed total sugar consumption from the total diet and from snacks. The median weekly sugar consumption was 865 g (range 102–5188 g). The median weekly consumption from snacks was 222 g (range 0–3190 g). Morgan et al. next examined whether sugar consumption differed in subsamples of different relative weight. They concluded that there was no relationship between sugar consumption and obesity, namely, that overall sugar consumption from snacks was not different for slight, average, heavy, and obese persons. Further, they noted that sugar consumption was higher for persons having higher portions of their eating as total meals, and that slight individuals consumed more of their sugar from snacks. These results certainly go against the layman's view that obese people eat too many sweets and that lighter people do not.

Kaufmann et al. (1975) found higher consumption of cake and of sugar for teenage Israeli boys than for girls. However, midweight girls (91–120% relative weight) reported consuming more sweets. While consumption of ice cream by boys (about once per week) appeared to be independent of their weight class, consumption by midweight girls was at about this level, but higher for lower weight and lower for higher weight girls.

Food intake over extended periods of time was measured in the two recent studies of military personnel noted above (Hirsch et al. 1984; Schnakenberg 1986,

Table 18.5. Estimated and weighed mean dessert intake (grams) (adapted from Hirsch et al. 1984)

	Experimental		Control	
	Estimated	Weighed	Estimated	Weighed
9/3/83	147.80	141.65	78.84	77.24
9/10/83	129.44	129.44	82.35	75.70
9/16/83	119.24	108.65	80.43	77.29
9/26/83	122.32	122.16	71.58	71.58

personal communication). Hirsch et al. measured individuals' intake of each food item for military rations and regular freshly prepared food. Dessert intake for the military rations is shown in Table 18.5. The researchers compared weighed and estimated dessert intake on 4 days over the test period. There was good agreement between weighed and estimated figures. There was better agreement between measures for the groups eating military rations, which are preportioned. From Table 18.5 one can also note that less military ration dessert is consumed by the group eating other foods during the day (control condition). There did not appear to be a tendency for dessert consumption to increase as the study proceeded. This same result was observed by Schnakenberg in a longer study of military rations.

Barr et al. (1983) reviewed studies on nutrient intakes of elderly women and presented new data. They compared intakes of institutionalized and free-living persons. The two studies of free-living elderly women reported higher energy intake than any of the seven studies of institutionalized women. This study should caution us not to generalize data from institutionalized elderly to the elderly population in general.

Cronin et al. (1982) reported on food usage in the United States from the 1977–78 Nationwide Food Consumption Survey of 15000 households containing 31000 people. The data were 3-day dietary records. Although there were some interesting trends for fruits and desserts, there were startling differences in the usage of the category of sweets by older people (age 51+). Thus if one looks at the overall category of "sweets" and the subcategories of "sugar and sweet spreads," "candy," "sugar-based beverages" and "carbonated beverages," people aged 51+ reported using this class of products less than any other age group. The differences were especially striking for beverages (Table 18.6). When the mean number of times of using foods per day was compared across age groups, the frequencies were lowest in the oldest group for "sweets", "sugar-based beverages," and "carbonated beverages" but not for "sugar and sweet spreads."

Food Waste of Sweet Foods

Food waste has been studied within school food service. Perhaps this is because of the sensitivity of that system to the possibility of waste and its resulting cost. Waste can also be used as an index of acceptability, and this is another possible reason for the focus on waste within school food service.

Generally, food waste can be measured in one of two ways, either through direct weighing or through visual estimation. Several recent papers have compared these two techniques within school food service. Comstock et al. (1981)

Table 18.6. Percent of persons who reported using sweets in 3-day dietary records by age (adapted from Cronin et al. 1982)

Food group	Age (years)					
	3–6	7–10	11–14	15–17	18–50	51 and over
Sweets	95	94	93	95	88	77
Sugar and sweet spreads	73	72	62	63	61	59
Candy	22	22	25	15	12	7
Sugar-based beverages	72	77	74	80	66	38
Carbonated beverages	66	69	65	73	62	33

used a six-point scale as follows: 5 – full portion remaining, 4 – nearly a full portion remaining, 3 – three-quarters of a full portion remaining, 2 – half of a full portion remaining, 1 – one-quarter of a full portion remaining, 0 – none remaining. After the visual estimations, detailed weighings were conducted on the same ten randomly drawn trays per day. Comstock et al. also compared a technique of child estimation using the following six-point scale: 1 – I ate none of it, 2 – I just tasted it, 3 – I ate a little, 4 – I ate half of it, 5 – I ate a lot, 6 – I ate all of it. Correlations between children's ratings and percent weighed waste varied from 0.69 to 0.77 for different food groups, indicating that, for example, correlations for sweet desserts were no different than for non-sweet meats or eggs. Similarly, corelations between visual estimations and percent weighed waste varied from 0.92 to 0.94 for various food groups. Visual estimation appears to be a less expensive, less intrusive means of collecting food waste data for sweet foods.

Stallings and McKibben (1982) studied visual estimation of plate waste using a five-point scale: 0, 1/4, 1/2, 3/4, 1. Correlation between visual estimations and weighed value were from 0.576 to 0.987. Sweet items included both higher and lower correlations. When plate waste was reported as the mean of the visual estimation scale data based on about 1200 plates studied, different food classes had different waste estimates. These ranged from the lowest plate waste for fruits (mean 0.50), closely followed by roll (0.54), milk (0.62), and meat (0.70), to starchy vegetable (1.14) and green vegetable (2.27). There was a significant gender difference for every food class, with males having less waste in every case (the difference approached significance for meats). For fruits, the average visual rating for 620 males was 0.41 and for 585 females, 0.57.

Jansen and Harper (1978) weighed the food waste from five randomly chosen trays per day for 10 days from 58 schools. The trays of five randomly selected students per day were weighed both before and after eating. Data were collected for both fifth and tenth graders. Desserts were generally highly consumed, from 61% to 93% for ice cream and sherbert. The most frequently served desserts (fruit desserts and juices) were consumed less (about 70%) than "cakes, pies, doughnuts, sweet rolls" (about 80%), "cookies and candies" (83%), or "ice cream and sherbert" (92%).

Harper et al. (1977) measured plate waste as a measure of acceptability. They calculated the percentage of serving consumed for ten food items, including chocolate pudding and sliced peaches. The vegetables had the highest waste (lowest percentage consumed), while the two desserts had among the lowest waste.

Comstock and Symington (1982) studied the distribution of serving sizes and plate waste in school lunch. This work defines the number of plates which must be observed in such studies. To determine serving sizes, 1907 serving sizes were measured on 204 different food items. The edible portion of each food was weighed to the nearest gram. To determine plate wastes, 13000 measurements were made on 214 different food items. The shapes of the distributions of serving sizes did not vary meaningfully for different food classes, including desserts, fruits, and fruit drinks. Comstock and Symington went on to analyze the waste data to provide measurement guidelines for other researchers. They analyzed the accuracy of waste sampling methods, providing data on how many waste samples should be drawn. Results for prepackaged foods showed that one sample was adequate. However, for on-site prepared foods using a sample of five foods in 200 produced 64% which failed to meet the criterion. With ten food samples in 200, 34% still failed to meet criterion. With populations of 1000 or 50 (compared with the 200 above), the percentage failing to meet the criterion would be 67% and 34%, and 61% and 26%

Comstock and Symington further determined that percent waste of a food item is best characterized by the proportion of individuals who consumed all or almost all of the serving, and the proportion who consumed none or almost none of the serving. These end zones were estimated at 10%. The percentage who consumed partial amounts was very evenly distributed across the midrange of waste measures. Since so many respondents had plate waste of either less than 11% or greater than 91%, this provides indirect support for the use of visual estimation since the ratings of "none or almost none wasted" and "all or almost all wasted" are easily made.

Conclusions

1. Sweet foods are not universally well liked and heavily consumed. Sweet foods contain a range of products with a range of appeal and a range of consumption. It is important for researchers and practitioners to realize this range of sweet products.

2. Several products deserve special mention because of their special appeal. Ice cream is especially well liked and highly accepted. Among many groups fresh fruit is highly accepted. Sweet beverages are well liked and heavily consumed.

3. Sweet beverages, including fruit drinks and carbonated beverages, are a key factor in differentiating the sweet food habits of many population subgroups, including racial groups and age groups.

4. Although there are many interesting and significant differences in the sweet food habits of different population subgroups, the similarities of the food habits are more striking. In general, individual variation is greater than group variation.

5. Sweet food habits do not appear to be related to relative body weight (overweight, underweight) in any clear way. There is no compelling pattern of high preference, acceptance, or consumption of sweet foods by the overweight.

References

Barr SI, Chrysomilides SA, Willis EJ, Beattie BL (1983) Nutrient intakes of the old elderly: a study of female residents of a long-term facility. Nutr Res 3: 417–431
Birch LL (1979) Dimensions of preschool children's food preferences. J Nutr Educ 11(2): 77–80
Camacho MM, Zallen EM, O'Brien KF (1984) Nutrient content of food in bagged lunches of children in kindergarten through third grade. School Food Service Res Rev 8(2): 122–125
Carlisle JC, Bass MA, Owsley DW (1980) Food preferences and connotative meaning of foods of Alabama teenagers. School Food Service Res Rev 4(1): 19–26
Comstock EM, Symington LE (1982) Distributions of serving sizes and plate waste in school lunches. J Am Diet Assoc 81: 413–422
Comstock EM, St Pierre RG, Mackiernan YD (1981) Measuring individual plate waste in school lunches. J Am Diet Assoc 79: 290–296
Cronin FJ, Krebs-Smith SM, Wyse BW, Light L (1982) Characterizing food usage by demographic variables. J Am Diet Assoc 81: 661–673
Einstein MA, Hornstein I (1970) Food preferences of college students and nutritional implications. J Food Sci 35: 429–436
Fromantin M (1985) L'Education nutritionelle lors du service national. Medicine et armees 13: 809–816
Halpern BP, Kelling ST, Meiselman HL (1986) An analysis of the role of stimulus removal in taste adaptation by means of simulated drinking. Physiol Behav 36: 925–928
Harper JM, Jansen GR, Shigetomi CT, Fallis LK (1977) Pilot study to evaluate food delivery systems used in school lunch programs. I. Menu item acceptability. School Food Service Res Rev 1(1): 20–23
Head MK, Giesbrecht FG, Johnson GN (1977) Food accepability research: comparative utility of three types of data from school children. J Food Sci 42: 246–251
Head MK, Giesbrecht FG, Johnson GN, Weeks RJ (1982) Acceptability of school-served foods III. Breads and desserts. School Food Service Res Rev 6(2): 98–101
Hirsch E, Meiselman HL, Popper RD, Smits G, Jezior B (1984) The effects of prolonged feeding meal, ready-to-eat (MRE) operational rations. Technical Report Natick/TR-85/035, United States Army Natick Research and Development Center, Natick, Massachusetts
Irrgang W (1984) Der Markt der Grossverbraucher. Anstalten, Kantinen, Gastronomie, Fast-Food- und Systemgastronomie. CMA Mafo-Jahrbuch
Jansen GR, Harper JM (1978) Consumption and plate waste of menu items served in the National School Lunch Program. J Am Diet Assoc 73: 395–400
Jansen GR, Harper JM, Frey AL, Crews RH, Shigetomi CT, Lough JB (1975) Comparison of type A and nutrient standard menus for school lunch. J Am Diet Assoc 66: 254–261
Kaufmann, NA, Poznanski R, Guggenheim K (1975) Eating habits and opinions of teen-agers on nutrition and obesity. J Am Diet Assoc 66: 264–268
Krantzler NJ, Mullen BJ, Comstock EM et al. (1982) Methods of food intake assessment – an annotated bibliography. J. Nutr Educ 14(3): 108–119
Lyman B (1982) Menu item preferences and emotions. School Food Service Res Rev 6(1): 32–35
McLaughlin B, Wickham C (1982) Nutritional quality of children's school pack lunches. Technical Bulletin No. 6, The Agricultural Institute, Dublin
Meiselman HL (1977) The role of sweetness in the food preference of young adults. In: Weiffenbach JM (ed) Taste and development: the genesis of sweet preference. DHEW Publication No. 77-1068, National Institutes of Health, Maryland, pp 269–281
Meiselman HL (1982) Design and evaluation of cards for customer feedback of food quality. J Food Service Systems 2: 7–21
Meiselman HL (1984) Consumer studies. In: Piggott JR (ed) Sensory analysis of foods. Applied Science Publishers, Barking, England
Meiselman HL (1986a) Measurement of food habits in taste and smell disorders. In: Meiselman HL, Rivlin RS (eds) Clinical measurements of taste and smell. MacMillan, New York
Meiselman HL (1986b) Use of sensory evaluation and food habits data in product development. In: Kawamura Y, Kare M (eds) Umami: physiology of its taste. Marcel Decker, New York
Meiselman HL, Waterman D (1978) Food preferences of enlisted personnel in the Armed Forces. J Am Diet Assoc 73: 621–629
Morgan KJ, Johnson SR, Stampley GL (1983) Children's frequency of eating, total sugar intake and weight/height stature. Nutr Res 3: 635–652

Moskowitz HR (1983) Product testing and sensory evaluation of foods. Food & Nutrition Press, Westport, Conn
Pangborn R (1981) Individual differences. In: Solms J, Hall R (eds) Criteria of food acceptance. Forster, Zurich
Peryam DR, Pilgrim (1957) Hedonic scale of measuring food preferences. Food Technol 11: 9–14
Piggott JR (1984) Sensory analysis of foods. Applied Science Publishers, Barking, England
Price DZ (1978) Food preferences of eight- to twelve-year-old Washington children. School Food Service Res Rev 2(2): 82–85
Restaurants and Institutions (1985) 1985 Tastes of America: Part I 95(25)
Stallings SF, McKibben GD Jr (1982) Validation of plate waste visual assessment technique in selected elementary schools. School Food Service Res Rev 6(1): 9–13
Stults VJ, Morgan KJ, Zabik ME (1982) Children's and teenagers' beverage consumption patterns. School Food Service Res Rev 6(1): 20–25
Tseng RY, Sakai L, Sun R, Smith E (1983) Eating patterns of California school children I. During school hours. School Food Service Res Rev 7(1): 19–28
Wyant K, Meiselman HL (1984) Sex and race differences in food preferences of military personnel. J Am Diet Assoc 84:169–175

Commentary

Beauchamp and Cowart: Meiselman's chapter raises two issues. First is his conclusion that individual variation is likely to be larger than group variation in sweet food habits. Since sweet food habits of very few ethnic groups have been studied, is it possible that this statement is premature? The material reviewed consisted mainly of studies of North American black and white individuals. A broader anthropological approach might reach different conclusions. Secondly, Meiselman rightly points out how difficulties may arise from different uses of a single term. Consider the meaning of the term acceptance or acceptability. Meiselman defines this as the hedonic judgment of a product which has been sampled and evaluated, and he criticizes those who use the term to refer to hedonic scales, amounts eaten, and/or amounts wasted as adding semantic confusion. In the developmental literature (and in some of the animal literature), acceptance typically refers to the amount consumed in a single choice situation. This is logically separate from hedonic judgments of the item, and it allows for the possibility that an item may be very acceptable without being very pleasant or vice versa (e.g., medicine when one is sick or ice cream when one is on a diet). This is a particularly important distinction in studies of human adults, a point Rozin has often made.

Meiselman: Concerning the semantics of food habits, I have simply proposed one possible scheme for organizing food habits and have tried to use my terms consistently. The current diversity of terms in the food habits field is confusing.

Bartoshuk: Meiselman raises the important point that sweet foods are not universally preferred by human subjects. The fact that this conclusion would startle many underlines how universally we consider sweetness to be highly desirable. Why then is not a highly desirable substance always highly desirable to everyone? Part of the answer is a sensory one. Sweeteners are present in the real world as components of complex mixtures. What we know about the laws of taste mixtures is discussed in Pangborn's chapter but we can go on to discuss higher order mixtures. For example, we might ask how taste and smell interact in mixtures. Studies on this have been done and the answer tends to be that they interact a lot less than one might expect. What about fattiness? This is especially interesting in the light of Drewnowski's lovely work on fat–sweet interactions. However, we might go even further and ask about the cognitive fusion of all of the sensations that go into "sweet foods". The Germans even had a term ("*Verschmelzung*") for the special way in which components of a complex mixture could seem to belong together so much that they are in some sense unitary. Can introspection reveal the components in such cases? The answer is — sometimes yes and sometimes no. Some mixtures are analytic, meaning that the components can be recognized. For example, auditory mixtures are analytic and I believe that taste mixtures are analytic. On the other hand, some mixtures are synthetic, meaning that the components fuse into qualitatively new sensations. Color mixing provides the classic example of synthetic mixing. In general, sensations from different modalities (e.g., taste and smell) are believed to "mix" analytically (although the perceived intensities of the identifiable components may be altered).

When we speak of "sweet foods" we might be speaking of foods that have a component (sweetness) that can be analyzed out and that meets some intensity criterion. On the other hand, perhaps we are really speaking of objects that meet some cognitive criterion. I'd bet that the meaning of "sweet" to the subjects in our studies varies a great deal from study to study. I'd also bet that the meaning to the subject will affect her/his behaviour.

Booth: With regard to the concept of sweetness, some food aromas and other smells are thought of and sensed as sweet or "sweetish". I believe that sugar on the tongue gives meaning to the word "sweet" (Chap. 10) but that does not give us the right to deny that those aromas really elicit a sort of sweet sensation.

Meiselman: Dr. Bartoshuk presents an interesting sensory basis for why not all sweet foods are well liked. Perhaps part of the reason also lies in the thinking and behavior of the researchers. When we consider sweet foods we think of chocolate cake or ice cream or cola. How often do we use raisin pie as our model sweet food? If we used a broader range of sweet foods, we might produce a broader range of results from sweet foods.

Rolls: Can this type of data be used to answer the following types of question?

Are particular types of meal more likely to be accompanied by sweet drinks or desserts?

Do we know whether inclusion of sweet foods or drinks increases intake in a meal? Does it increase satisfaction with a meal? Does the type of sweetener or sweetness intensity affect acceptance of the sweet foods in the meal?

Does repeat consumption ever decrease the acceptance of sweet foods?

Meiselman: We can answer one of your questions. The data from Hirsch et al. (1984) showed no change in acceptance or consumption of sweet desserts over a lengthy study in which items were served repetitively. Unfortunately, food habits research in either laboratory or field has not yet addressed the other questions.

Drewnowski: I agree that there is no single best method for measuring dietary intakes. Assessments of sugar consumption by children or adults have been especially difficult to perform. However, some methods and some studies are clearly more thorough and reliable than others. For example, Kaufmann et al. (1975) asked a few basic questions regarding nutrition and diet of a limited number of Israeli teenagers. In contrast, studies by Morgan and Zabik (1981) are based on a cross-sectional sample (n = 657) of children aged 5–12 years and 7-day food studies. These last two authors typically work with large-scale nutritional survey data, such as the National Food Consumption Survey (NFCS) or the National Health and Nutrition Survey (NHANES), and have access to the Michigan State University data base, which includes the precise sucrose content of most commercial food products, including snacks, soft drinks, candies and confectionery products as provided by the manufacturers. Did Tseng et al. (1983) have access to a comparable data base in calculating sugar intake during school hours?
What often happens with dietary assessment procedures is that the deeper you probe and the more questions you ask, the more answers you get. Tseng et al. (1983) studied snack intake "by questionnaire" and obtained intake estimates of 14–18 g sucrose/daily. Morgan and Zabik (1981) estimate the average daily total sugar consumption at 134.3 g, with a few children consuming over twice as much per average day. These discrepancies are clearly due to methodological differences between the studies, and additional work on dietary intake methods is required.

References

Morgan KJ, Zabik ME (1981) Amount and food sources of total sugar intake by children ages 5 to 12 years. Am J Clin Nutr 34: 404–413
(For remaining references, see preceding reference list.)

Meiselman: Unfortunately, a view of food intake methods was beyond the scope of this paper. Drewnowski correctly cautions the reader to consider methodological details in evaluating or generalizing the data from experiments cited.

Blass: If not already in practice, Meiselman's food evaluation system (as presented in Fig. 18.1) holds some therapeutic promise. It can be utilized with eating disorder patients. At the least, the data will objectively confirm clinical impressions of return to health during institutionalization. At best, Meiselman's evaluation system might indicate asymmetries between the patient's behavior and real attitudes toward food. Such asymmetries might predict recovery and stability upon discharge.

Index

Acceptability tests 240, 244
Acceptance psychophysics
  parameter relations 153
  rating anchors 151–3
Adapting concentration, 'water taste' 37
Adenosine, receptor activity 4
Affective responses, obesity 180
Affective tests 51
Age
  salt taste acceptance 135
  sweet acceptability effects 132–5
  sweet taste intensity 128–9
  sweetness thresholds 44
  taste modifications 233
AH,B glucophores 4–6, 9–10
Alliesthesia 73, 161–2
  negative 186, 197
Alloxan, taste receptor effects 34
Anorexia nervosa
  diagnostic criteria (DSM III) 193–4
  dietary patterns 195–6
  glucose tolerance 198–200
  sweet sensitivity and hedonic valuation 196–7
  taste psychobiology 198
  taste responsiveness 186
  trace metals and taste sensitivity 196
Antisocial behaviour, dietary studies 218–21
  test population 212, 220
Anxiety
  disorders, sugar challenge 215–16
  eating disorders, sweet preference effects 200
Appetite
  psychophysics, sensory affect quantification, free descriptors 151
  transient suppression 155–6
Apple juice, sweetness/acidity rating 255–6
Aqueous medium type, taste interaction effects 53
Aqueous solutions
  ethyl alcohol and sweetness enhancement 56
  taste perception, temperature effects 59–60
Aroma, sweetness interaction 57
Artichoke, taste-modifying effects 42
Artificial sweeteners
  food intake and body weight 167–9
  and marketing 241–3
  UK-permitted 6
  world bulk output 248, 249
Aspartame 6
  governments' acceptance 242
  hyperactivity, sucrose differences 214–15
  intake/energy incomplete compensation 156
  marketing 242
  sensory-specific satiety 162–3
  storage temperature effects 254
Athletic performance
  carbohydrate intake
    intra-exercise 217–18
    pre-exercise 217
  long-term dietary studies 216–17
  research strategies 206–9
  test population 212
Attention-deficit disorder 206, 214
  sugar challenge 215–16
Australia, consumer attitudes 92
Auxoglucs (Oertly and Myers) 4

Bacteria, sugar chemotaxis 115
Behavioral studies
  acute and chronic 207–9
  correlational 208
  independent variable definition 209–11
  research strategies 206–9
  test population 211–12
Beverages
  children's consumption patterns 268–9
  diet 254
  preferences of blacks 263, 264
  preload, fluid or food? 166
  sugar pairing 100, 107
  taste interactions 55–6

see also Carbonated drinks
Binge eating *see* Bulimia nervosa
Body weight
artificial sweeteners and food intake 167–9
control problems 154–6
sweetness pleasantness 70–3
see also Obesity; Overweight
Bombesin, food satiety, sucrose solution effects 167
Brain, electrical self-stimulus, morphine/morphine antagonist effects 118
Brand names, marketing 244
Bulimia nervosa
binge sweet intake 195
diagnostic criteria (DSM III) 193–4
dietary patterns 195–6
glucose tolerance 199–200
metabolic and behavioural aspects 199–200
sweet sensitivity and hedonic valuation 196–7
taste psychobiology 198
Bulk sweeteners 6

Cabanac reponse alliesthesia theory 73, 161–2
Caloric dilution, and compensation 169–70
Caloric load, behavioural aspects 211
Calorific need, sweet taste acceptability 134–5
see also Energy status; Satiety
Cancer, sweet sensitivity 234, 238
Carbonated drinks
high fructose corn syrup 253
invert sucrose 252
psychological attribute ratings 75
Carboxymethylcellulose viscosity, sweetness perception 58
CCK, food satiety, sucrose solution effects 167
Chemical classes, sweetness 5–7
Chemoreception 3
mechanism 9–10
new approaches 10–12
studies 10
Child behaviour, sugar effects
challenge studies 213–16
correlational studies 212–13
Children
food attitudes 263–5
normal, sugar challenge studies 215
nutrition education 232–3
sweet food preferences and familiarity 187
upbringing and education 228–32
Chlorodeoxy sugars 6–7
Chocolate 251
Chorda tympani, taste mediation 121
Coca Cola
consumer resistance to change 244
marketing 244
Cognitively coded sensation 74–5
Colour, sweetness interaction 56
Compliance, extrinsic motivation 105
Compounds, molecular configuration changes, taste effects 6–9
Concentration, perceived sweetness relationships 68
Concept tests 243
Conditioned taste aversions (CTAs) 22–3
Conduct disorders, sugar challenge 215–16
Consumer attitudes
behaviour relationships 91–2
fluctuations and trends 91–3
food nutritional content 92
marketing effects 239–40
scientific/medical community effects 93, 95–6
sex differences 76
sweetness and sweet foods 74–7, 239–40
Contagion law (Frazer; Mauss) 102
Context-dependent effect, sweet taste acceptability 135
Context-dependent responses 49–50
Contexts and meanings, acquisition 108–9
Corn syrup, high fructose *see* High fructose corn syrup (HFCS)
Cornstarch viscosity, sweetness perception 58
Crime
dietary studies, test population 212
sugar intake 218–21
Crohn's disease 235
Cross-adaptation
asymmetric 41
reciprocal 55
Cross-adaptation studies 37–41
artifacts 38–41
research reports 40–1
unequal sweetness 38–9
Cyclamates, banning 242

Dairy products, frozen 252
Depressive states, taste perception changes 234
Diabetes mellitus 43
non-insulin dependent, carbohydrate/fat/sugar intake 177
sugar/starch warnings 104
Diencephalon, taste afferents 21–2
Diet, carbohydrate/protein ratio, behavioural effects 210
Diet drinks 254
Dietary experience, sweet taste acceptability 130–2
Dietary intake, obesity 177–8
Dieting (slimming) 91
Discrimination ratio 149
Disorders, 'sugar-related' 88–9
Distress vocalizations, sucrose/endorphin effects 120, 121–2
Dose-effect reversal belief 102
Dufty, 'Sugar Blues' 88
Dulcin, species response differences 36

Eating disorders 193–4
see also Anorexia nervosa; Bulimia nervosa; Obesity

Eating intimacy, sugar associations 100–1
Education, sweetness implication 227–38
Elderly, sweetness thresholds 44
Endearment terms, sweetness-related 228
Energy status, sweet taste preferences 185–6
Ethyl alcohol
  aqueous solution sweetness enhancement 56
  sugar as substitute 89
Evaluative conditioning 106–7
Excitability, dietary aspects 208
Exposure, food liking effects 106

Facial expressions
  child decoding 230
  evoked 230–1
  gustatory 230–1
Fat
  intake, adult-onset diabetes 177
  sugar combinations, obesity 181, 182–5
  sweet taste effects 251
Feeding, stress-induced 121
Feingold elimination diet 206
Fishbein multi-attribute model 74
Flavour, sweetness combination, acceptance 100
Food
  acceptance/rejection
    categories 105–6
    social-affective context 107
    sweet foods 265–7
  anticipation, cephalic insulin release 200
  deprivation, sweetness pleasantness effects 73, 161–2
  intake behaviour, sweetness effects 67–90
  interactions, palatability effects 164
  likes/dislikes
    acquisition 105–6
    exposure effects 106
    overjustification effects 107–8
  plate wastage 270–2
  rating cards 266
  taste interactions 55–6
Food attitudes
  food service systems 262–5
  sex and racial aspects 263–4
Food selection 144–5
  food complex and sweet taste 146–7
  intake and 267–70
  sweetener effects, development 145–6
  sweetness objective influences 145
  sweetness role, rapid objective measurement 154
Frazer and Mauss, laws of sympathetic magic 102, 103
French press, attitudes 93–6
ß-Fructofuranose 253
D-Fructopyranose, AH,B system 10
Fructose
  fruit flavour accentuation 253
  gastric emptying 166
  plasma insulin levels 165
  preloads, hunger and food intake 165–6
  saccharin synergy 253
  squirrel monkey response 34
  see also High fructose corn syrup (HFCS)
Fruit flavour accentuation, fructose-associated 253
Fruit juices
  children's consumption 269
  preferences 263

Gastric emptying, various sugar effects 166
Gelatin viscosity, sweetness perception 57
Glucophores 9–10
Glucose
  gastric emptying 166
  metabolism, hedonic response relationships 184–5
  plasma insulin levels 165
  preloads, hunger and food intake 165–6
D-Glucose, sweetener synergisms 55
Glycosuria, discovery (Willis) 86
Gustofacial reflex (GFR) 230–1
Gymnema sylvestre, taste receptor effects 35–6

Health, sweetness perception 234–5
Hedonic value 69
Hedonics, neurophysiology 21–8
High fructose corn syrup (HFCS)
  carbonated drinks 253
  home/industrial uses per capita 250
High school students, food attitudes 263–4
Home consumption, sugar 248–50
Honey 104
  'humoral qualities' 85
Hyperactivity
  dietary aspects 205–6
  sucrose/aspartame differences 214–15
  sugar association 90
  sugar challenge 213–15
Hypoglycaemia 90
  exercise-induced 199
  insulin-induced 185

Ice cream, psychological attribute ratings 75
Ideal-rejection points
  perceived sweetener differences 153
  sweet taste 147, 152
Ideal-relative intensity rating 152
Industrial consumption, sugar 248–50
Infants, obesity and sweet taste 186–7
  see also Newborn infants
Insulin, plasma levels
  carbohydrate intake 198
  glucose/fructose ingestion 165
Intake/energy, incomplete compensation, aspartame 156
Intense sweeteners 6
Intensity
  perceived 50–1, 68–9

rating 69
and thresholds 50

Juvenile offenders, dietary studies 218–21

Labeled-line taste coding theory (Pfaffmann) 34
Labelling information, product acceptance 103
Law of contagion (Frazer; Mauss) 102
Law of similarity (Frazer; Mauss) 103
Laws of sympathetic magic (Frazer; Mauss) 102
Laxatives, sugar preference effects 200
Lipid metabolism, hedonic response relationships 184–5
Lipophilicity, sweetness enhancement 4–5, 10

Macaque, taste cortex gustatory projections 25–8
McBurney gustometer, adaptation measurement 39
Maltodextrins, transient appetite suppression 155
Marketing 239–46
contemporary 240–3
difficulties 242–3
methodologies 243–5
Masturbation, sugar consumption analogy 90
Mayonnaise
hedonic scale rating 244–5
product optimization 244–5
Media
consumer attitude moulding 93–6
French, attitudes to sugar 93–6
issues discussed 94–5
types, attitudes to sugar 94
Methylcellulose viscosity, sweetness perception 57
Military, food intake studies 269–70
Milk, children's consumption 269
Miracle fruit, taste receptor effects 36
Mixtures
component substitution/synergism 43
pairing discrimination, reaction time 44
synergism 8, 43
taste suppression 55
cross-adaptation artifacts 39
see also Taste interactions
Molar volume, sugar, water volume displacement 11
Monellin 39–41
Moral attitudes, sugar and sweetness 89–90
Morphine, saccharin ingestion effects 118

NaCl
bitterness side taste 53
sweetness side taste 53
Naloxone
stress/opioid-enhanced pain threshold effects 118–19
sugar ingestion effects 118
Naltrexone, sugar ingestion effects 118
Nausea, food distaste acquisition 106
Neohesperidin dihydrochalcone (neo DHC) 39–41
Neural code, sweetness 15–32
alterations 22–8
experience effects 22–3
physiological needs 23–8
Neurons, sweet stimuli response 15–21
see also Taste neurons
Neurophysiology 15–32
Newborn infants 232–3
forehead stroking/sucrose delivery conditioning 117
sweet taste development 128
Non-verbal communication, social behaviour and sweetness 230–1
Nucleus tractus solitarius (NTS)
glucose response 15
taste responses 18–20
taste-evoked activity, satiety effects 24–8
Nutrition, and taste 101
Nutrition education 232–5
adults 234–5
children 232–3

Obesity
affective responses 180
individual differences 181
dietary intake 177–8
fat/sugar combinations 181, 182–5
satiety deficit 155
sugar consumption relationships 269
sweetness perception intensity 179–80
taste over-responsiveness 178–9
see also Overweight
Opioid antagonists, sugar ingestion effects 118
Opioids, endogenous
affective system effects 121–2
sugar ingestion control 118
Orocline (Booth) 144
Overweight, sweetness pleasantness 70
see also Body weight; Obesity

Paediatricians, consumer attitude moulding 96
Pain thresholds, stress/opioid-enhanced, naloxone effects 118–19
Palatability
food interaction effects 164
repeated consumption effects 164–5
Parabrachial nuclei, taste neurons, hamster 16–17
Parachors 4
Parental attitudes 90
Pearson product-moment correlation coefficient, neuron functional similarity 16–17
Pediatricians, *see* Paediatricians
Pepsi Cola, marketing 244

Persistence 7
Pharmacophores 3
Phenylthiocarbamide (PTC), detection thresholds 42
Physicians, consumer attitude moulding 96
Pleasantness, experienced 69–73
Polydipsia, sweet taste solutions 167
Preference function 69, 71
Preference tests 244
Product acceptance and use, effect of semantics 103–4
Product development
  sweetener choice 252–3
  sweetness role 247–60
Product optimization 254–8
  marketing methodology 244–5
  sensory characteristic measuring 255–6
  trials 256–8
    factorial design method 257
    sequential approach 256
    simplex method 257
    single factor method 256–7
    surface response method 257–8
Propylthiouracil (PROP), detection thresholds 42
Protein/carbohydrate ratio, dietary, behavioral effects 210
Proteins, sugar behavioral effect suppression 212
Proteolytic enzymes, taste receptor effects 34
Protocepts, consumer testing 243
Psychohedonic function 149
Psychohedonicity (Frijters) 71
Puritan values, sugar as sinful 100

Quality effects, temporal relationships 7
Quinine, saltiness side taste 53

Rats
  sucrose ingestion 115–17
  sweet system 116–17
Reaction time
  delayed 7
  pairing discrimination 44
Receptor mechanisms
  competitors 34
  multiple
    neurophysiology 33–6
    psychophysical aspects 36–44
Receptors 11
Rewards, food liking 107–8
  see also Sugar tender

Saccharin
  banning 242
  ingestion, morphine effects 118
  species response differences 36
Saccharin/sucrose mixtures, receptor stimulation 36
Saccharophobia 87–9
Safety beliefs 101–2
Salivation
  product-specific 72
  sugar concentration decrease 72–3
Salt taste acceptance, age effects 135
Saltiness/sweetness, within-modality, aqueous mixtures 54
Sapophores 3, 4
Satiety
  deficit, obesity 155
  fluid-related, saccharin solution intake 167
  food liking 106
  food selection effects 155–6
  NTS taste-evoked activity effects 24–8
  sensory-specific 162–3, 186
  sugar type comparisons 165–6
  sweetness pleasantness effects 73, 161–2
  transient 155–6
  see also Calorific need; Energy status
School food service, food plate wastage 270–2
Scientific/medical community, consumer attitude moulding 93, 95–6, 97
Scurvy, sugar as cause (Willis) 86
Self-adaptation 37
  failure 39
Serotonin, brain levels
  behavioral changes 207
  carbohydrate ingestion 206
Sex steroids, sweet preference modulation 135
Side tastes 52–3
Similarity law (Frazer; Mauss) 103
Sinful beverages, sugar association 100
Social-affective context, food acceptance 107
Socialization
  intrinsic and extrinsic motivation 104–5
  sweetness as agent 106–7
Sources of sweetness 248–9
Specific apparent volume, taste studies 11
Stress, eating disorders, sweet preference effects 200
Stroking, sucrose delivery conditioning 117
Structure:activity studies 4
Sucrose
  adaptation, water taste 39
  average intake, students 267–8
  endogenous opioid release 120–1
  home/industrial uses per capita 250
  hyperactivity, aspartame differences 214–15
  pain threshold effects 119–21
  product levels 251
  psychophysical functions 71
  sensory-specific satiety 162–3
  solution concentration, rat ingestion 115–17
  squirrel monkey response 34
  stress reponse effects 119–21
  thaumatin negative correlation 43
Sugar
  ambivalent attitudes 89–90
  attitude scale 75–6
  consumption 99, 248–50

early status 84–5, 240
emergence of reservations 85–6
ethical-political aspects 86
historical aspects 84–7
moral attitudes 89–90
nineteenth century social attitudes 86–7
as spice and medicine 84–5, 240
whiteness, early views 85
Sugar tender (Fischler) 229, 233
see also Rewards
Sugars
chlorination, bitter products 6
transient appetite suppression 155–6
Sweet taste
acceptability
individual differences 153–4
measurement stability 153–4
sweetness levels and context 146
development, newborn infants 128
dietary experience effects 130–2
effects of fat 251
familial aspects 130
ideal and rejection points 147, 152
maturation and sensitivity effects 128–9
preferences within food complex 146–7
tolerance triangle 148–9
Sweeteners
correlations across 42–3
home consumption 248–50
industrial consumption 248–50
properties and functional uses 252–3
stability 254
synergy 253
world bulk output 248, 249
Sweetness, theories 3–5
Synergy, sweeteners 253

Taste
coding, labeled-line theory (Pfaffmann) 34
cultural aspects 233–4, 235
nutritional aspects 101
weaning and 233
see also Sweet taste
Taste interactions 49, 51
aqueous solutions 51–6
Taste neurons
cluster analysis (Everitt) 17–18
response
spatial distribution 15–20
temporal distribution 20–1
Taste/odour, between modality, aqueous mixtures 54
Taste perception 143–4
affecting factors 250–1
Taste responsiveness
hormonal/metabolic variables 185
individual differences 181
obesity 178–9, 181
Telencephalon, taste afferents 21–2
Texture, sweetness acceptance 250–1
Thaumatin 39–41
species response differences 36
sucrose negative correlation 43
Thresholds, and intensity 50
Tolerance functions, sweet taste 149–53
Tolerance triangle, sweet taste 148–9
Tongue loci, sweetness 41
Tooth decay 91
Trace metals, anorexic taste sensitivity 196
Tryptophan levels, carbohydrate ingestion 206
Twinkie defence 205
Twins, sweet taste correlations 130

Ulcer treatment, spices and laws of similarity 104
Undimensional ideal-point scale (McBride) 70
Upbringing, sweetness implications 227–38

Viscosity, sweetness interaction 57–9

Water structure, sugar molecule effects 8
Water taste 8
adapting concentration and 37
bitter 39
sweet 39
Water volume displacement, sugar molar volume 11
Weaning, taste role 233
Weber ratio 147–8
White wines, tartaric acid effects 56
World bulk output 248, 249
Wundt curve 69

Xanthines, beverages 100, 107
Xylose, gastric emptying 166

Yoghurt, psychological attribute ratings 75

Zinc deficiency 235